PLANT LIPID
BIOCHEMISTRY

EXPERIMENTAL BOTANY
An International Series of Monographs

CONSULTING EDITORS

J. F. Sutcliffe

School of Biological Sciences, University of Sussex, England

AND

P. Mahlberg

Department of Botany, Indiana University, Bloomington, Indiana, U.S.A.

Volume 1. D. C. SPANNER, Introduction to Thermodynamics. 1964
Volume 2. R. O. SLATYER, Plant-Water Relationships. 1967
Volume 3. B. M. SWEENEY, Rhythmic Phenomena in Plants. 1969
Volume 4. C. HITCHCOCK and B. W. NICHOLS, Plant Lipid Biochemistry. 1971

IN PREPARATION

J. F. LONERAGEN, Trace Elements in Plants
R. CLELAND, The Cell Wall

PLANT LIPID BIOCHEMISTRY

THE BIOCHEMISTRY OF
FATTY ACIDS AND ACYL LIPIDS
WITH PARTICULAR REFERENCE TO
HIGHER PLANTS AND ALGAE

C. HITCHCOCK and B. W. NICHOLS

Unilever Research Laboratory, Colworth House,
Sharnbrook, Bedford, England

1971

ACADEMIC PRESS · LONDON AND NEW YORK

ACADEMIC PRESS INC. (LONDON) LTD
Berkeley Square House
Berkeley Square
London, W1X 6BA

U.S. Edition published by
ACADEMIC PRESS INC.
111 Fifth Avenue
New York, New York 10003

Library of Congress Catalog Card Number: 77-170752
ISBN: 0-12-349650-0

PRINTED IN GREAT BRITAIN BY
WILLIAM CLOWES AND SONS LIMITED, LONDON, COLCHESTER AND BECCLES

Preface

The development of new analytical techniques over the past 15 years has given great impetus to research in lipid chemistry and biochemistry generally, and the increase in our knowledge of lipid and fatty acid metabolism in plants has been particularly dramatic. During this period a few books and monographs have appeared which are concerned fairly specifically with mammalian systems, but no attempt to survey current understanding of lipid metabolism in plants has yet been made. This book attempts to fulfil that function with regard to the structure, distribution, metabolism and function of acyl lipids and fatty acids in algae and higher plants. We have also described aspects of mammalian and microbial metabolism where the comparison appears relevant.

The preparation of the text has required reference to over 1000 original papers and while it is possibly too much to hope that all relevant work has been considered we believe that most of the literature published before October, 1970 has been covered. In an attempt to keep the book up to date, we also include a supplementary list of original papers which came to our attention between completion of the manuscript, and going to print in July, 1971. We have assumed that the book will seldom be read from cover to cover but will more likely be employed to retrieve information regarding a specific feature or area of plant lipid metabolism. Accordingly we have prepared what we hope will prove to be a comprehensive index, and have also allowed a small degree of repetition to occur in the text when it seemed justified.

Finally and especially, we wish to thank all those authors and publishers who have generously allowed us to reproduce figures and tables from their publications, and our colleagues Drs M. I. Gurr, A. T. James and L. J. Morris for commenting on parts of the manuscript.

Any errors of fact or interpretation are ours alone and we welcome all comments and suggestions, constructive or destructive.

C. Hitchcock
B. W. Nichols

June, 1971

Introductory Note

The nomenclature, abbreviations, conventions and symbols used in this book need little explanation; they are familiar to chemists and lipid biochemists and are in general based on the recommendations of the Biochemical Society, and therefore usually follow the principles of the International Union of Pure and Applied Chemistry and of the International Union of Biochemistry. Accepted abbreviations are used without definition, and include AMP, adenosine 5'-phosphate; ADP, adenosine 5'-pyrophosphate; ATP, adenosine 5'-triphosphate; CDP, cytidine 5'-pyrophosphate; CTP, cytidine 5'-triphosphate; UDP, uridine 5'-pyrophosphate; UTP, uridine 5'-triphosphate; FAD, flavin-adenine dinucleotide; FMN, flavin mononucleotide; NAD, nicotinamide-adenine dinucleotide (NAD^+, oxidized form; NADH, reduced form); NADP, nicotinamide-adenine dinucleotide phosphate ($NADP^+$, oxidized form; NADPH, reduced form); DAF, days after flowering; ACP, acyl carrier protein; CoA, coenzyme A; acyl-CoA (e.g. stearoyl-CoA), the acyl derivative (thiolester) of CoA. In figures, ACP and CoA are often written ACPSH and CoASH, to illustrate actions at their thiol groups.

Optically active fatty acids are distinguished by the prefix D- or L-. Natural unsaturated fatty acids usually contain olefinic (ethylenic) bonds (.CH: CH.) with the two hydrogen atoms on the same side of the molecule; such geometry is called *cis*, while the less usual opposite situation is described as the *trans*-configuration. In formulae, the bonds may be distinguished by *c* or *t* respectively; the letter R represents an alkyl group and E (or enz) an enzyme.

Trivial names of fatty acids and lipids are often used, but these are defined in the first two chapters. For convenience, and especially in tables, a popular system of symbols for fatty acids has been adopted in which two numbers are used, separated by a colon. The first is the number of carbon atoms in the acid and the second is the number of unsaturated centres. The position of unsaturation may be denoted by additional numbers after the main symbol; after each additional number the letters c, t, a or e may respectively indicate *cis*-olefinic, *trans*-olefinic, acetylenic or ethylenic bonds of unknown or irrelevant configuration. The absence of a letter implies the usual *cis*-olefinic bond. Substituents may be described as prefixes to the main symbol: hydroxy-derivatives, keto-derivatives (i.e. oxo-derivatives) and hydroperoxy-derivatives are designated by the letters h, k or p respectively, preceded by a number indicating the position of substitution. For example, a pentadecenoic acid is represented by 15 : 1; oleic acid (*cis*-9-octadecenoic acid) by 18 : 1 (9) or 18 : 1 (9c); ximenynic acid (*trans*-11-octadecen-9-ynoic acid) by 18 : 2 (9a 11t); coriolic acid (13-hydroxy-*cis*-9,*trans*-11-octadecadienoic acid) by 13h-18 : 2 (9c 11t).

The symbol C_n denotes a molecule containing n carbon atoms, while C-m and $C_{(m)}$ refer to the carbon atom numbered m. Two systems of numbering are used; normally the accepted convention is followed in which fatty acids are numbered from the carboxyl group (C-1). Here C-2 is also known as the α-carbon atom and C-3 is β, etc.; the last position is denoted by ω, so that 2-hydroxypalmitate and α-hydroxypalmitate are synonymous, while 16-hydroxypalmitate is an ω-hydroxy acid. It is occasionally expedient to number from the ω end of the fatty acid chain, and the numbers so generated are always distinguished by the prefix ω (see p. 5). Hence linoleic acid (*cis*-9,*cis*-12-octadecadienoic acid) usually appears as 18 : 2 (9,12) but also as 18 : 2 (ω6,9).

Unless specified otherwise, fatty acid compositions of lipids are given as the percentage weight of each acid in a sample relative to the total weight of acids present; these data are normally determined by gas-liquid chromatography and are subject to experimental error of perhaps about 5%.

In general, the systematic names of plants used in experiments are given as reported in the original literature. Those appearing in Chapter 1 are indexed (Tables 1.14 and 1.15, pp. 36–42). Introductory and supplementary books are listed in a short bibliography at the end of this volume.

Contents

PREFACE v

INTRODUCTORY NOTE vii

Chapter 1

Structure and Distribution of Plant Fatty Acids 1

   1. Introduction 1

   2. Major Fatty Acids 2

   3. Minor Fatty Acids 2
      A. Saturated Minor Acids 3
      B. Unsaturated Minor Acids 5

   4. Unusual Fatty Acids 9
      A. Non-Conjugated Ethylenic Acids 11
      B. Conjugated Ethylenic Acids 14
      C. Acetylenic Acids 17
      D. Substituted Acids 22
      E. Branched-chain Acids 28

   5. Taxonomy 30
      A. Lower Plants 31
      B. Higher Plants 33

Chapter 2

Plant Acyl Lipids 43

   1. The Glycerides 43
      A. Phosphoglycerides 44
      B. Glycosyl Diglycerides 48
      C. Other Glycerides 50

   2. Diol Lipids 50

   3. Other Acyl Esters 51
      A. Sterol Esters and Sterol Glycoside Esters 51
      B. Wax Esters 51
      C. Carotenoid Esters 53
      D. Terpenoid Esters 53

   4. Sphingolipids 53
      A. Cerebrosides (Ceramide Monoglycosides) 54
      B. Phytoglycolipid and Related Substances 55

   5. Cutin 57

Chapter 3

The Lipid and Fatty Acid Composition of Specific Tissues   59

   1. Higher Plants   59
      A. Fruits   59
      B. Corms, Tubers and Bulbs   64
      C. Flowers   65
      D. Mitochondria   65
      E. Leaves   66

   2. Mosses and Ferns   73

   3. Algae   73
      A. Higher Algae   73
      B. Blue-green Algae (Cyanophyceae)   76

Chapter 4

Distribution of Individual Fatty Acids between Lipid Classes   81

   1. Seed Lipids   81
      A. Triglycerides   81
      B. Phospholipids and Glycolipids   89

   2. Corms, Tubers and Bulbs   90

   3. Photosynthetic Tissues   90
      A. Distribution of Fatty Acids between Lipid Classes . . .   90
      B. Positional Distribution of Fatty Acids in Individual Phospholipids and Glycolipids   93

Chapter 5

Biosynthesis of Plant Fatty Acids   96

   1. Introduction   96

   2. Activation of Carboxylic Acids   97

   3. Biosynthesis of Saturated Fatty Acids   102
      A. Acetyl-CoA Carboxylase   102
      B. Fatty Acid Synthetase   104
      C. Fatty Acid Elongation   119
      D. Biosynthesis of Saturated Fatty Acids in Higher Plants .   124

   4. Biosynthesis of Unsaturated Fatty Acids   130
      A. Anaerobic Pathway of Monoenoic Acid Biosynthesis . .   131
      B. Aerobic Pathway of Monoenoic Acid Biosynthesis . . .   134
      C. Biosynthesis of Polyenoic Acids   140
      D. Mechanism of Desaturation   146
      E. Control of Desaturation   152
   5. Biosynthesis of Unusual Fatty Acids   156
      A. Non-conjugated Ethylenic Acids   157

B. Conjugated Ethylenic Acids 160
C. Acetylenic Acids. 164
D. Substituted Acids 168
E. Branched-chain Acids 173

Chapter 6

The Biosynthesis of Acyl Lipids 176

1. Glyceride Biosynthesis 176
   A. The Initial Acyl Acceptor—Biosynthesis of Phosphatidic
      Acid 176
   B. Di- and Tri-glycerides 178
   C. Phosphoglycerides 179
   D. Glycosyl Glycerides 183
   E. Factors Affecting the Fatty Acid Composition of Individual
      Lipids. 185

2. Biosynthesis of Other Acyl Lipids 190
   A. Wax Esters 190
   B. Steryl Glucoside Esters 191
   C. Sphingolipids. 191
   D. Cutin 191

Chapter 7

Lipolytic Enzymes 193

1. Lipase (Glycerol Ester Hydrolase) 193
   A. Castor Bean Lipase 194

2. Phospholipases 195
   A. Phospholipase A 196
   B. Phospholipase B 196
   C. Phospholipase D 197

3. Glycolipid Hydrolases 199
   A. Galactosyl Glyceride Acyl Hydrolases 199
   B. Sulphoquinovosyl Diglyceride Acyl Hydrolases 200

Chapter 8

Biological Degradation of Plant Fatty Acids 201

1. Beta-oxidation of Fatty Acids 201
   A. Introduction 201
   B. Beta-oxidation in Plants 205

2. Alpha-oxidation 213
   A. Alpha-oxidation in Germinating Seeds 213
   B. Alpha-oxidation in Leaves 214
   C. Other Alpha-oxidation Systems 217
   D. Function of Alpha-oxidation 219

3. Omega-oxidation of Fatty Acids 221

4. Oxidation by Lipoxygenase 223
   A. Autoxidation 223
   B. Lipoxygenase 226

Chapter 9

Lipid and Fatty Acid Metabolism During Organogenesis and Senescence . 236

1. Seed and Fruit Maturation 236

2. Seed Germination and Seedling Development 241

3. Chloroplast Development 245

4. Maturation of Green Tissue 248

5. Lipid and Fatty Acid Metabolism during "Steady-state" photo-
   synthesis 248
   A. General Metabolism in Leaves and Algae 248
   B. Cuticular Waxes 252

6. Lipid Metabolism in Plant Mitochondria. Ageing 255

7. Senescence 258

8. Effect of Environment on Fatty Acid Synthesis and Metabolism . 259
   A. Leaves 260
   B. Oil-seeds 260

9. Genetic Control 261

Chapter 10

The Role of Lipids in Plant Metabolism 263

1. Plant Lipids as Components of Membranes 263
   A. Structural Role 263
   B. Involvement in Membrane Function 269
2. Other Functions 274
   A. Energy Storage 274
   B. Cellular Repair Mechanisms 276
   C. Cutin Formation 277
   D. Ethylene Production 278

Chapter 11

The Analysis of Plant Lipids 279

1. Extraction and Storage 279

2. Fractionation and Analysis of Plant Lipids 281
   A. Preliminary Separations 281
   B. Separation of Neutral Lipid Classes 283

C. Separation of Phospholipids and Glycolipids 284
D. Isolation of Specific Molecular Species of Individual Lipid
   Classes 289
E. Detection of Lipids on Thin Layer Chromatograms . . 292
F. Stability of Lipids During Chromatography 293
G. Quantitative Determination of Plant Lipids 294

3. Fractionation and Analysis of Component Fatty Acids . . . 295
   A. Gas Chromatography 295
   B. Argentation Thin Layer Chromatography 298
   C. Determination of Double Bond Position 299
   D. Specific Methods of Fatty Acid Analysis 299

4. Fractionation and Analysis of Sphingolipid Bases (Sphingosines) 300

5. Identification of Isolated Plant Lipids 300
   A. Ultraviolet Spectroscopy 301
   B. Infrared Spectroscopy 301
   C. Mass Spectroscopy 302
   D. Nuclear Magnetic Resonance Spectroscopy 302

6. Developments in Lipid Methodology 303
   A. Gas Chromatography 303
   B. Liquid Chromatography 304

References 306

More Recent Publications Not Considered in the Text 333

Supplementary Reading 337

Author Index 339

Subject Index 353

CHAPTER 1

Structure and Distribution of Plant Fatty Acids

1. Introduction

The acyl lipids owe their characteristic properties to the fact that their molecules are mainly composed of long chains of carbon atoms; these properties may be modified by the presence of a small number of more reactive and polar groups in the molecule. The hydrocarbon chains are provided by the higher saturated and unsaturated monocarboxylic aliphatic acids, and some similar substituted acids, all of which are generally termed "fatty acids". These acids are chemically combined, usually as esters, but occasionally as amides or ethers, to form complex lipids which may also contain alcohols (especially glycerol), bases, phosphate esters, sugars or sterols (Chapter 2). A wide variety of fatty acids occur naturally, particularly in the vegetable kingdom; thus different fatty acids are found on the same lipid, and the same acids are distributed throughout many lipids, with the result that each acyl lipid is represented by a large number of individual molecular species. Not all fatty acids are equally important in nature, and three types may be distinguished: those which are widespread and are major constituents of natural lipids; those of related structure which are generally minor constituents; and those of unusual structure whose occurrence is limited to a few sources, where they may account for a large proportion of the fatty acids present. These "major", "minor" and "unusual" acids are important in the sense that they represent end-products of metabolic processes, and so accumulate; they can therefore be extracted, investigated and utilized by man. The present chapter deals with their structure and occurrence in plants, a subject covered in some detail by Hilditch and Williams (1964) and Markley (1960, 1968). During biosynthesis and degradation, other intermediate acids are involved that do not normally accumulate to any extent; these are discussed in Chapters 5 and 8.

Acyl lipids occur in all organisms. In higher plants, fatty acids accumulate mainly as triglyceride in the seed and/or in the fleshy part of the fruit, which act as food stores. While fruit coat fats and seed oils provide a huge commercial source of lipid, the rest of the plant also contains fatty acids, though in smaller quantity. In the leaf, stem and roots they tend to be combined as glycolipids and phospholipids; these polar lipids are

1

associated with metabolic and structural functions (Chapter 10, p. 263). The fatty acid composition of these functional lipids varies somewhat from plant to plant, and similarly from species to species; this composition can be markedly different from the storage triglyceride, which may contain unusual fatty acids absent from other parts of the same plant and from the lipids of other species. In general, botanically related species often produce similar seed oils, which has led to a general classification of higher plants on the basis of their chief seed acids (p. 33; Hilditch and Williams, 1964; Shorland, 1963) though in a minority of cases this is somewhat ill-defined. The lipid composition of other parts of the plant, such as leaves, are not so characteristic; the fatty acids of specific tissues are discussed in Chapter 3 (p. 59).

2. Major Fatty Acids

The major fatty acids are all saturated or unsaturated monocarboxylic acids with a straight even-numbered carbon chain; the reason for this structure rests on the mechanism of their biosynthesis (Chapter 5, p. 96). The saturated homologues lauric (dodecanoic), myristic (tetradecanoic), palmitic (hexadecanoic) and stearic (octadecanoic) acids all occur in plants, but even more abundant are the unsaturated analogues oleic (*cis*-9-octadecenoic), linoleic (*cis*-9,*cis*-12-octadecadienoic) and linolenic (*cis*-9,*cis*-12,*cis*-15-octadecatrienoic) acids; structures are given in Table 1.1. These seven acids alone accounted for 94% of those in the world's commercial vegetable fats in 1969; the breakdown of that figure (Table 1.1) illustrates their importance in the total industrial production from all commercial crops. In general, they are also widely distributed in the lipids throughout all parts of all plants, often with palmitate, oleate and linoleate predominating (Chapter 3, p. 59). The unsaturated fatty acids illustrate the general ubiquity of the *cis*-double bond in the 9-position and of the 1,4-diene or "methylene-interrupted" structure $(.CH:CH.CH_2.CH:CH.)$ in the polyenoic acids. This pattern is also described as "divinylmethane" or "skipped methylene". Linolenic acid is sometimes called α-linolenic acid (9,12,15-octadecatrienoate) to distinguish it from the isomeric γ-linolenic acid (6,9,12-octadecatrienoate).

3. Minor Fatty Acids

The allocation of fatty acids between the "minor" and "unusual" classes is necessarily somewhat arbitrary. This classification is convenient

TABLE 1.1

Structures of the major fatty acids[a]

World distribution[b] %	Common name	Symbol	Structure
4	Lauric	12:0	$CH_3.[CH_2]_{10}.CO_2H$
2	Myristic	14:0	$CH_3.[CH_2]_{12}.CO_2H$
11	Palmitic	16:0	$CH_3.[CH_2]_{14}.CO_2H$
4	Stearic	18:0	$CH_3.[CH_2]_{16}.CO_2H$
34	Oleic	18:1(9c)	$CH_3.[CH_2]_7.CH\overset{c}{:}CH.[CH_2]_7.CO_2H$
34	Linoleic	18:2(9c 12c)	$CH_3.[CH_2]_3.[CH_2.CH\overset{c}{:}CH]_2.[CH_2]_7.CO_2H$
5	Linolenic[c]	18:3(9c 12c 15c)	$CH_3.[CH_2.CH\overset{c}{:}CH]_3.[CH_2]_7.CO_2H$

[a] The "major acids" are those responsible for a large proportion of fatty acids present in most plant lipids; other acids are ubiquitous but usually present in small quantities ("minor acids", Table 1.2), or are found in a few sources only ("unusual acids", Table 1.5).

[b] Estimation of the percentage distribution of each acid in all commercial vegetable fats grown throughout the world in 1969–1970. Other acids (Tables 1.2 and 1.5) accounted for some 6%, including 3% erucic acid (22:1,13c; p. 7), 1% ricinoleic acid (12h-18:1,9c; p. 24), 0·7% arachidic acid (20:0; p. 4), 0·3% eleostearic acid (18:3,9c 11t 13t; p. 15). The total annual production of these fats was about thirty million tons.

[c] Also called α-linolenic acid to distinguish it from its "minor" isomer γ-linolenic acid (18:3,6c 9c 12c).

to distinguish broadly between acids which are present in minor quantities in most fats, and those which occur in a few unusual sources where they may be present in large amounts. Minor acids are often disregarded in analytical reports, and their detection often depends on the sensitivity of the methods employed; it is therefore difficult to judge how widespread is their distribution. Here, the "minor" class is taken to include all saturated acids and those unsaturated analogues which are directly related to the major unsaturated acids (Table 1.2). All other natural fatty acids are classed as "unusual".

A. Saturated Minor Acids

The carbon numbers of the minor saturated fatty acids may be odd or even, and may be smaller or greater than those of their "major" counterparts. Small quantities of hexanoic, octanoic and decanoic acids are

TABLE 1.2

Structures of some minor fatty acids[a]

Class	Common name	Symbol
Saturated		
	Caproic	6:0
	Caprylic	8:0
	Capric	10:0
	Margaric	17:0
	Arachidic	20:0
	Behenic	22:0
	Lignoceric	24:0
Unsaturated[b]		
$\Delta 9$-family	Palmitoleic	16:1(9c)
$\omega 9$-family	Erucic	22:1(13c)
$\omega 6$-family	γ-Linolenic	18:3(6c 9c 12c)
	Arachidonic	20:4(5c 8c 11c 14c)
$\omega 3$-family	—	16:3(7c 10c 13c)
	—	18:4(6c 9c 12c 15c)
	—	22:6(4c 7c 10c 13c 16c 19c)

[a] The "minor" fatty acids are of widespread occurrence but usually account for only a small proportion of the acids present.

[b] See Table 1.3.

probably ubiquitous, though coconut oil from *Cocus nucifera* contains abnormally high concentrations of these lower fatty acids (about 1%, 8% and 7% respectively). The higher saturated acids are not often specifically looked for, and may not be very widespread; they have been detected in some vegetable fats of unusually high melting point. For instance, the wax isolated by Kleiman *et al.* (1969c) from the hulls of sunflower seeds (pericarp of *Helianthus* species) contained all the even-numbered saturated acids from palmitate (1%) to tricontanoate (3%) including eicosanoate (44%), docosanoate (22%) and octacosanoate (8%). Here they were present as monoesters, being randomly esterified to a series of saturated monohydric long-chain alcohols.

Fatty acids of odd carbon number are rare. Heptadecanoic acid seems to be present in trace amounts in many vegetable oils, though early reports of its ubiquity (leading to its gaining the trivial name "margaric acid") were unfounded, resulting from analyses of unresolved mixtures of other acids. However, traces of pentadecanoate and heptadecanoate are synthesized in leaves (Stumpf and James, 1963), and this capacity is probably general.

B. Unsaturated Minor Acids

The minor unsaturated acids are related to the major unsaturated acids in one of two ways: the carbon chain is extended (or shortened) either at the methyl or at the carboxyl end of the molecule (Table 1.3). In the former case the positions of the double bonds as described in the systematic Δ-notation remain the same: oleic (9-octadecenoic) acid is related to palmitoleic (9-hexadecenoic) acid. In the latter case, this position relative to the carboxyl group is altered: oleic acid is also related to erucic (13-docosenoic) acid. Here it is convenient to count the double bond position from the methyl terminus, using the ω-notation: for example oleic acid ($18:1, \omega 9c$) belongs to the same $\omega 9$ family as erucic ($22:1, \omega 9c$) and nervonic ($24:1, \omega 9c$) acids. Two other series of unsaturated natural fatty acids are recognizable from the position of the double bond nearest to the terminal methyl group. The members of the linoleic ($\omega 6$) family are related through their $\omega 6, 9$-double bonds, and include γ-linolenic acid ($18:3, \omega 6c\,9c\,12c$). The linolenic ($\omega 3$) family is characterized by the $\omega 3c\,6c\,9c$-double bond trio present in α-linolenic acid. Methylene-interrupted all-*cis* polyenoic acids may be unambiguously described by the position of the first double bond only: α-linolenic acid is therefore represented by $18:3(\Delta 9)$ or $18:3(\omega 3)$ as well as by the standard symbol $18:3(9c\,12c\,15c)$.

The natural unsaturated fatty acids are generally members of the $\Delta 9$, $\omega 9$, $\omega 6$ or $\omega 3$ families with even carbon numbers and double bonds in relative 1,4-positions. These structures cover the "major" and "minor" acids; any deviation in family or in this specific methylene-interrupted spacing is considered to give rise to an "unusual" acid. Occasionally, a tissue is found in which "minor" acids accumulate to such an extent that they become principal constituents.

1. Δ9-Family of unsaturated acids

According to Markley (1960) small quantities of *cis*-9-hexadecenoic (palmitoleic) acid ($16:1, 9c$) are present in all plants (and animals) from highest to lowest; an unusually abundant source is the seed oil of *Doxantha unguis-cati*, where 64% of the fatty acid is palmitoleate (Chisholm and Hopkins, 1965). Myristoleic acid ($14:1, 9c$) is also a minor plant component. The seed oil of *Asclepias syriaca* contains 9,12-hexadecadienoate (Chisholm and Hopkins, 1960a).

2. ω9-Family of unsaturated acids

11-Eicosenoic acid ($20:1, 11c$) is present in small amounts in some vegetable oils; the seed of *Cardiospermum halicacabum* is unusual in

TABLE 1.3

Structural relationships between some major and minor unsaturated fatty acids

Family	Class	Common name	Structure
$\Delta 9$	Major	Oleic	$CH_3 \cdot [CH_2]_7 \cdot \overset{c}{CH:CH} \cdot [CH_2]_7 \cdot CO_2H$
	Minor	Palmitoleic	$CH_3 \cdot [CH_2]_5 \cdot \overset{c}{CH:CH} \cdot [CH_2]_7 \cdot CO_2H$
$\omega 9$	Major	Oleic	$CH_3 \cdot [CH_2]_7 \cdot \overset{c}{CH:CH} \cdot [CH_2]_7 \cdot CO_2H$
	Minor	Erucic	$CH_3 \cdot [CH_2]_7 \cdot \overset{c}{CH:CH} \cdot [CH_2]_{11} \cdot CO_2H$
	Minor	Nervonic	$CH_3 \cdot [CH_2]_7 \cdot \overset{c}{CH:CH} \cdot [CH_2]_{13} \cdot CO_2H$
$\omega 6$	Major	Linoleic	$CH_3 \cdot [CH_2]_3 \cdot [CH_2 \cdot \overset{c}{CH:CH}]_2 \cdot [CH_2]_7 \cdot CO_2H$
	Minor	γ-Linolenic	$CH_3 \cdot [CH_2]_3 \cdot [CH_2 \cdot \overset{c}{CH:CH}]_2 \cdot CH_2 \cdot \overset{c}{CH:CH} \cdot [CH_2]_4 \cdot CO_2H$
	Minor	Arachidonic	$CH_3 \cdot [CH_2]_3 \cdot [CH_2 \cdot \overset{c}{CH:CH}]_2 \cdot [CH_2 \cdot \overset{c}{CH:CH}]_2 \cdot [CH_2]_3 \cdot CO_2H$
$\omega 3$	Major	α-Linolenic	$CH_3 \cdot [CH_2 \cdot \overset{c}{CH:CH}]_3 \cdot [CH_2]_7 \cdot CO_2H$
	Minor	—	$CH_3 \cdot [CH_2 \cdot \overset{c}{CH:CH}]_3 \cdot [CH_2]_5 \cdot CO_2H$
	Minor	—	$CH_3 \cdot [CH_2 \cdot \overset{c}{CH:CH}]_3 \cdot CH_2 \cdot \overset{c}{CH:CH} \cdot [CH_2]_4 \cdot CO_2H$
	Minor	—	$CH_3 \cdot [CH_2 \cdot \overset{c}{CH:CH}]_3 \cdot [CH_2 \cdot \overset{c}{CH:CH}]_2 \cdot [CH_2]_2 \cdot CO_2H$

having 42% (Chisholm and Hopkins, 1958). Erucic acid (22:1,13c) also does not occur to any great extent except in some seed oils from the families Cruciferae and Tropaeolaceae. Rapeseed (*Brassica napus*) is so important in commerce that erucic acid, here classed as a "minor" vegetable acid, in fact accounted for about 3% of the world production in 1969 (Table 1.1). However, no erucic acid could be detected in a quarter of over 100 species of Cruciferae examined by Miller *et al.* (1965); the proportion in the remainder varied up to 55%. Higher homologues of the ω9-family have also been sometimes detected: in particular, *Ximenia caffra* yields seed oil containing minor amounts of 13-docosenoic, 15-tetracosenoic, 17-hexacosenoic, 19-octacosenoic, and 21-tricontenoic acids (Ligthelm *et al.*, 1954). *Tropaeolum speciosum* seed fat is an unusually rich source of *cis*-15-tetracosenoate (42%) and of *cis*-17-hexacosenoate (8%); this plant is related to the common nasturtium (*Tropaeolum majus*), which contains the highest level of erucic acid (79 mole %) found in any seed (Litchfield, 1970).

3. ω6-*Family of unsaturated acids*

Lower plants such as phytoplankton are a source of ω6-polyenoic acids (Table 1.4), which are also occasionally present in higher plants: for instance the seed oil of *Ephedra campylopoda* contains 1% *cis*-11,*cis*-14-eicosadienoate (Kleiman *et al.*, 1967). When sixty-two species of the family Boraginaceae were analysed by Kleiman *et al.* (1964) and Miller *et al.* (1968), γ-linolenic acid (18:3,6c9c12c) was sometimes found to be an important constituent (0–27%). This minor acid is also present (up to 21%) in the blue-green alga *Spirulina platensis* (Nichols and Wood, 1968) but it is more typical of animals than of plants. Similarly, arachidonic acid (20:4,5c8c11c14c) is a vital metabolite in animals (p. 141); it has not been found in higher plants but has been detected in unicellular algae, mosses and ferns. In the lipids of mosses (mainly *Brachythecium* and *Mnium*) it accounts for up to 34% of the total fatty acids (Schlenk and Gellerman, 1965). Its presence in fern fronds has been reported by Haigh *et al.* (1969) who found 26% in lipids from hart's-tongue (*Scolopendrium vulgare*).

4. ω3-*Family of unsaturated acids*

The C_{16}-analogue of α-linolenic acid was isolated from rape leaves (*Brassica napus*) by Heyes and Shorland (1951), and has been identified as *cis*-7,*cis*-10,*cis*-13-hexadecatrienoic acid. This acid is accompanied by the corresponding 4,7,10,13-tetraene in various algae such as *Chlorella pyrenoidosa* (Paschke and Wheeler, 1954) and *Scenedesmus obliquus* (Klenk *et al.*, 1963). The related *cis*-11,*cis*-14,*cis*-17-eicosatrienoate is

TABLE 1.4

Some fatty acids present in lower plants[a]

	9	10	11	12	13	14	15	16	17	18	19	20	22	24
Saturated														
Even carbon number		10:0		12:0[b]		14:0[b]		16:0[b]		18:0[b]		20:0	22:0	24:0
Odd carbon number	9:0		11:0		13:0		15:0		17:0		19:0			
Δ9-Family														
Δ9-Monoenoates						14:1		16:1		18:1[b]				
					13:1		15:1		17:1					
Δ9-Dienoates										18:2[b]				
									17:2		19:2			
Δ9-Trienoates										18:3[b]				
									17:3					
ω9-Family														
ω9-Monoenoates						14:1		16:1		18:1[b]		20:1	22:1	
ω9-Dienoates										18:2		20:2		
ω6-Family														
ω6-Dienoates								16:2		18:2[b]		20:2	22:2	
ω6-Trienoates								16:3		18:3		20:3	22:3	
ω6-Tetraenoates												20:4	22:4	
ω6-Pentaenoates													22:5	
ω3-Family														
ω3-Trienoates								16:3		18:3[b]		20:3		
ω3-Tetraenoates								16:4		18:4		20:4	22:4	
ω3-Pentaenoates												20:5	22:5	
ω3-Hexaenoates													22:6	

[a] Identified in various species of algae. The fatty acid composition of lower plants is discussed and illustrated in Chapter 3 (Tables 3.13 to 3.16).

[b] "Major" fatty acids; all others are "minor" acids.

present in *Ephedra campylopoda* seed oil to the extent of 2% (Kleiman *et al.*, 1967) and similar amounts of *cis*-8,*cis*-11,*cis*-14-heptadecatrienoate have been found in the seeds of *Thymus vulgaris* (Smith and Wolff, 1969). Other members of the ω3-family are to be found in lower plants, where minor unsaturated fatty acids abound. Between and within species, large quantitative variations in the content of C_{16}, C_{18}, C_{20} and C_{22} polyunsaturated acids may occur; chain lengths greater than C_{22} are rare.

The fatty acid compositions of a range of marine phytoplankton species cultured under standard conditions have been reported by Ackman *et al.* (1968) and by Chuecas and Riley (1969); they identified many "minor" fatty acids, the more important including 6,9,12,15-octadecatetraenoate, 5,8,11,14,17-eicosapentaenoate, 7,10,13,16,19-docosapentaenoate and 4,7,10,13,16,19-docosahexaenoate. These planktonic algae represented Bacillariophyceae, Chrysophyceae, Chlorophyceae and other classes; Korn (1964a) characterized fifty-one of the constituent acids of the phytoflagellate *Euglena gracilis*. Table 1.4 illustrates the variety of "minor" acids sometimes present in these lower plants, which may also contain unusual acids such as *trans*-3-hexadecenoate, *cis*-7-hexadecenoate and 4,7,10,13,16-heneicosapentaenoate. The fatty acid composition of specific tissues and organisms is discussed and illustrated in Chapter 3.

4. Unusual Fatty Acids

Some natural oils contain fatty acids with structures that seem unrelated to those classified as "major" or "minor". These "unusual" acids are often found only in the seeds of related plants, and may be restricted to a few individual species, a genus or a whole family; however, within this narrow distribution they may represent the principal acid of the oil. Examples of their occurrence in seeds are given in Table 3.4 (p. 64). They are sometimes also found in specific lower forms of plant life, such as yeasts. The unusual feature is generally a substituent and/or an unsaturated bond in an abnormal position (Table 1.5); thus they comprise all natural fatty acids other than the unsubstituted straight-chain saturated series, the monoenoic acids of the $\Delta 9$ and $\omega 9$ series, and their polyunsaturated analogues with methylene-interrupted *cis*-double bonds in the linoleic and linolenic families. Because of the interest in the biochemistry and industrial use of acyl lipids, and thanks to the availability of sophisticated analytical methods (p. 279), a great number of unusual fatty acids have been isolated from vegetable sources. Most of these have been shown to have combinations of *cis*- and *trans*-ethylenic bonds, acetylenic bonds, and/or oxygenated (or hydrocarbon) substituents, but not all structural combinations of these features are found in nature.

The unusual acids may be conveniently divided into five groups (Table 1.5). Those unsaturated acids whose double bonds do not correspond with those of the "major" or "minor" acids may be classed either as "non-conjugated ethylenic acids" (Group A) or "conjugated ethylenic acids" (Group B). Any acid with one or more triple bond falls into Group C ("acetylenic acids") whether double bonds are also present or not. Substituted derivatives of members of Groups A, B and C are themselves members of the same Group, but the substituted "major" and "minor"

TABLE 1.5

Structures of some unusual fatty acids[a]

Group	Common name	Symbol	Structure
A. Non-Conjugated Ethylenic[b]	Trans-3-hexadecenoic	16:1(3t)	$CH_3 . [CH_2]_{11} . CH : CH . CH_2 . CO_2H$ (t)
	Petroselinic	18:1(6c)	$CH_3 . [CH_2]_{10} . CH : CH . [CH_2]_4 . CO_2H$ (c)
B. Conjugated Ethylenic[c]	α-Eleostearic	18:3(9c 11t 13t)	$CH_3 . [CH_2]_3 . [CH : CH]_2 . CH : CH . [CH_2]_7 . CO_2H$ (t)(c)
C. Acetylenic[d]	Crepenynic	18:2(9c 12a)	$CH_3 . [CH_2]_4 . C : C . CH_2 . CH : CH . [CH_2]_7 . CO_2H$ (c)
D. Substituted[e]	Ricinoleic	12h-18:1(9c)	$CH_3 . [CH_2]_5 . CH(OH) . CH_2 . CH : CH . [CH_2]_7 . CO_2H$ (c)
	Vernolic	12,13-epoxy:18:1(9c)	$CH_3 . [CH_2]_4 . CH . CH . CH_2 . CH : CH . [CH_2]_7 . CO_2H$ (c) O
E. Branched-Chain	Sterculic	9,10-methylene-18:1(9)	$CH_3 . [CH_2]_7 . C : C . [CH_2]_7 . CO_2H$ CH_2
	Hydnocarpic	11-(2-cyclopentenyl)-11:0	$CH_2 . CH_2 . CH . [CH_2]_{10} . CO_2H$ CH=CH

[a] Distribution illustrated in Table 3.4 (p. 64).
[b] See Table 1.6.
[c] See Table 1.7.
[d] See Tables 1.8 and 1.9.
[e] See Tables 1.10 and 1.11.

acids form Group D ("substituted acids"). There remain a number of acids whose skeleton is branched and these constitute Group E (branched-chain acids). In Group E, hydrogen of the hydrocarbon chain has been substituted by carbon; in Group D by another element, usually oxygen. This classification is somewhat arbitrary, but consideration of structural and distributional similarities within Groups A to E, described in the following Sections A to E, sometimes imply metabolic relationships; these are discussed in corresponding Sections A to E of Chapter 5 (p. 157). The occurrence of unusual fatty acids in plants has been reviewed by Markley (1960, 1968), Hilditch and Williams (1964), Wolff (1966), Hopkins and Chisholm (1968) and Smith (1970).

A. Non-conjugated Ethylenic Acids

First, we consider a number of unsubstituted ethylenic fatty acids which are unusual only in the positions or geometry of their double bonds, which are neither allenic nor conjugated: we shall consider these in an order depending on the position of the unusual bond (Table 1.6).

TABLE 1.6

Structures of unusual non-conjugated ethylenic fatty acids found in plants[a]

18:3(2t 9c 12c)
16:1(3t); 18:1(3t); 18:3(3t 9c 12c); 18:4(3t 9c 12c 15c)
10:1(4c); 12:1(4c); 14:1(4c)
20:1(5c); 22:1(5c); 22:2(5c 13c); 16:1(5c); 18:1(5c); 20:3(5c 11c 14c)
   20:4(5c 11c 14c 17c); 16:2(5c 9c); 18:2(5c 9c); 18:2(5c 11c); 17:2(5c 9c)
16:1(5t); 18:1(5t); 18:2(5t 9c); 18:3(5t 9c 12c)
18:1(6c); 18:1(6t); 16:1(6c); 16:1(6t)
18:1(7e)
18:1(11c)
20:1(13c)
18:2(9t 12t); 18:2(9c 12t)

---

[a] The distribution of these acids is described in the text in this order. Conjugated ethylenic and acetylenic acids are not included (see Tables 1.7, 1.8 and 1.9).

Mixed pollen gathered by honey bees contains a free acid identified by Hopkins et al. (1969b) as trans-2,cis-9,cis-12-octadecatrienoic acid, which is apparently synthesized in the plant and acts as a food marker for the bee. Other insect attractants possess the unusual trans-2 double bond, and some trans-2-enoic acids are important metabolites in bees.

Trans-3-hexadecenoic acid is a constituent of photosynthetic tissue, where it occurs exclusively esterified at the 2-position of phosphatidyl glycerol (Haverkate and van Deenen, 1965; Nichols, 1965); its metabolism is discussed on p. 157. It is also found in some seeds: one-tenth of the fatty acids in oil from *Helenium bigelowii* is accounted for by this *trans*-acid, but *H. hoopesii* contains none (Hopkins and Chisholm, 1964c). Other seeds of the Compositae family are good sources: *Grindelia oxylepsis* has both *trans*-3-hexadecenoate (14%) and *trans*-3-octa-decenoate (2%) (Kleiman *et al.*, 1966), while *trans*-3,*cis*-9,*cis*-12-octadecatrienoic acid is also present in *Calea urticaefolia* (31%) (Bagby *et al.*, 1965a), *Aster alpinus* (14%) and *Arctium minus* (10%) (Morris *et al.*, 1968). The seed oil of *Tecoma stans* contains 19% *trans*-3, *cis*-9,*cis*-12,*cis*-15-octadecatetraenoic acid (Hopkins and Chisholm, 1965a).

A series of *cis*-4-monoenoates was discovered by Hopkins *et al.* (1966) in the seed oil of *Lindera umbellata*: these were *cis*-4-decenoate (4%) *cis*-4-dodecenoate (47%) and *cis*-4-tetradecenoate (5%).

The principal fatty acids in the seeds of meadow foam (*Limnanthes douglasii*) are *cis*-5-eicosenoate (65%) and *cis*-5-docosenoate (7%) (Smith *et al.*, 1960b); also present is *cis*-13-docosenoate (13%), accompanied by *cis*-5,*cis*-13-docosadienoate (10%) (Bagby *et al.*, 1961). Other *cis*-5-unsaturated analogues have also been found in seed oils: *cis*-5-hexadecenoate (2%) and *cis*-5-octadecenoate (21%) in *Carlina acaulis* (Spencer *et al.*, 1969); *cis*-5-octadecenoate (18%), *cis*-5-eico-senoate (3%), *cis*-5,*cis*-11,*cis*-14-eicosatrienoate (23%) and *cis*-5,*cis*-11,*cis*-14,*cis*-17-eicosatetraenoate (1%) in *Caltha palustris* (Smith *et al.*, 1968); and *cis*-5,*cis*-11,*cis*-14,*cis*-17-eicosatetraenoate (22%) in *Ephedra campylopoda* (Kleiman *et al.*, 1967). Some *cis*-5-enoic acids are also present in the slime mould *Dictyostelium discoideum* (16:2,5c9c; 18:2,5c9c; 18:2,5c11c; 17:2,5c9c; Davidoff and Korn, 1963) and in bacteria (16:1,5c; 18:1,5c; Fulco *et al.*, 1964). The natural occurrence of fatty acids containing a *trans*-5 double bond has also been reported by Bhatty and Craig (1966). *Thalictrum venulosum* yields a seed oil which contains 3% *trans*-5-hexadecenoate, 11% *trans*-5-octadecenoate, 5% *trans*-5,*cis*-9-octadecadienoate, and 44% *trans*-5,*cis*-9,*cis*-12-octadeca-trienoate, as well as traces of *cis*-5-monoenoates.

Petroselinic acid (18:1,6c) is the characteristic fatty acid of the seeds of plants belonging to the Umbelliferae family, and of the closely related Araliaceae family of the order Umbellales (Placek, 1963). The seed oils of parsley, fennel, carrot, celery, parsnip, caraway, dill, chervil, and ivy are rich in this unusual acid (50–90%), which may be distinguished from its ubiquitous isomer oleic acid by analysis of its cleavage products or by silver-ion thin-layer chromatography (Kleiman *et al.*, 1967a). The seeds of *Picramnia sellowii* are a rare source of tariric acid (18:1,6a; 85%); Spencer *et al.* (1970) report that they also contain small quantities of 6-monoenoates (18:1,6t; 18:1,6c; 16:1,6c; 16:1,6t; 20:1,6t). An indirect analytical method indicates the presence of 7-octadecenoate (11–30%) and 5-octadecenoate (0–5%) as well as 6-octadecenoate (11–42%) in ten Indian species of the family Umbelliferae (Kartha and Khan, 1969; Kartha and Selvaraj, 1970).

Traces of *cis*-7-hexadecenoate, *cis*-7, *cis*-10-hexadecadienoate and *cis*-7,*cis*-10,*cis*-13-hexadecatrienoate appear to be quite widespread in lower plants, where they are classed as "minor" acids of the ω9-, ω6- and ω3-families respectively. Nor-linolenic acid (*cis*-8,*cis*-11,*cis*-14-heptadecatrienoate) is an odd-numbered acid of the ω3-family; it has been discovered in the seed oil of *Thymus vulgaris* by Smith and Wolff (1969). Unsaturated acids with a *cis*-double bond in the Δ9-position are the most widespread in nature; these have also been considered in previous sections.

Vaccenic acid (18:1,11c) is the monoenoic acid characteristic of those bacteria which synthesize their unsaturated acids by an anaerobic mechanism (p. 131); they often also contain *cis*-9-hexadecenoate and *cis*-7-tetradecenoate (Scheuerbrandt and Bloch, 1962). While the biosynthesis of these ω7 acids in plant tissue may therefore indicate bacterial contamination, vaccenic acid is also a plant product, being present at the 15% level in the seed oil of *Doxantha unguis-cati* and of *Asclepias syriaca* (Chisholm and Hopkins, 1965). It is also a normal constituent of animal tissues (Holloway and Wakil, 1964). Rapeseed contains *cis*-13-eicosenoic acid (Haeffner, 1970).

The *trans*-analogues of linoleic acid have also been isolated from seed oils: 15% *trans*-9,*trans*-12-octadecadienoic acid was found in *Chilopsis linearis* (Chisholm and Hopkins, 1963). *Dimorphotheca sinuata* (syn. *D. aurantiaca*) does not contain this acid, but 1·3% *cis*-9,*trans*-12-octadecadienoic acid in addition to 65% β-dimorphecolic (9-hydroxy-*trans*-10,*trans*-12-octadecadienoic) acid (Morris and Marshall, 1966). *Crepis rubra* seed oil, whose major component is crepenynic (*cis*-9-octadecen-12-ynoic) acid, also contains 3% *cis*-9,*trans*-12-octadecadienoic acid. *Trans*-9,*cis*-12-octadecadienoate has not yet been detected in natural lipids.

Unusual unsaturated fatty acids with conjugated systems are considered together in the next section; non-conjugated acetylenic fatty acids are discussed with their conjugated analogues: these include stearolic acid (18:1,9a) on p. 18, crepenynic acid (18:2,9c 12a) on p. 19 and tariric acid (18:1,6a) on p. 20.

B. Conjugated Ethylenic Acids

We now consider a group of unusual polyenoic acids clearly recognizable by their conjugated (1,3-spaced) double bonds from the usual methylene-interrupted (1,4-spaced) pattern. The conjugated fatty acids of seed oils have been surveyed by Hopkins and Chisholm (1968), who point out that their occurrence in nine of the forty-eight orders of plants in eleven families ranging from the primitive Santalaceae to the relatively recent Compositae suggests a random distribution in nature with little evidence for any pattern. Wherever they are found, the conjugated acids show some general structural characteristics, which are typically near the middle of the chain. Those which have so far been isolated and investigated have ethylenic and/or acetylenic bonds and sometimes hydroxyl substitution in characteristic positions. We shall consider all acetylenic acids separately (p. 17).

Conjugated ethylenic fatty acids fall into four families recognizable by common cis-9, cis-12,- trans-9- and trans-12-double bond positions.

TABLE 1.7

Some conjugated ethylenic fatty acids of biochemical interest[a]

	Family	Common name	Symbol
(i)	cis-9	Coriolic	13h-18:2(9c 11t)
		α-Eleostearic	18:3(9c 11t 13t)[b]
		Punicic	18:3(9c 11t 13c)
		α-Parinaric	18:4(9c 11t 13t 15c)[b]
		α-Kamlolenic	18h-18:3(9c 11t 13t)
(ii)	cis-12	α-Dimorphecolic	9h-18:2(10t 12c)
		Calendic	18:3(8t 10t 12c)
		Jacaric	18:3(8c 10t 12c)
(iii)	trans-9	—	13h-18:2(9t 11t)[c, d]
		β-Eleostearic	18:3(9t 11t 13t)[c]
		Catalpic	18:3(9t 11t 13c)
(iv)	trans-12	β-Dimorphecolic	9h-18:2(10t 12t)[e]
		—	18:3(8t 10t 12t)[c]
		—	18:3(8c 10t 12t)[c]
		—	18:2(10t 12t)

[a] The distribution of these acids is described in the text by families in this order.
[b] The 4-keto-derivatives 4k-18:3(9c 11t 13t) and 4k-18:4(9c 11t 13t 15c) also exist.
[c] Postulated intermediates not yet isolated from natural sources.
[d] The keto-analogue 13k-18:2(9t 11t) exists.
[e] The keto-analogue 9k-18:2(10t 12t) exists.

Each of these families (Table 1.7) contains a hydroxydienoic acid and two trienoic acids, of which some have not yet been found in nature.

1. Cis-9-family

The principal conjugated fatty acids having a *cis*-9-double bond are 13-hydroxy-*cis*-9,*trans*-11-octadecadienoic acid (coriolic or α-artemesic acid; 13h-18:2,9c11t); *cis*-9,*trans*-11,*trans*-13-octadecatrienoic acid (α-eleostearic acid; 18:3,9c11t13t); and *cis*-9,*trans*-11,*cis*-13-octadecatrienoic acid (punicic acid; 18:3,9c11t13c).

Coriolic (13h-18:2,9c11t) is the predominant fatty acid of coriaria seed oils; in *Coriaria nepalensis* it accounts for 66–68% of the fatty acids present (Tallent *et al.*, 1968) where it has the D-configuration and is not accompanied by the 9-hydroxydienoic acid analogues (Tallent *et al.*, 1966b). Coriolic acid is also present in the seeds of *Dimorphotheca sinuata* (syn. *D. aurantiaca*; Morris *et al.*, 1960a) and of *Xeranthemum annuum* (Powell *et al.*, 1967b), where it exists in minor amounts with α-dimorphecolic acid (9h-18:2,10t12c; p. 16). In *Monnina emarginata* the dienolic acid present (30%) is the enantiomer of coriolic acid, i.e. L-13-hydroxy-*cis*-9,*trans*-11-octadecadienoic acid (Phillips *et al.*, 1970).

α-Eleostearic acid (18:3,9c11t13t) is the principal component (72–82%) of tung oil, from the seeds of *Aleurites fordii* and *A. montana* (Hilditch and Mendelowitz, 1951); it is also present (50%) in the seed oil of *Kentranthus macrosiphon* (Hopkins and Chisholm, 1965b). Its isomer punicic acid (18:3,9c11t13c) accounts for 71% of the acids from pomegranate (*Punica granatum*) seeds (Crombie and Jacklin, 1957). When six different species of the family Cucurbitaceae were analysed, three contained α-eleostearic acid (49–65%), and the other three punicic acid (32–56%); the two acids seemed mutually exclusive (Hopkins *et al.*, 1969a).

The structure of α-kamlolenic acid was proved to be 18-hydroxy-*cis*-9,*trans*-11,*trans*-13-octadecatrienoic acid (18h-18:3,9c11t13t); it is thus the 18-hydroxyderivative of α-eleostearic acid and not of catalpic acid (p. 16) (Hopkins *et al.*, 1969). It occurs in Kamala oil from *Mallotus phillipensis* (Ahlers and Gunstone, 1954). The seed oil of *Impatiens edgeworthii* contains 48% α-parinaric acid (*cis*-9,*trans*-11,*trans*-13,*cis*-15-octadecatetraenoic acid; 18:4,9c11t13t15c) (Bagby *et al.*, 1966). The 4-keto-derivatives of α-eleostearic acid and of α-parinaric acid have been reported by Gunstone and Subbarao (1967b) to occur in the seed oil of *Chrysobalanus icaco*. The oil contains 10% α-licanic acid (4k-18:3, 9c11t13t) and 18% of 4-oxo-*cis*-9,*trans*-11,*trans*-13,*cis*-15-octadeca-tetraenoate (4k-18:4,9c11t13t15c).

2. Cis-*12-family*

The hydroxydiene of this family is 9-hydroxy-*trans*-10,*cis*-12-octadecadienoic acid (α-dimorphecolic acid; 9h-18:2,10t12c) which occurs in several oils (Morris *et al.*, 1960a) including *Xeranthemum annuum*, where it has the D-configuration (Powell *et al.*, 1967b). In these sources it coexists with coriolic acid (D-13h-18:2,9c11t; p. 15), but it has been found without any accompanying 13-hydroxydienes in *Calendula officinalis* (Badami and Morris, 1965). Seed oil from this plant contains calendic acid (*trans*-8,*trans*-10,*cis*-12-octadecatrienoic acid; 18:3, 8t10t12c) as major component (56%; McLean and Clark, 1956). Calendic acid was characterized by Chisholm and Hopkins (1960b) and also found to account for 36% of the fatty acids of another seed oil, *Osteospermum hyoseroides* (Hopkins and Chisholm, 1965b). The isomeric jacaric acid (*cis*-8,*trans*-10,*cis*-12-octadecadienoic acid; 18:3, 8c10t12c) has been found as a major component of the seed oil of *Jacaranda mimosifolia* by Chisholm and Hopkins (1962).

3. Trans-*9-family*

The hydroxydiene of this series, which would logically be 13-hydroxy-*trans*-9,*trans*-11-octadecadienoic acid, has not been found in nature. The only natural member of the series is catalpic acid (*trans*-9,*trans*-11,*cis*-13-octadecatrienoic acid; 18:3,9t11t13c), which was isolated by Hopkins and Chisholm (1962a) from seeds of *Catalpa ovata*. Traces of ketoacids (13k-18:2,9t11t; 13k-18:1,11t) have been reported in *Monnina emarginata* seeds (Phillips *et al.*, 1970). β-Eleostearic acid (18:3,9t11t13t) has been synthesized chemically, and when necessary to distinguish it from its natural *cis*-analogue (18:3,9c11t13t), the latter is regarded as α-eleostearic acid which is a member of the *cis*-9-family of conjugated acids.

4. Trans-*12-family*

Dimorphecolic acid was isolated from dimorphotheca oil by Smith *et al.* (1960); it comprises 65% of the fatty acids of *Dimorphotheca sinuata* (syn. *D. aurantiaca*) seeds and its structure proved to be 9-hydroxy-*trans*-10,*trans*-12-octadecadienoic acid (9h-18:2,10t12t). It is regarded as β-dimorphecolic acid when necessary to distinguish it from the α-acid (9h-18:2,10t12c). A minor constituent of dimorphotheca oil is 2·5% of 9-oxo-*trans*-10,*trans*-12-octadecadienoate (Binder *et al.*, 1964). The two expected trienes of this series (*cis*-8,*trans*-10,*trans*-12- and *trans*-8,*trans*-10,*trans*-12-octadecatrienoic acids) have not been isolated from natural sources. However the related diene *trans*-10,*trans*-12-octadecadienoic acid was discovered by Hopkins and Chisholm

(1962b, 1964a) as a component (9%) of the seed oil of *Chilopsis linearis*.
Of all the possible combinations of conjugated ethylenic groupings with or without extra hydroxysubstituents, only those belonging to the four series described above have been found in nature; this selective occurrence of conjugated fatty acids must be explained in terms of the mechanism of their biosynthesis and accumulation (p. 160). A fatty acid having a furanoid structure, which may be regarded as the 9,12-epoxide of the conjugated 9,11-octadecadienoic acid is considered on p. 121.

The only exceptions to this classification of conjugated fatty acids are apparently the 2,4-dienoic acids, whose proposed mechanism of biosynthesis (p. 164) would nevertheless relate them to the *cis*-9-family or *trans*-9-family. Deca-2,4-dienoate is a constituent of stillingia oil, from the seeds of *Sapium sebiferum* (syn. *Stillingia sebifera*; Crossley and Hilditch 1949); it occurs there as the ester with an unusual hydroxy-allenic acid (p. 21). The 5% of conjugated acid was thought to account for the abnormally good drying properties of the oil, whose principal fatty acids are 40% linolenate and 25–30% linoleate (Devine, 1950). Its geometry was found to be *trans*-2,*cis*-4-decadienoate; it was also identified with a flavour component isolated from pears (Jennings *et al.*, 1964), and later, the *trans*,*trans*-isomer was also found to be present (Heinz and Jennings, 1966). A 2,4-dodecadienoic acid has been identified in *Sebastiana lingustrina* seed oil (Holman and Hanks, 1955).

C. Acetylenic Acids

Several hundred compounds with conjugated acetylenic bonds have been identified in fungi (particularly Basidiomycetes) and in plants of some nine families including Compositae, Umbelliferae and Leguminosae (Johnson, 1965; Jones, 1966). Some of these are fatty acids and it seems likely that others are formed from fatty acids by various biochemical processes (Bu'Lock and Smith, 1967). Many belong to the class of compounds known as polyketides, having structures which may be derived from a polyketone skeleton of alternate oxidized and reduced carbon atoms ($.CH_2.CO.CH_2.CO.CH_2.$) by hypothetical but chemically rational transformations. The chemistry and biochemistry of these natural acetylenes have been studied particularly by F. Bohlman, J. D. Bu'Lock, E. R. H. Jones, N. A. Sørenson and their various co-workers (Sørenson, 1963; Bu'Lock, 1964; Johnson, 1965; Jones, 1966; Bohlmann, 1967). We are here concerned particularly with the structure and distribution in higher plants of fatty acids with one triple bond or more, and of acids related to these.

1. Stearolic family

One distinct group of these is characterized by eighteen (or seventeen) carbon atoms per molecule with an acetylenic bond in the ω9-position; this is accompanied by acetylenic or ethylenic bonds (usually in conjugated positions) further from the carboxyl group (Table 1.8). Their monohydroxyderivatives sometimes occur, the hydroxyl group always appearing in the ω10-position. They are found as seed-oil components, but are restricted to only two plant families (Santalaceae and Olacaceae, both in the order Santales), though fairly widely distributed within these families. The monoynoic acid which is the structural parent of this family is stearolic acid (octadec-9-ynoic acid; 18:1,9a). It has been isolated from the seeds of four species of the family Santalaceae, where it coexists with ximenynic acid (i.e. santalbic acid or *trans*-11-octadecen-9-ynoic acid; 18:2,9a11t) (Hopkins and Chisholm, 1964b; Morris and Marshall, 1966a). Ximenynic acid (24%) was originally found in the seed oils of *Ximenia caffra* (Ligthelm *et al.*, 1954) where it coexists with 3% 8-hydroxyximenynic acid (Ligthelm, 1954). These two acids are present in *Santalum album* seed oil (Gunstone and Sealy, 1963). Their lower homologues are also known: pyrulic acid (*trans*-10-heptadecen-8-ynoate; 17:2, 8a10t) was discovered in the seed oil of *Pyrularia pubera*,

TABLE 1.8

Acetylenic fatty acids of the stearolic family isolated from plant species of the families Olacaceae and Santalaceae[a]

	18:1	18:2	18:3	18:4
Monoynes	9a	9a11t[b,c] 9a17e	9a11e13e 9a11t17e[b,c]	
Diynes		9a11a[b]	9a11a13t[b] 9a11a13c[b] 9a11a17e[b]	9a11a13c15e 9a11a13c17e[b]
Triynes				9a11a13a15e 9a11a13a17e

[a] After Gunstone and Sealy (1963), Powell *et al.* (1966) and Hopkins *et al.* (1968). Common names of the acetylenic acids are as follows: 18:1(9a), stearolic; 18:2(9a11t), ximenynic; 18:3(9a11a17e), isanic; 18:3(9a11a13t), exocarpic; 18:4(9a11a13c17e), bolekic; 8h-18:3(9a11a17e), isanolic; 17:2(8a10t), pyrulic.

[b] The corresponding 8-hydroxyacid also occurs. There are 7 cases: 8h-18:2(9a11t); 8h-18:2(9a11a); 8h-18:3(9a11t17e); 8h-18:3(9a11a13t); 8h-18:3(9a11a13c); 8h-18:3(9a11a17e); 8h-18:4(9a11a13c17e).

[c] The analogous C_{17} acids also occur. There are 4 cases: 17:2(8a10t); 7h-17:2(8a10t); 17:3(8a10t16e); 7h-17:3(8a10t16e).

where stearolic acid is the major component (Hopkins *et al.*, 1967, 1968b); 7-hydroxypyrulic acid (6%) is present in the seed oil of *Acanthosyris spinescens* (Powell *et al.*, 1966) which also contains the unconjugated analogue 17-octadecen-9-ynoic acid (18%; 18:2,9a 17e) (Powell and Smith, 1966). This plant is also a source of *trans*-11,17-octadecadien-9-ynoic acid (4%; 18:3,9a 11t 17e) and its 8-hydroxyderivative (4%), together with their lower homologues *trans*-10,16-heptadecadien-8-ynoic acid (10%; 17:3,8a 10t 16e) and its 7-hydroxyderivative (9%).

Isano (or boleko) oil is expressed from the seeds of *Onguekoa* (or *Ongokea*) *gore*: it contains at least nine acetylenic fatty acids including four 8-hydroxyacids (Gunstone and Sealy, 1963; Morris, 1963). Of the total fatty acid content, 10% is 9,11-octadecadiynoic acid (18:2,9a 11a), and this is accompanied by its 8-hydroxyderivative (4–14%; 8h-18:2, 9a 11a) (Morris, 1963). Isano oil also contains 32% isanic acid (17-octadecen-9,11-diynoic acid; 18:3,9a 11a 17e) with 15–20% of its 8-hydroxyderivative, isanolic acid; 2% *cis*-13-octadecen-9,11-diynoic acid (18:3,9a 11a 13c; the *cis*-analogue of exocarpic acid described below) with 1% of its 8-hydroxyderivative; and 6% *cis*-13,17-octadecadien-9,11-diynoic acid (bolekic acid; 18:4,9a 11a 13c 17e) with1% of 8-hydroxybolekic acid. Ximenynic acid (18:2,9a 11t), or its *cis*-analogue, is also present at a 1% level. Exocarpic acid (*trans*-13-octadecen-9,11-diynoic acid; 18:3,9a 11a 13t) has been isolated from *Buckleya distichophylla* where it comprises 29% of the seed oil (Hopkins and Chisholm, 1966b). It was discovered in the roots of *Exocarpus cupressiformis* and *E. stricta* by Hatt *et al.* (1959); the seeds of these plants contain ximenynic acid but not exocarpic acid. In the somatic lipids of mature plants of the Santalaceae family, Bu'Lock and Smith (1963) and Hatt *et al.* (1959, 1960) have identified a number of related acids (18:3,9a 11t 13t; 18:3,9a11a 13t; 18:4,9a 11a 13e 15e; 18:4, 9a 11a 13a 15e; 18:5,9a 11a 13a 15e 17e).

2. *Crepenynic family*

The natural acetylenic fatty acids which do not contain the ω9-triple bond are by no means unimportant; they are not however structurally related to those which do. The ω6-triple bond is a feature of the non-conjugated crepenynic acid (*cis*-9-octadecen-12-ynoic acid, 18:2,9c 12a). This represents 60% of the fatty acids in the seeds of *Crepis foetida* (Mikolajczak *et al.*, 1964). It is present in other *Crepis* species, sometimes with vernolic acid; however some species contain no detectable crepenynate (Earle *et al.*, 1966). The enynoic acid also occurs in the fungus *Tricholoma grammopodium* (Bu'Lock and Smith, 1967), and the legume *Afzelia cuanzensis* (Gunstone *et al.*, 1967); in both cases it coexists with its dehydro-derivative *cis*-9,*cis*-14-octadecadien-12-ynoic acid (18:3, 9c 12a 14c). Helenynolic acid, extracted from *Helichrysum bracteatum* seeds by Powell *et al.* (1965), has the structure 9-hydroxy-*trans*-10-octadecen-12-ynoic acid (9h-18:2, 10t 12a). Also present in this seed oil is *cis*-9,10-epoxyoctadec-12-ynoic acid, as well as coronaric and crepenynic acids (Conacher and Gunstone, 1970).

Higher plants contain other groups of acetylenic compounds, and although few of them are fatty acids, their structures suggest a close relationship. One such group of C_{10} acetylenes is associated with plants of the family Compositae, sub-family Tubuliflorae (Sørensen, 1963). Crystals of dehydromatricaria ester (methyl *cis*-2-decen-4,6,8-triynoate) were observed in essential oil from the root of *Artemisia vulgaris* as long ago as 1826; the C_{10} polyacetylenic acids identified in Compositae are listed in Table 1.9. Their triple bond at the ω6-position corresponds to that in crepenynic acid. They occur together with many related polyacetylenes which are hydrocarbons, alcohols, glycols, chlorhydrins, aldehydes or ketones having between ten and seventeen carbon atoms per molecule; benzene, furan, oxiran or thiophen rings are occasionally present. They are found in most parts of the plant, the root usually being the richest source; the seed oils of Compositae are mainly normal glycerides containing commonplace fatty acids.

TABLE 1.9

C_{10} and C_{14} acetylenic fatty acids isolated from species of the family Compositae (sub-family Tubuliflorae)[a]

Present as methyl esters	Present as lactones	Present as isobutylamides
10:3(2c 4a 6a)	4h-10:3(2e 4e 6a)	10:4(2e 4a 6a 8a)
10:3(2t 4a 6a)	4h-10:4(2e 4e 6a 8e)	14:4(2e 4e 8a 10a)
8h-10:3(2e 4a 6a)		14:5(2e 4e 8a 10a 12e)
10:3(4a 6a 8c)		
10:4(2c 4a 6a 8c)		
10:4(2t 4a 6a 8c)		
10:4(2c 4a 6a 8a)		
10:4(2t 4a 6a 8a)		

[a] After Sørenson (1963) and Johnson (1965).

3. Other acetylenic acids

The simplest natural acetylenic fatty acid was also the first to be isolated (Arnaud, 1892) and proves to be the most unusual. Tariric acid (octadec-6-ynoic acid; 18:1,6a), found in the seed oil of several species of *Picramnia* is unique among acids so far discovered in the position of its triple bond. It coexists with *cis*- and *trans*-6-monoenoic acids (Spencer *et al.*, 1970).

Groups of polyacetylenes occur in plant families other than the Compositae, particularly the Umbelliferae and Araliaceae. They are hydrocarbons, alcohols and ketones, often with C_{17} or C_{13} chains. The corresponding acids seldom accumulate, though they may be involved in the metabolic processes.

4. Allenic acids

Fatty acids with vicinal double bonds are considered here, since the allenic grouping (.CH:C:CH.) is isomeric with the acetylenic grouping (.C:C.CH$_2$.); indeed chemical interconversion by movement of a proton is a possibility. Most known natural allenes have been isolated from fungi which also contain conjugated polyacetylenes; they have chains of intermediate length (C_8 to C_{13}) and the allenic function is usually also conjugated (Johnson, 1965). In higher plants, the characteristic allenic acid is labellenic acid, which probably has general occurrence within the subfamily Stachydoideae (Labiatae). It was isolated from the seed oil of *Leonotis hepetaefolia* (16%) and shown by Bagby *et al.* (1965b) to have the structure (−)5,6-octadecadienoic acid. *Lamium purpureum* seed oil contains 16% of the analogue 5,6-*trans*-16-octadecatrienoic acid (Mikolajczak *et al.*, 1967). The only other allene identified in higher plants is 8-hydroxy-5,6-octadienoate from *Sapium sebiferum* seeds (syn. *Stillingia sebifera*), where it is esterified at the hydroxyl group with *trans*-2,*cis*-4-decadienoic acid (p. 44; Sprecher *et al.*, 1965).

5. Furanoid acids

Disubstituted derivatives of furan are very rare in nature, but there are at least two unrelated sources of fatty acids incorporating this heterocyclic system. The shoots of the broad bean *Vicia faba* contain an antifungal acetylenic furanoid ketoester christened wyerone and characterized by Fawcett *et al.* (1968). Its systematic name is methyl 3-[5-(hept-*cis*-4-en-2-ynoyl)-2-furyl]prop-*trans*-2-enoate, and it can be regarded as the 4,7-epoxide of 8k-14:5(2t 4c 6c 9a 11c). Wyerone coexists with small proportions of the same derivative of 8h-14:5(2t 4c 6c 9a 11c) and of 8k-14:4(2t 4c 6c 9a). Another simpler furanoid acid is 8-(5-hexyl-2-furyl)octanoic acid, i.e. the 9,12-epoxide of 18:2(9c 11c). This was isolated from the seed oil of *Exocarpus cupressiformis* and identified by Morris *et al.* (1966). The structure was confirmed by synthesis (Elix and Sargent, 1968).

D. Substituted Acids

A number of hydroxyderivatives of conjugated and acetylenic fatty acids have already been described; here we consider other substituted fatty acids, with substitution of hydrogen by an oxygen function (e.g. hydroxyl group or oxiran ring). Substitution by a carbon function (e.g. methyl group or cyclopropene ring) leads to "branched-chain" acids (p. 28). Other substitutions are very uncommon; one such rarity occurs in the seeds of ratsbane (*Dichapetalum toxicarum*), which owe their

TABLE 1.10

Structures of monohydroxy-, keto-, polyhydroxy- and epoxy-derivatives of fatty acids found in plants[a]

1. *Monohydroxyacids*	2. *Ketoacids*
2h-acids[b]	15k-24:1(18c)
3h-16:0	17k-26:1(20c)
3h-18:0	19k-28:1(22c)
3h-20:0	8k-16:0
	13k-18:1(9t)
12h-18:1(9c)	
9h-18:1(12c)	3. *Polyhydroxyacids*
14h-20:1(11c)	
12h-18:2(9c 15c)	10h 16h-16:0
12h-16:1(9c)	10h 18h-18:0
	9h 10h 18h-18:0
8h-16:0	9h 10h-18:0
9h-16:0	11h 12h-20:0
9h-18:0	13h 14h-22:0
7h-10:0	15h 16h-24:0
11h-14:0	9h 10h 18h-18:1(12e)
11h-16:0	8h 9h 13h-22:0
	15h 16h-16:0
17h-18:0	2h 15h 16h-16:0
17h-18:1(9c)	9h 10h 12h 13h-22:0
16h-16:0	
15h-16:0	4. *Epoxyacids*
18h-18:0	
18h-18:1(9c)	9,10-epoxy-18:0
14h-14:0	12,13-epoxy-18:1(9c)
12h-12:0	15,16-epoxy-18:2(9c 12c)
	9,10-epoxy-18:1(12c)
	9,10-epoxy-18:1(12a)

[a] The distribution of these acids is described in the text in this order. Derivatives of conjugated ethylenic and acetylenic acids are not included (see Tables 1.7 and 1.8).
[b] See Table 1.11.

poisonous properties to the presence of ω-fluoroacids. 18-Fluoro-oleic acid is the principal unusual acid, and 16-fluoropalmitic acid is also present (Ward et al., 1964).

When long-chain oxyacids accumulate in plants, the lipids involved are significantly different in chemical and physical properties from their unsubstituted analogues. The hydroxyl group imparts considerable polarity to the molecule, and probably optical activity as well, and moreover is a reactive centre at which further biochemical transformations can take place. While classed as unusual structures, hydroxyacids are not altogether uncommon. They are bound through the carboxyl group as the triglycerides of a few seeds or as the amide in leaf cerebrosides; through the hydroxyl group in some plant glycosides; and through both functional groups in some waxes. Polyhydroxyacids, epoxyacids and ketoacids are also known to occur in plant lipids. We first deal with monohydroxyacids, starting with substitutions near the carboxyl group; then ketoacids, polyhydroxyacids and epoxyacids are considered (Table 1.10). The hydroxyderivatives of conjugated ethylenic (p. 14) and acetylenic (p. 17) acids have been discussed with their unsubstituted analogues.

1. Monohydroxyacids

The long-chain 2-hydroxyacids have a wide distribution in nature, being found in small quantities in animals, plants and micro-organisms; they could be regarded as "minor" rather than "unusual" fatty acids. The 2-hydroxyderivatives of commonplace fatty acids are associated with specific sphingolipids, particularly cerebroside, ceramide and phytoglycolipid, where they usually coexist with their homologues. Hence while they may be present in trace amounts they can still represent a significant proportion of the acids of an isolated lipid fraction. Though the 2-hydroxyderivatives of palmitate and stearate exist, longer chain derivatives are more common, including those of docosanoate, tetracosanoate and hexacosanoate together with lesser amounts of the homologues of odd carbon number. Table 1.11 illustrates their occurrence. Sastry and Kates (1964b) isolated from runner-bean leaves (Phaseolus multiflorus) a glucocerebroside fraction which contained only 2-hydroxyacids. Carter and Koob (1969) have analysed P. vulgaris leaves; their data on the fatty acid composition of the phytoglycolipid, cerebroside and ceramide are also given in Table 1.11, together with the composition of phytoglycolipid from flax seed (Linum usitatissimum). The 2-hydroxy-palmitate isolated from pea leaves (Pisum sativum) was optically active, and, like all other known natural long-chain 2-hydroxyacids, had the D-configuration (Hitchcock et al., 1968b).

TABLE 1.11

Fatty 2-hydroxyacids. Percentage composition of fatty acid mixtures obtained from some plant sphingolipids

Fatty acid	Phytoglycolipid		Ceramide	Cerebroside	
	Flax seed[a]	Bean leaf[b]	Bean leaf[b]	Bean leaf[b]	Bean leaf[c]
2h-16:0	4	—	—	1	30
2h-18:0	—	—	—	—	1
2h-20:0	—	—	—	—	2
20:0	—	—	—	3	—
2h-22:0	7	20	13	15	20
22:0	—	1	—	4	—
2h-23:0	7	6	6	4	5
2h-24:0	44	48	59	42	38
24:0	2	4	—	—	—
2h-25:0	17	6	7	4	2
2h-26:0	11	10	15	12	3
Others	8	6	—	14	—

[a] *Linum usitatissimum* (Carter and Koob, 1969).
[b] *Phaseolus vulgaris* (Carter and Koob, 1969).
[c] *Phaseolus multiflorus* (Sastry and Kates, 1964a,b,c).

Other more unusual D-2-hydroxyacids have been isolated from seed oils, where they occur as ordinary triglyceride. The seeds of *Pachira insignis* and *Bombacopsis glabra* contain 2-hydroxysterculic acid at about the 20% and 10% levels respectively (Morris and Hall, 1967). The oil of thyme seeds (*Thymus vulgaris*) contains 13% 2-hydroxy-linolenic acid (Smith and Wolff, 1969).

The 3-hydroxyacids are important metabolites in the biosynthesis and degradation of fatty acids, but rarely accumulate in the ubiquitous systems involving synthetase (p. 104) or β-oxidation (p. 201). Several species of the red yeast *Rhodotorula* produce extracellular glycolipids which consist of a mixture of mannitol and pentitol mono-esters of D-3-hydroxypalmitic and D-3-hydroxystearic acids, with the remaining hydroxyl groups acetylated (Tulloch and Spencer, 1964). *Nocardia asteroides* contains a peptolipid whose lipid moiety is D(−)-3-hydroxy-eicosanoate (Guinand and Michel, 1966).

Ricinoleic acid (12-hydroxyoleic acid) is a well-characterized component of commercial castor oil, accounting for up to 90% of the fatty acids of the seed oil of *Ricinus communis* (Canvin, 1963) where it occurs exclusively as ordinary triglyceride (Morris, 1967). It has the D(+)

configuration (Serck-Hanssen, 1958). Its only other major source is the parasitic fungus *Claviceps purpurea* (ergot), where it is present (up to 44%) as estolide triglyceride (Morris and Hall, 1966). An isomer of ricinoleate, 9-hydroxy-12-octadecenoate, occurs (7%) in *Strophanthus sarmentosus* seed oil (Gunstone, 1952); a homologue of ricinoleate, lesquerolate (14-hydroxy-*cis*-11-eicosenoate) is the predominant acid in *Lesquerella* oil. While *L. lasiocarpa* contains 40–45% and *L. lindheimerii* 51–72% (Smith *et al.*, 1961), *L. densipila* contains 38% of 12-hydroxy-*cis*-9,*cis*-15-octadecadienoic (densipolic) acid instead (Smith *et al.*, 1962), together with ricinoleic acid and small quantities of 12-hydroxypalmitoleic acid (Binder and Lee, 1966).

Oil from the spores of the clubmoss *Lycopodium complanatum* contains 7% of (+)8-hydroxypalmitate (Tulloch, 1965); 8-oxopalmitate is also present in small amounts (0·4%). The yeast *Pityrosporum ovale* requires long-chain fatty acids for growth; in the medium, a polar material accumulates which contains 9-hydroxypalmitate and 9-hydroxystearate. The former is a major product, and both may be present as glycolipid (Wilde and Stewart, 1968). Some hydroxyacids indeed occur as glycosides, with the hydroxy group carrying a sugar moiety. Such glycosides occur in various parts of plants belonging to the family Convolvulaceae; the acids involved include 7-hydroxydecanoate, 11-hydroxytetradecanoate and 11-hydroxyhexadecanoate (Legler, 1965).

Similar acidic glycosides are excreted by some yeasts (Stodola *et al.*, 1967); here the carboxyl group may also be linked to the sugar, forming cyclic lactonic glycosides (Tulloch *et al.*, 1968). The nature and amount of the hydroxyacids present depends on what additions are made to the medium: *Torulopsis apicola* (syn. *T. magnoliae*) normally excretes mainly L-17-hydroxystearate and L-17-hydroxyoleate, but when long-chain nutrients are available, products can include a range of homologous hydroxyacids substituted at the ultimate ω-position as well as at the penultimate (ω–1) position (Tulloch *et al.*, 1962). *Torulopsis gropengiesseri*, a similar if not identical yeast, converts added long-chain alkanes partly to ω-hydroxyacid, ω-carboxyacid and (ω–1)hydroxyacid; for instance fermentation on hexadecane leads to accumulation of 16-hydroxypalmitate, 1,16-hexadecandioate and 15-hydroxypalmitate in glycolipid, the yields from hexadecane being 34%, 7% and 27% respectively (Jones and Howe, 1968). Plant cutin contains 18-hydroxystearate and 18-hydroxyoleate as well as polyhydroxyacids (Matic, 1956). The wax from the leaves of the Colorado spruce (*Picea pungens*) hydrolyses to yield mainly ω-hydroxyacids: the major acid component is 14-hydroxymyristate, and it is accompanied by 12-hydroxylaurate and 16-hydroxypalmitate (von Rudloff, 1959).

2. Ketoacids

A few ketoacids have been reported to occur in plants; those which also have conjugated ethylenic bonds (4k-18:3,9c11t13t; 4k-18:4, 9c11t13t15c; 9k-18:2,10t12t; 13k-18:2,9t11t) are considered in a previous section (p. 14) with their unsubstituted analogues. One group which does not fall into this category is present in the seed oil of *Cuspidaria pterocarpa*, and was identified by Smith (1966) as 15-keto-*cis*-18-tetracosenoate (5%), 17-keto-*cis*-20-hexacosenoate (13%) and 19-keto-*cis*-22-octacosenoate (3%). The occurrence of 8-ketopalmitate is mentioned on p. 25, and of 13-keto-*trans*-octadecenoate on p. 16.

3. Polyhydroxyacids

The waxy outer surfaces of most higher plants display a wide diversity of chemical constituents, including fatty acids and primary alcohols, with their ω-hydroxy- and ω-carboxy-derivatives all mainly of even carbon number; also present are usually odd-numbered alkanes, secondary alcohols and ketones (Eglinton and Hamilton, 1967). Cutin (p. 57) contains 16-hydroxypalmitate, 10,16-dihydroxypalmitate, 18-hydroxystearate, 10,18-dihydroxystearate, 9,10,18-trihydroxystearate, 9,10 epoxy,18-hydroxystearate, and 18-hydroxyoleate (Matic, 1956; Baker and Martin, 1963).

A minor component of isano oil was identified as *threo*-9,10-dihydroxystearate (Morris, 1963); the *erythro*-isomer is present in castor oil (King, 1942). Saturated *erythro*-dihydroxyacids are also present in considerable amounts in the seed oil of *Cardamine impatiens*: this contains 1% 9,10-dihydroxystearate, 1% 11,12-dihydroxyeicosanoate, 16% 13,14-dihydroxydocosanoate and 6% 15,16-dihydroxytetracosanoate (Mikolajczak *et al.*, 1965). The unusual acids are found in the triglycerides at the 1- or 3-position of glycerol, with one hydroxyl group free; the other is acetylated (Mikolajczak *et al.*, 1968). Acetoxyacids have also been identified in the seed oil of *Chamaepeuce afra* by Mikolajczak and Smith (1967) who report the parent acids are (+)*threo*-9,10,18-trihydroxystearate (9%) and (+)*threo*-9,10,18-trihydroxy-12-octadecenoate (14%).

An extracellular lipid produced by a yeast has been identified as the triacetate of 8,9,13-trihydroxydocosanoic acid (Stodola *et al.*, 1965). When corn smut *Ustilago zeae* is cultured, extracellular crystals of a glucolipid are formed; this contains two ustilic acids A and B, characterized by Lemieux (1953) as D-15,16-dihydroxypalmitate and D-2,D-15,16-trihydroxypalmitate respectively.

Three tetrahydroxyacids have been reported to occur in lichens (Solberg, 1960): the principal one is 9,10,12,13-tetrahydroxydocosanoic acid, and it is sometimes accompanied by similarly substituted heneicosanoic and tricosanoic acids.

4. Epoxyacids

Epoxyacids have been isolated from the seed oil of plants from several families, including Compositae, Euphorbiaceae, Onagraceae, Dipsacaceae, Olacaceae and Valerianaceae. Their structures proved to resemble the more usual fatty acids, with the oxiran ring replacing one double bond. The seed oil of *Tragopogon porrifolius* contains *cis*-9,10-epoxystearate (Chisholm and Hopkins, 1959) as does the wheat stem rust (*Puccinia graminis*) uredospore (Tulloch, 1960). The configuration of the 3% in *Xeranthemum annuum* is L-9,L-10-epoxystearate (Powell *et al.*, 1967a). Isano oil (seed oil of the tree *Onguekoa gore*) contains 1·4% of this epoxyacid (Morris, 1963), and it accounts for 2% in the lipids of the clubmoss *Lycopodium complanatum*, where it occurs with 4% *threo*-9,10-dihydroxystearate (Tulloch, 1965).

Vernolic acid (*cis*-12,13-epoxyoleic acid) is found in the seeds of *Vernonia anthelmintica* at the 72% level (Gunstone, 1954; Tallent *et al.*, 1966a), and of other plants belonging to the families Compositae, Euphorbiaceae, Onagraceae, Valerianaceae and Dipsacaceae (Wolff, 1966). When eight *Crepis* species were examined by Earle *et al.* (1966), the vernolate content varied between 68% and zero. *Cis*-15,16-epoxylinoleic acid also occurs naturally in *Camelina sativa* seed oil (Gunstone and Morris, 1959). The absolute configuration of (+)vernolic acid is *cis*-D-12,D-13-epoxyoleic acid (Morris and Wharry, 1966). The (−)enantiomorph is present (1% to 7%) in six species of the family Malvaceae, though not in four others (Hopkins and Chisholm, 1960), nor has it been found elsewhere.

The isomer, coronaric acid (*cis*-9,10-epoxy-*cis*-12-octadecenoic acid) was isolated from *Chrysanthemum coronarium* (Smith *et al.*, 1960a), from *Aster alpinus* (Morris *et al.*, 1968), and shown to have the L-9,L-10-configuration by Powell *et al.* (1967) who found it at the 8% level in *Xeranthemum annuum*. The acetylenic analogue of coronaric acid, *cis*-9,10-epoxy-12-octadecynoic acid (p. 20) occurs in *Helichrysum bracteatum* seed oil, along with coronaric and other unusual acids (Conacher and Gunstone, 1970).

The fact that fresh samples of oil from *Helianthus annuus* contain only traces of oxygenated fatty acids, suggests that specific oxidation can take place on prolonged storage in air, since the oil then contained significant quantities, including 2% coronaric acid (Mikolajczak *et al.*, 1968a). This disturbing report emphasizes the necessity of using fresh or carefully stored samples in order to prevent the formation of artifacts.

5. Dicarboxylic acids

The fruit-coat fat of the sumach (*Rhus succedanea*) is apparently unique in containing a significant (6%) proportion of straight-chain dicarboxylic acids of general formula $HO_2C.[CH_2]_n.CO_2H$, esterified to glycerol at both ends (Hilditch and Williams, 1964). Hexadecandioic and similar acids occur in some yeasts (p. 25). Traumatic acid (2-dodecendioic acid) is associated with damaged plant tissue (p. 276).

E. Branched-chain Acids

While most natural fatty acids have a normal ("straight") carbon skeleton, there are several groups of natural fatty acids in which the carbon chain is branched or cyclic. Some methyl-branched fatty acids and those with long-chain branches are characteristic of bacteria, as are those containing a cyclopropane ring; however, a few plants contain acids with methyl branches or cyclopropene or cyclopentene rings. The distribution and biochemistry of such fatty acids has been reviewed by Abrahamsson *et al.* (1963), Hilditch and Williams (1964), Lennarz (1966) and Asselineau (1966).

1. Branched and cyclic acids from bacteria

Tubercle bacilli contain D(−)-10-methylstearic (tuberculostearic) acid, which represents about 10% of the total. Also present in this and other mycobacteria are smaller proportions of many branched fatty acids of greater or lesser complexity: three types can be recognized. Those with a single methyl substituent near the middle of the carbon chain include tuberculostearate, 10-methylmargarate, 10-methyl-palmitate and 8-methylpalmitate. Another type has two, three or four methyl substituents near the carboxyl group: examples are 2,4-dimethyl-docosanoate, 2,4,6-trimethyl-2-tetracosenoate and 2,4,6,8-tetramethyl-octacosanoate. Finally, the mycolic acids are 2-alkyl-3-hydroxyacids usually with 70 to 90 carbon atoms per molecule; a typical structure is $C_{53}H_{103}.CH(OH).CH(C_{24}H_{49}).CO_2H$.

Some eubacteria contain branched fatty acids, but here a single methyl substituent occurs near the hydrocarbon end of the molecule. The iso-acids, which have the isopropyl terminus, are (ω−1)-methyl-substituted acids typified by 13-methyltetradecanoate; the methyl group is also found in the anteiso-position, as with the (ω−2)-substituted 12-methyltetradecanoate. The iso- and anteiso-acids are characteristic of Gram-positive organisms. The unusual fatty acids of Gram-negative bacteria carry a cyclopropane ring; they are saturated, principally cis-11,12-methyleneoctadecanoic (lactobacillic) acid, cis-9,10-methyl-eneoctadecanoic (dihydrosterculic) acid and cis-9,10-methylenehexa-decanoic acid. The Gram-positive lactobacilli are exceptional in having alicyclic acids rather than iso-acids.

The presence of these branched acids in plants or animals is only significant if this is not due to contaminating bacteria. The iso- and anteiso-acids occur in sheep wool wax and human sebum; trace quantities are widely distributed in other animal fats. The uropygial or "preen" gland of geese, ducks and swans (but not chickens) excretes lipid contain-ing 2,4,6-trimethylundecanoate and similar branched acids (Haahti and Fales, 1967).

2. Branched and cyclic acids from plants

In plants, the occurrence of branched fatty acids is apparently rare. The simplest case is the accumulation of 2-methyl-2-butenoic acid: the cis-isomer (angelic acid) is found in the roots of Angelica archangelica, while the trans-form (tiglic acid) is present in root and seeds of Croton tiglium. Breakdown of terpenoids can lead to the formation of polymethyl isoprenoid fatty acids: thus phytanic acid (3,7,11,15-tetramethyl-palmitate) and pristanic acid (2,6,10,14-tetramethylpentadecanoate) are probably derived from the phytol present in chlorophyll. Radunz (1965) has reported the unexpected presence of a series of seven iso- and five anteiso-acids in Antirrhinum majus and mutants: 16-methylhepta-decanoate, 12-methyltridecanoate and optically active 14-methyl-palmitate were identified. The branched acids were found in normal cotyledons (3–11%), petals (7–9%), seeds (0·7%) and discoloured leaves of a mutant plant (4–6%); only traces were detected in mature green leaves.

Alicyclic fatty acids are components of the oils from certain trees and plants. Cyclopropanoid acids are present in Euphoria longana, where the major seed acid (17%) is 8-(2-octylcyclopropyl)octanoate (i.e. dihydrosterculate); traces (1%) of homologues were indicated (Kleiman et al., 1969b). In this case no cyclopropenoid acids were detected, but in Hibiscus syriacus dihydrosterculic coexists with sterculic and malvalic acids (Wilson et al., 1961), whose systematic names are 8-(2-octyl-cycloprop-1-enyl)octanoic and 7-(2-octylcycloprop-1-enyl)heptanoic acids respectively. The structure of sterculic acid is illustrated in Table 1.5. These three alicyclic acids are associated with seed and leaf oils of Malvaceae, Sterculiaceae, Bombacaceae and Tiliaceae, four families belonging to the order Malvales.

Sterculia foetida is exceptional in having a seed oil of high cyclopropene content in which sterculic acid predominates (53%) over malvalic acid

(10%); dihydrosterculic acid accounts for 0·4%. In *Hibiscus syriacus* these three components constitute 2%, 14% and 1% respectively; homologues are occasionally observed (Raju and Reiser, 1966). Dihydromalvalic acid is often absent, but has been detected in young seeds (Johnson *et al.*, 1967a). The cotton plant (*Gossypium hirsutum*) contains smaller quantities of alicyclic acids. Sterculynic acid (8,9-methylene-8-octadecen-17-ynoic acid) has been identified by Jevans and Hopkins (1968) at the 8% level in *Sterculia alata*. Substantial amounts of D-2-hydroxysterculate occur in the seed oils of *Pachira insignis* and *Bombacopsis glabra*; the latter oil contains about 10% of this hydroxyacid (Morris and Hall, 1967) together with 34% sterculate, 3% malvalate and 3% dihydrosterculate (Raju and Reiser, 1966).

The occurrence of cyclopentene-fatty acids seems to be confined to the family Flacourtiaceae where they are major constituents of the seed oil. The alicyclic ring is always at the methyl terminus of the molecule, and the principal examples are 11-(2-cyclopentenyl)-undecanoic (hydnocarpic) acid, 13-(2-cyclopentenyl)-tridecanoic (chaulmoogric) acid, and 13-(2-cyclopentenyl)-6-tridecenoic (gorlic) acid. These structures are those of palmitic, stearic and petroselinic acids respectively, with cyclization of the five terminal carbon atoms; hydnocarpic acid is illustrated in Table 1.5. Chaulmoogra oil from the seeds of *Hydnocarpus wightiana* is optically active and contains 49% hydnocarpic, 27% chaulmoogric and 12% gorlic acids, together with 3% of lower cyclic homologues (Cole and Cardoso, 1939).

5. Taxonomy

Over a quarter of a million different species of flowering plants have been described, as well as over 100,000 lower plants including about 10,000 algae. The classification of such a range is a science in itself, the classical approach to which is based on morphology and has led to the naming of plant species, each name covering all individual plants which appear identical or nearly so, and which interbreed freely to produce fertile offspring; varieties within species are recognized when small but consistent differences become evident. Similar species may be arranged in larger groups (genera), which are assembled into families, and so on, in orders, classes and divisions of the plant kingdom. Higher plants are identified in this chapter by the accepted botanical system involving the first three ranks. Such classification is possible because of the long evolutionary history of the plant kingdom, during which the different plant forms evolved in sequence and in parallel; primitive biochemical systems gave way to more complex metabolism, with consequent changes in metabolites as well as in the morphology of the plant. There is therefore an important interaction between phylogeny and taxonomy, and the chemical composition of a plant can be relevant to its phylogenetic and taxonomic relationships. Now that accumulated metabolites can be conveniently identified, plants can be rapidly screened and analytical data can contribute to the solution of taxonomic problems. Sometimes the presence of certain compounds is closely correlated with accepted botanical classification, as with the cyclopentenoid acids in the seeds of Flacourtiaceae species (p. 30): here chemical and morphological characters have evolved together. On the other hand, some plants from very different families contain similar compounds, such as the conjugated ethylenic fatty acids (p. 14), where the ability to synthesize these acids has apparently been developed at different stages of evolution.

Some plant materials are of such wide occurrence that their taxonomic value is negligible; moreover the composition of a particular plant can depend on genetic history as well as on external environment (p. 261). For example in trials using a single variety of safflower (*Carthamus tinctorius*) carried out by Horowitz and Winter (1957), most seed oils contained 72–79% linoleic and 22–15% oleic acids. However two individual plants gave oils with 11–19% linoleic and 79–74% oleic acids, and the progeny of one of these continued to give similar seed oil for four generations. The influence of temperature is discussed on p. 153; in general, oil from plants grown at higher temperatures is more saturated. Lack of oxygen has a similar effect, which is more obvious in lower non-photosynthetic organisms which cannot generate oxygen for themselves. The taxonomic significance of fatty acid composition must therefore be treated with some reserve.

In this chapter, we have described the sources of each fatty acid in an order depending on its structure and biochemistry. We now briefly consider whether this distribution can lead to a classification of the sources.

A. Lower Plants

Algal lipids usually contain ordinary fatty acids of the "major" and "minor" classes, and are sometimes a particularly good source of "minor" polyunsaturated acids. The classification of algae on this basis suffers from similar disadvantages to those encountered with some seeds: reliance must be placed on differences in the quantitative concentrations of the same acids, which may vary considerably according to culture conditions. However some generalizations are apparent (Nichols, 1969; Nichols and James, 1968).

Green freshwater algae grown photoautotrophically exhibit a fatty acid composition similar to that of many leaves, except that the proportion of α-linolenic acid is somewhat lower (Tables 3.8 and 3.13; p. 76). Heterotrophic growth leads to a great reduction in the accumulation of polyenoic acids. Marine algae, the green Chlorophyceae, the red Rhodophyceae and the brown Phaeophyceae and diatoms (Bacillariophyceae) differ from freshwater algae in synthesizing C_{20} and C_{22} polyenoic acids (Table 1.4). Blue-green algae (Cyanophyceae) unlike other cells which perform the Hill reaction, frequently contain little or no polyenoic acids and no *trans*-3-hexadecenoate.

Polyunsaturated fatty acids may be biosynthesized by the α-linolenic or γ-linolenic pathways (p. 141; Fig. 5.18), and this provides a basis for a distinction between plant classes, since the pathways involve different intermediate acids which accumulate. This classification depends on

TABLE 1.12

Classification of algae based on their fatty acid composition and lipid metabolism[a]

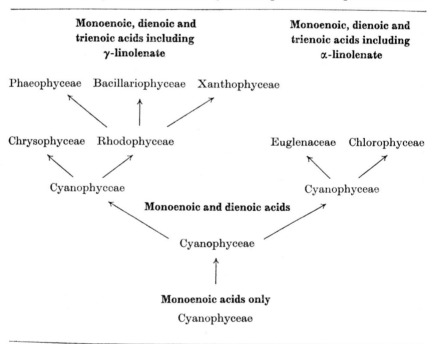

[a] After Nichols (1969).

whether α-linolenic acid or γ-linolenic acid is present; the former is typical of higher plants and algae, while γ-linolenate and arachidonate are associated with higher animals, ciliates and amoebae (p. 141). Some organisms accumulate acids of both pathways, the relative importance of which may be affected by growth conditions; other organisms contain acids of neither pathway. When the trienoic acid content of three blue-green algae is examined (Nichols, 1969), *Anabaena cylindrica* is found to contain 11% α-linolenate, while *Spirulina platensis* contains 21% γ-linolenate; *Anacystis nidulans* contains neither. Thus a general classification based on fatty acid composition is possible (Table 1.12).

A possible evolutionary significance of the polyunsaturated fatty acids in blue-green algae has been suggested by Kenyon and Stanier (1970). Such considerations imply a pattern of plant evolution which corresponds fairly well to conventional views. Many bacteria contain only monoenoic acids, synthesized by the "anaerobic" pathway; later, an "aerobic" pathway evolved (p. 130). If bacteria were the precursors of the plant kingdom, it seems likely the most primitive algae were those blue-greens which contain only saturated and monoenoic acids. It is possible to arrange certain species (or groups of species) in order of their capacity to synthesize fatty acids of increasing degrees of unsaturation, and it is not unreasonable to suppose that such an arrangement is also consistent with their order of evolutionary development.

B. Higher Plants

Somatic lipids are generally composed of commonplace fatty acids, but the acid of stored lipid sometimes contains unusual components which invite classification. The fatty acid composition of whole leaves or their chloroplasts show a remarkably consistent pattern over a wide variety of plant families, but the leaf cuticular waxes show variations with taxonomic possibilities (Mazliak, 1968). Fruit coat acids may be useful when more analytical results are available; seed oils provide the taxonomist with most data.

Hilditch and Williams (1964) base their classification of all seed oils primarily on the principal component acids although they point out that this does not always lead to clearly defined groups nor to a unique grouping for a particular botanical family. However, if analytical data are available, most families can be conveniently classified in this way (Shorland, 1963). The usefulness of such a classification (Table 1.13) is limited by two considerations. Firstly, the large number of seeds in which the "major" acids predominate makes it necessary to group them

2

TABLE 1.13

Classification of seed oils based on their fatty acid composition[a]

Class[b]	Group[c]	Principal acids[d]	Numbers known	Representative families[a,e] Examples
1		MAJOR ACIDS ONLY		
1.1	Ia	Linolenate-rich	15	Labiatae, Linaceae, etc.
1.2	Ib	Linoleate-rich	35	Compositae, Dipsacaceae, Olacaceae, Onagraceae, Ranunculaceae, Scrophulariaceae, etc.
1.3	—	Oleate-rich	0[f]	
1.4	IVh	Stearate-rich	10	Convolvulaceae, Sterculiaceae, etc.
1.5	III	Palmitate-rich	28	Acanthaceae, Apocynaceae, Bombacaceae, Caprifoliaceae, Malvaceae, Tiliaceae, etc.
1.6	IVj	Myristate/Laurate-rich	6	Lauraceae, Palmae, etc.
2		MAJOR ACIDS WITH CHARACTERISTIC MINOR ACIDS		
2.1		*Saturated acids*		
2.1.1	IVj	10:0	3	Lauraceae, Ulmaceae, etc.
2.1.2	IVg	20:0, 22:0, etc.	5	Leguminosae, Sapindaceae, etc.
2.2		*Unsaturated acids*		
2.2.1	—	Δ9-Family (16:1)	2	Bignoniaceae, Asclepidaceae
2.2.2	—	ω9-Family	6	
	IVd	20:1		Buxaceae, Cruciferae, Olacaceae, Sapindaceae
	IVe	22:1		Cruciferae, Tropaeolaceae
	IVf	24:1, 26:1, etc.		Cruciferae, Leguminosae, Olacaceae, Tropaeolaceae
2.2.3	—	ω6-Family (18:3)	1	Boraginaceae
2.2.4	—	ω3-Family	0	
3		MAJOR ACIDS WITH CHARACTERISTIC UNUSUAL ACIDS		
3.1		*Non-conjugated ethylenic acids*		
3.1.1	—	16:1(3t), etc.	2	Bignoniaceae, Compositae

		Group[c]	[d]	[d]
3.1.2	12:1(4c)	—	1	Lauraceae
3.1.3	18:1(5c), 20:1(5c), 18:3(5t,9c,12c), etc.	—	4	Compositae, Ephedraceae, Limnanthaceae, Ranunculaceae
3.1.4	18:1(6c)	—	3	Araliaceae, Simaroubaceae, Umbelliferae
3.1.5	18:1(11c), etc.	—	2	Asclepidaceae, Bignoniaceae
3.2	*Conjugated ethylenic acids*			
3.2.1	9c-Family	II	10	Balsaminaceae, Chrysobalanaceae, Compositae, Coriariaceae, Cucurbitaceae, Euphorbiaceae, Polygalaceae, Punicaceae, Rosaceae, Valerianaceae
3.2.2	12c-Family	II	2	Bignoniaceae, Compositae
3.2.3	9t-Family	II	1	Bignoniaceae
3.2.4	12t-Family	II	2	Bignoniaceae, Compositae
3.3	*Acetylenic acids*			
3.3.1	Stearolic family	IVb	2	Olacaceae, Santalaceae
3.3.2	Crepenynic family	IVb	1	Compositae
3.3.3	18:1(6a)	IVb	1	Simaroubaceae
3.3.4	Allenic acids	—	1	Labiatae
3.4	*Substituted acids*			
3.4.1	12h-18:1(9c), 14h-18:1(11c), 12h-18:2(9c,15c), etc.	—	2	Cruciferae, Euphorbiaceae
3.4.2	Epoxy acids	—	5	Compositae, Euphorbiaceae, Dipsacaceae, Onagraceae, Valerianaceae
3.5	*Branched-chain acids*			
3.5.1	Cyclopropanoid acids	IVc	1	Sapindaceae
3.5.2	Cyclopropenoid acids	IVc	4	Bombacaceae, Malvaceae, Sterculiaceae, Tiliaceae
3.5.3	Cyclopentenoid acids	IVc	1	Flacourtiaceae

[a] Re-classification adapted and extended from Hilditch and Williams (1964) and Shorland (1963).

[b] The Classes are based on the characteristic fatty acids present, and correspond to the classification of acids followed in this chapter.

[c] The Groups according to Hilditch's original classification (also based on fatty acid composition) are given; Groups I, II and III correspond to drying, semi-drying and non-drying oils respectively.

[d] Characteristic acids are usually present in quantities greater than about 10% of the total seed oil acids of at least one species of the given family.

[e] See Table 1.14 for an index.

[f] Some fruit-coat oils are oleate-rich: e.g. olive oil (65–86%); palm oil (40–52%).

according to quantitative data on the relative importance of these acids, which may be open to uncertainty. Secondly, some species that are closely related by their botany give seed oils which are not related by their chemistry and *vice versa*. For instance, some rapeseed oils contain no erucic acid, which is typical of others (Table 3.3, p.63); conversely the Lauraceae and Palmae are unrelated families whose species have similar seed fats rich in lauric acid. Evidently the nature of the plant is not absolutely correlated to the biochemical system which yields a particular seed lipid. The classification rests on the composition of a biological fuel whose role is to be stored and degraded, and whose structure is not critical. These limitations may well become more significant as new strains of plants are developed for novel seed-oils.

Tables 1.14 and 1.15 comprise an index of higher plants whose fatty acids are considered in this chapter; they are listed in alphabetic order of genera and of families.

TABLE 1.14

Index of plants by families[a]

Acanthaceae 1.5 (acanthus family) spp.: 34
Anacardiaceae 1.5 (cashew family)
  Rhus succedanea (sumach): 28
Apocynaceae 1.5 (dogbane family) spp.: 34
  Strophanthus sarmentosus: 25
Araliaceae 3.1.4 (aralia or ginseng family) spp.: 12, 21, 35
  Hedera helix (English ivy): 12
Asclepidaceae 2.2.1, 3.1.5 (milkweed family) spp.: 34, 35
  Asclepias syriaca (common milkweed or silkweed): 5, 13
Balsaminaceae 3.2.1 (touch-me-not family) spp.: 35
Bignoniaceae 2.2.1, 3.1.1, 3.1.5, 3.2.2, 3.2.3, 3.2.4 (bignonia family) spp.: 34, 35
  Catalpa ovata (Indian bean): 16
  Chilopsis linearis (desert-willow or flowering willow): 13, 17
  Cuspidaria pterocarpa: 26
  Doxantha unguis-cati (cat's claw): 5, 13
  Jacaranda mimosifolia (syn. *J. acutifolia*, *J. ovalifolia*; jacaranda): 16
  Tecoma stans (syn. *Stenolobium stans*; yellow-bells, trumpet flower): 12
Bombacaceae 1.5, 3.5.2 (bombax family) spp.: 29, 34, 35
  Bombacopsis glabra: 24, 30
  Pachira insignis: 24, 30
Boraginaceae 2.2.3 (borage family) spp.: 7, 34
Buxaceae 2.2.2 (box family) spp.: 34
Caprifoliaceae 1.5 (honeysuckle family) spp.: 34
Chailletaceae *see* Dichapetalaceae
Chrysobalanaceae 3.2.1 spp.: 35
  Chrysobalanus icaco (coco plum); 15

Compositae 1.2, 3.1.1, 3.1.3, 3.2.1, 3.2.2, 3.2.4, 3.3.2, 3.4.2 (daisy family) spp.:
   17, 20, 27, 34, 35
   Arctium minus (common burdock): 12
   Artemisia vulgaris (common mugwort): 20
   Aster alpinus (aster, starwort, Michaelmas daisy): 12, 27
   Calea urticaefolia: 12
   Calendula officinalis (pot-marigold): 16
   Carlina acaulis (weather-thistle, silver-thistle): 12
   Carthamus tinctorius (safflower, false saffron): 31
   Chamaepeuce afra: 26
   Chrysanthemum coronarium (garland chrysanthemum, crown daisy): 27
   Crepis spp.: 19, 27
   Crepis foetida: 19, 64
   Crepis rubra: 13
   Dimorphotheca sinuata, (syn. *D. aurantiaca D. calendulacea*; cape-marigold):
      13, 15, 16
   Grindelia oxylepsis (gumweed, tarweed): 12
   Helenium bigelowii (sneezeweed): 12, 64
   Helenium hoopesii (sneezeweed): 12
   Helianthus spp. (sunflower): 4
   Helianthus annuus (common sunflower): 27
   Helichrysum bracteatum (strawflower): 20, 27
   Osteospermum hyoseroides: 16
   Tragopogon porrifolius (salsify, vegetable-oyster, oyster-plant): 27
   Vernonia anthelmintica (ironweed): 27, 64
   Xeranthemum annuum (common immortelle): 15, 16, 27, 27
Convolvulaceae 1.4 (convolvulus family) spp.: 25, 34
Coriariaceae 3.2.1 spp.: 35
   Coriaria nepalensis: 15
Cruciferae 2.2.2, 3.4.1 (mustard family) spp.: 7, 34, 35
   Brassica napus (rape, colza): 7, 7, 13, 36, 63
   Camelina sativa (cf. cameline, dodder or German sesame oil): 27
   Cardamine impatiens: 26
   Lesquerella lasiocarpa: 25
   Lesquerella densipila: 25
   Lesquerella lindheimerii: 25
Cucurbitaceae 3.2.1 (gourd family) spp.: 15, 35
Dichapetalaceae (syn. Chailletaceae)
   Dichapetalum toxicarum (ratsbane): 22
Dipsacaceae 1.2, 3.4.2 (teasel family) spp.: 27, 34, 35
Ephedraceae 3.1.3 spp.: 35
   Ephedra campylopoda: 7, 8, 12
Euphorbiaceae 3.2.1, 3.4.1, 3.4.2 (spurge family) spp.: 27, 35
   Aleurites fordii (China wood-oil tree, tung-oil tree): 15
   Aleurites montana (China wood-oil tree, tung-oil tree): 15, 64
   Croton tiglium: 29
   Mallotus phillipensis (Kamala tree): 15
   Ricinus communis (castor bean, castor-oil plant): 24, 64
   Sapium sebiferum (syn. *Stillingia sebifera*; Chinese tallow-tree): 17, 21
   Sebastiana lingustrina: 17

Euphorbiaceae—*continued*
  Stillingia sebifera: see *Sapium sebiferum*
Flacourtiaceae 3.5.3 (flacourtia family) spp. 30, 31, 35
  Hydnocarpus wightiana (cf. Chaulmoogra oil): 30, 64
Gramineae 1.5 (grass family)
  Triticum vulgare (wheat): 63
  Zea mays (maize): 63
Labiatae 1.1, 3.3.4 (mint family) spp.: 21, 34, 35
  Lamium purpureum (deadnettle): 21
  Leonotis hepetaefolia (lion's-ear): 21
  Thymus vulgaris (common thyme): 13, 24
Lauraceae 1.6, 2.1.1, 3.1.2 (laurel family) spp.: 34, 35, 36
  Lindera umbellata (syn. *Benzoin umbellatum*): 12
Leguminosae 2.1.2, 2.2.2 (pulse family) spp.: 17, 34
  Afzelia cuanzensis: 19
  Arachis hypogaea (peanut, groundnut): 63
  Phaselous multiflorus (syn. *P. coccineus*; runner bean, multiflora bean, scarlet runner): 23, 24
  Phaseolus vulgaris (kidney bean, French bean): 23, 24
  Pisum sativum (garden pea): 23
  Soja hispida (syn. *Glyine hispida*; soya bean, soy bean): 63
  Vicia faba (broad bean): 21
Limnanthaceae 3.1.3 (limnanthes or false mermaid family) spp.: 35
  Limnanthes douglasii (meadow foam): 12
Linaceae 1.1 (flax family) spp.: 34
  Linum usitatissimum (flax): 23
Malvaceae 1.5, 3.5.2 (mallow family) spp.: 29, 34, 35
  Gossypium hirsutum (upland cotton): 30
  Hibiscus syriacus (bush mallow or shrubby althea): 29, 30
Olacaceae 1.2, 2.2.2, 3.3.1 spp.: 12, 27, 34, 35
  Onguekoa gore (syn. *Ongokea gore*; cf. isano or boleko oil): 19, 27
  Ximenia caffra: 7, 18
Onagraceae 1.2, 3.4.2 (evening primrose family) spp.: 27, 34, 35
Palmae 1.6 (palm family) spp.: 34, 36
  Cocus nucifera (coconut): 4
  Elaeis guineensis (oil palm): 61, 62
Pinaceae 1.1 (pine family)
  Picea pungens (Colorado spruce): 25
Polygalaceae 3.2.1 (milkwort family) spp.: 35
  Monnina emarginata: 15, 16
Punicaceae 3.2.1 (pomegranate family) spp.: 35
  Punica granatum (pomegranate): 15
Ranunculaceae 1.2, 3.1.3 (buttercup family) spp.: 34, 35
  Caltha palustris (marsh marigold, kingcup): 12
  Thalictrum venulosum (meadow rue): 12
Rosaceae 3.2.1 (rose family) spp. 35
Santalaceae 3.3.1 (sandalwood family) spp.: 18, 19, 35
  Acanthosyris spinescens: 19
  Buckleya distichophylla: 19
  Exocarpus cupressiformis: 19, 21

Santalaceae—*continued*
  Exocarpus stricta: 19
  Pyrulia pubera (oil-nut, buffalo-nut): 18
  Santalum album: 18
Sapindaceae 2.1.2, 2.2.2, 3.5.1 (soapberry family) spp.: 34, 35
  Cardiospermum halicacabum (balloon-vine): 5
  Euphoria longana (longan, lungan): 29
Scrophulariaceae 1.2 (figwort family) spp.: 34
  Antirrhinum majus (snapdragon): 29
Simaroubaceae syn. Simarubaceae 3.1.4, 3.3.3 (quassia family) spp.: 35
  Picramnia spp.: 13, 21
  Picramnia sellowii: 13
Stachydoideae (sub-family) *see* Labiatae
Sterculiaceae 1.4, 3.5.2 (sterculia family) spp.: 29, 34, 35
  Sterculia alata: 30
  Sterculia foetida: 29, 64
Tiliaceae 1.5, 3.5.2 (linden family) spp. :29, 34, 35
Tropaeolaceae 2.2.2 (tropaeolum family) spp.: 7, 34
  Tropaeolum majus (nasturtium): 7
  Tropaeolum speciosum (nasturtium): 7
Tubuliflorae (sub-family) *see* Compositae spp.: 20
Ulmaceae 2.1.1 (elm family) spp.: 34
Umbelliferae 3.1.4 (parsley family) spp.: 12, 13, 17, 21, 35,
  Anethum graveolens (dill): 12
  Angelica archangelica: 29
  Apium graveolens (celery): 12
  Carum carvi (caraway): 12
  Chaerophyllum sativum (chervil): 12
  Daucus carota (carrot): 12
  Foeniculum vulgare (fennel): 12
  Pastinaca sativa (parsnip): 12
  Petroselinum sativum (parsley): 12, 64
Valerianaceae 3.2.1, 3.4.2 (valerian family) spp.: 27, 35
  Centranthus macrosiphon see *Kentranthus*
  Kentranthus macrosiphon: 15

---

[a] Included are all plants whose seed fatty acids are considered in Chapter 1 (and in Tables 3.2, 3.3 and 3.4). Plant families are listed in alphabetic order followed by the class according to Table 1.13. Named species are listed in alphabetic order under the appropriate family, preceded by any unnamed species; common names are given in brackets. References are page numbers.

TABLE 1.15

Index of plants by genera[a]

Acanthosyris spinescens (Santalaceae): 19
Afzelia cuanzensis (Leguminosae): 19
Aleurites fordii (Euphorbiaceae): 15
Aleurites montana (Euphorbiaceae): 15, 64
Anethum graveolens (Umbelliferae): 12
Angelica archangelica (Umbelliferae): 29
Antirrhinum majus (Scrophulariaceae): 29
Apium graveolens (Umbelliferae): 12
Arachis hypogaea (Leguminosae): 63
Arctium minus (Compositae): 12
Artemisia vulgaris (Compositae): 20
Asclepias syriaca (Asclepidaceae): 5, 13
Aster alpinus (Compositae): 27, 12
Bombacopsis glabra (Bombacaceae): 24, 30
Brassica napus (Cruciferae): 7, 7, 13, 36, 63
Buckleya distichophylla (Santalaceae): 19
Calea urticaefolia (Compositae): 12
Calendula officinalis (Compositae): 16
Caltha palustris (Ranunculaceae): 12
Camelina sativa (Cruciferae): 27
Cardamine impatiens (Cruciferae): 26
Cardiospermum halicacabum (Sapindaceae): 5
Carlina acaulis (Compositae): 12
Carthamus tinctorius (Compositae): 31
Carum carvi (Umbelliferae): 12
Catalpa ovata (Bignoniaceae): 16
Centranthus macrosiphon see *Kentranthus*
Chaerophyllum sativum (Umbelliferae): 12
Chamaepeuce afra (Compositae): 26
Chilopsis linearis (Bignoniaceae): 13,17
Chrysanthemum coronarium (Compositae): 27
Chrysobalanus icaco (Chrysobalanaceae): 15
Cocus nucifera (Palmae): 4
Coriaria nepalensis (Coriariaceae): 15
Crepis spp. (Compositae): 19, 27
Crepis foetida (Compositae): 19,64
Crepis rubra (Compositae): 13
Croton tiglium (Euphorbiaceae): 29
Cuspidaria pterocarpa (Bignoniaceae): 26
Daucus carota (Umbelliferae): 12
Dichapetalum toxicarum (Dichapetalaceae): 22
Dimorphotheca sinuata syn. *D. aurantiaca* (Compositae): 13, 15, 16
Doxantha unguis-cati (Bignoniaceae): 5, 13
Elaeis guineensis (Palmae): 61, 62
Ephedra campylopoda (Ephedraceae): 7, 8, 12

Euphoria longana   (Sapindaceae): 29
Exocarpus cupressiformis   (Santalaceae): 19, 21
Exocarpus stricta   (Santalaceae): 19
Foeniculum vulgare   (Umbelliferae): 12
Gossypium hirsutum   (Malvaceae): 30
Grindelia oxylepsis   (Compositae): 12
Hedera helix   (Araliaceae): 12
Helenium bigelowii   (Compositae): 12, 64
Helenium hoopesii   (Compositae): 12
Helianthus spp.   (Compositae): 4
Helianthus annuus   (Compositae): 27
Helichrysum bracteatum   (Compositae): 20, 27
Hibiscus syriacus   (Malvaceae): 29, 30
Hydnocarpus wightiana   (Flacourtiaceae): 30, 64
Jacaranda mimosifolia syn. *J. acutifolia* or *J. ovalifolia*
    (Bignoniaceae): 16
Kentranthus macrosiphon   (Valerianaceae): 15
Lamium purpureum   (Labiatae): 21
Leonotis hepetaefolia   (Labiatae): 21
Lesquerella lasiocarpa   (Cruciferae): 25
Lesquerella densipila   (Cruciferae): 25
Lesquerella lindheimerii   (Cruciferae): 25
Limnanthes douglasii   (Limnanthaceae): 12
Lindera umbellata   (Lauraceae): 12
Linum usitatissimum   (Linaceae): 23, 24
Mallotus phillipensis   (Euphorbiaceae): 15
Monnina emarginata   (Polygalaceae): 15, 16
Onguekoa gore syn. *Ongokea gore*   (Olacaceae): 19, 27
Osteospermum hyoseroides   (Compositae): 16
Pachira insignis   (Bombacaceae): 24, 30
Pastinaca sativa   (Umbelliferae): 12
Petroselinum sativum   (Umbelliferae): 12, 64
Phaseolus multiflorus syn. *P. coccineus*   (Leguminosae): 23, 24
Phaseolus vulgaris   (Leguminosae): 23, 24
Picea pungens   (Pinaceae): 25
Picramnia spp.   (Simaroubaceae): 13, 21
Picramnia sellowii   (Simaroubaceae): 13
Pisum sativum   (Leguminosae): 23
Punica granatum   (Punicaceae): 15
Pyrulia pubera   (Santalaceae): 18
Rhus succedanea   (Anacardiaceae): 28
Ricinus communis   (Euphorbiaceae): 24, 64
Santalum album   (Santalaceae): 18
Sapium sebiferum syn. *Stillingia sebifera*   (Euphorbiaceae):
    17, 21
Sebestiana lingustrina   (Euphorbiaceae): 17
Soja hispida syn. *Glycine hispida*   (Leguminosae): 16
Sterculia alata   (Sterculiaceae): 30
Sterculia foetida   (Sterculiaceae): 29, 64
Stillingia sebifera see *Sapium sebiferum*

Strophanthus sarmentosus (Apocynaceae): 25
Tecoma stans (Bignoniaceae): 12
Thalictrum venulosum (Ranunculaceae): 12
Thymus vulgaris (Labiatae): 13, 24
Tragopogon porrifolius (Compositae): 27
Triticum vulgare (Graminae): 63
Tropaeolum majus (Tropaeolaceae): 7
Tropaeolum speciosum (Tropaeolaceae): 7
Vernonia anthelmintica (Compositae): 27, 64
Vicia faba (Leguminosae): 21
Xeranthemum annuum (Compositae): 15, 16, 27, 27
Ximenia caffra (Olacaceae): 7, 18
Zea mays (Gramineae): 63

---

a Included are all named species whose seed fatty acids are considered in Chapter 1 (and in Tables 3.2, 3.3 and 3.4). Plant genera are listed in alphabetic order, with family names in brackets. References are page numbers. Common names are given in Table 1.14.

CHAPTER 2

Plant Acyl Lipids

The acyl lipids of plants comprise a wide variety of different molecular structures of which the majority contain fatty acid ester groups, although in a few cases the acids are present as amide derivatives. It is the purpose of this chapter to describe the general structure of these fatty acid-containing lipids, leaving the detailed consideration of the nature and arrangement of their constituent acyl groups for subsequent sections.

1. THE GLYCERIDES

By far the largest group of plant lipids, the glycerides, are based on the trihydric alcohol, glycerol. Of these the structurally simplest member of importance is triglyceride (I, Fig. 2.1), in which all three hydroxyl groups of the alcohol are esterified with fatty acids. This arrangement permits both structural and optical isomerism, and these aspects of triglyceride chemistry are discussed in a later section.

$$\begin{array}{lll}
CH_2O.CO.R^1 & CH_2O.CO.R^1 & CH_2O.CO.R^1 \\
| & | & | \\
R^2CO.O.CH & R^2CO.O.CH & CH_2OH \\
| & | & | \\
CH_2O.CO.R^3 & CH_2OH & CH_2O.CO.R^2 \\
\text{I. Triglyceride} & \text{II. 1,2-Diglyceride} & \text{III. 1,3-Diglyceride}
\end{array}$$

$CH_2O.CO.[CH_2]_7CH{=}CH.CH_2CH{=}CH.CH_2CH{=}CH.CH_2CH_2$
|
$CH.O.CO.[CH_2]_7CH{=}CH.CH_2CH{=}CH[CH_2]_4CH_3$
|
$CH_2O.CO.[CH_2]_3CH{=}C{=}CH.CH_2O.CO.CH{=}CH.CH{=}CH.CH{=}CH.[CH_2]_4CH_3$

IV. Tetraacid triglyceride from *Sapium sebiferum* oil (Sprecher *et al.*, 1965)

FIG. 2.1 (I–IV) Neutral glycerides [R^1, R^2, R^3, etc. indicate alkyl groups].

An additional variation in triglyceride structure can occur in those molecular species which contain a hydroxy acid, since the free hydroxyl group of the acid permits interesterification with further fatty acids. This type of glyceride occurs only rarely in plants, but Sprecher *et al.*

(1965) have shown that the seed oil of *Sapium sebiferum* contains a tetraester glyceride in which 8-hydroxy-5,6-octadienoic acid is esterified through its carboxyl group to glycerol, and to 2,4-decadienoic acid through the ω-hydroxyl group (IV, Fig. 2.1).

In diglycerides (II, III, Fig. 2.1) only two of the hydroxyl groups of glycerol are esterified, and although this class of compound seldom occurs in large quantities in plant cells, 1,2-diglycerides are important intermediates in the biosynthesis of other classes of glyceride (p. 176).

A. Phosphoglycerides

Plant phosphoglycerides are structurally related to 1.2-diacyl-*sn*-glycero-3-phosphoric acid (I, Fig. 2.2), more commonly known as phosphatidic acid, and are optically active. Their optical antipodes do not occur naturally. Phosphatidic acid itself is seldom a major component of lipid extracts from plant tissues although it is, like 1,2-diglyceride, an important intermediate in lipid biosynthesis (p. 176).

1. Phosphatidyl choline (lecithin)

Phosphatidyl choline, the choline ester of phosphatidic acid (II, Fig. 2.2) was the first plant phospholipid to be completely characterized. The universally employed trivial name for this lipid, lecithin, is derived from the Greek word for egg-yolk (lekithes) the original source from which this lipid was characterized, although in the past "lecithin" has also been used to denote the soluble fraction, rich in phosphatidyl choline, which is obtained from crude plant phospholipids by precipitation with ethanol.

2. Phosphatidyl ethanolamine

The widely distributed ethanolamine ester of phosphatidic acid (III, Fig. 2.2) has often been called "cephalin" because of its preponderance in the alcohol-insoluble fraction of oil-seed phosphatides (see above), although this usage is now uncommon. More recently N-acyl derivatives of this lipid (IV, Fig. 2.2) have been isolated from wheatflour (Bomstein, 1965), from a variety of seeds and beans (Dawson *et al.*, 1969) and from crude soyabean phosphatides (Aneja *et al.*, 1969), so that this derivative is probably a minor constituent of many plant tissues.

3. Phosphatidyl serine

The presence of serine in the hydrolysates of peanut phospholipids prompted Hutt and coworkers (1950) to propose the existence of phosphatidyl serine (V, Fig. 2.2) in plant tissues. Subsequently, Benson and

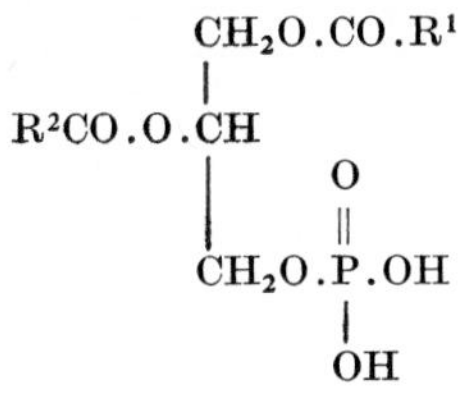

$$CH_2O.CO.R^1$$
$$|$$
$$R^2CO.O.CH$$
$$|\qquad O$$
$$\qquad ||$$
$$CH_2O.P.OH$$
$$|$$
$$OH$$

I. Phosphatidic acid

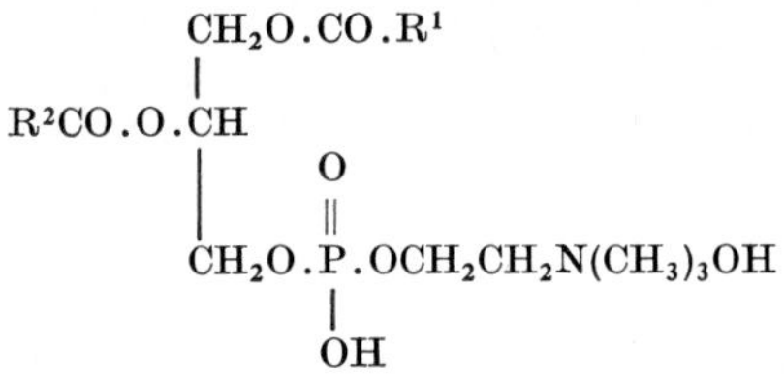

$$CH_2O.CO.R^1$$
$$|$$
$$R^2CO.O.CH$$
$$|\qquad O$$
$$\qquad ||$$
$$CH_2O.P.OCH_2CH_2N(CH_3)_3OH$$
$$|$$
$$OH$$

II. Phosphatidyl choline (lecithin)

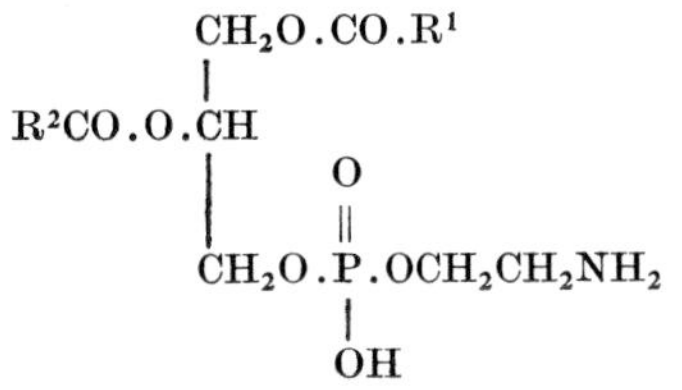

$$CH_2O.CO.R^1$$
$$|$$
$$R^2CO.O.CH$$
$$|\qquad O$$
$$\qquad ||$$
$$CH_2O.P.OCH_2CH_2NH_2$$
$$|$$
$$OH$$

III. Phosphatidyl ethanolamine

$$CH_2O.CO.R^1$$
$$|$$
$$R^2CO.O.CH$$
$$|\qquad O$$
$$\qquad ||$$
$$CH_2O.P.OCH_2CH_2NH.CO.R^3$$
$$|$$
$$OH$$

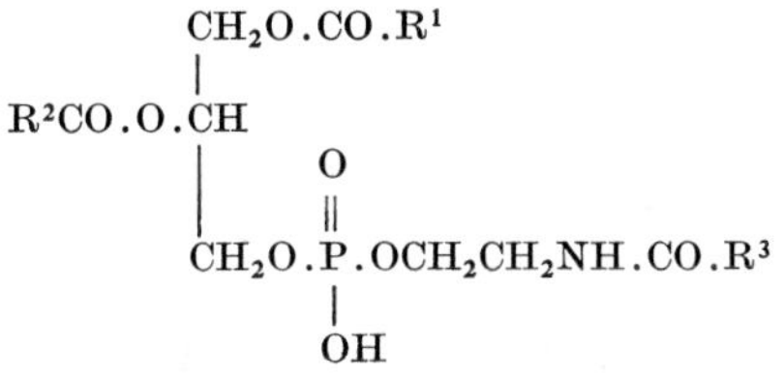

IV. N-acyl phosphatidyl ethanolamine

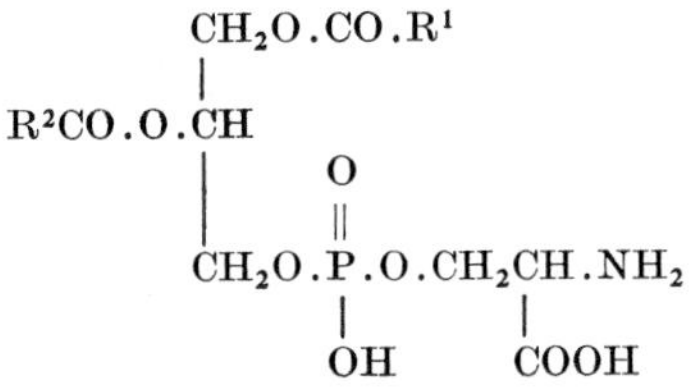

$$CH_2O.CO.R^1$$
$$|$$
$$R^2CO.O.CH$$
$$|\qquad O$$
$$\qquad ||$$
$$CH_2O.P.O.CH_2CH.NH_2$$
$$|\qquad\qquad |$$
$$OH\qquad COOH$$

V. Phosphatidyl serine

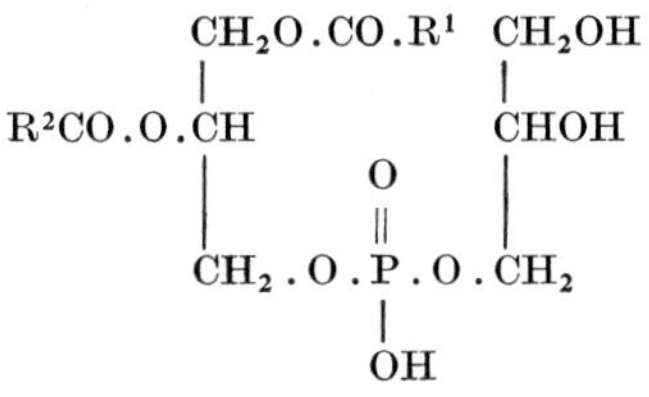

$$CH_2O.CO.R^1 \quad CH_2OH$$
$$|\qquad\qquad\qquad |$$
$$R^2CO.O.CH \qquad\qquad CHOH$$
$$|\qquad O \qquad\quad |$$
$$\qquad ||$$
$$CH_2.O.P.O.CH_2$$
$$|$$
$$OH$$

VI. Phosphatidyl glycerol

$$OH$$
$$|$$
$$CH_2O.CO.R^1 \quad CH_2O.P.O.CH_2$$
$$|\qquad\qquad\qquad |\quad ||\quad |$$
$$\qquad\qquad\qquad\qquad O$$
$$R^2CO.O.C \qquad CHOH \qquad HC.O.CO.R^3$$
$$|\qquad O \qquad |\qquad\qquad |$$
$$\quad ||$$
$$CH_2O.P.O.CH_2 \quad R^4CO.O.CH_2$$
$$|$$
$$OH$$

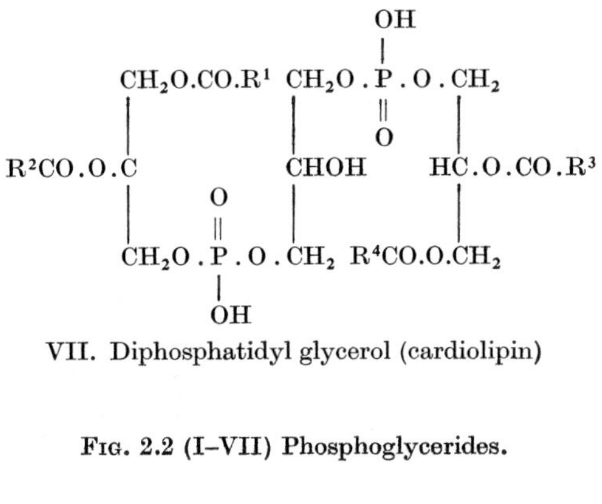

VII. Diphosphatidyl glycerol (cardiolipin)

FIG. 2.2 (I–VII) Phosphoglycerides.

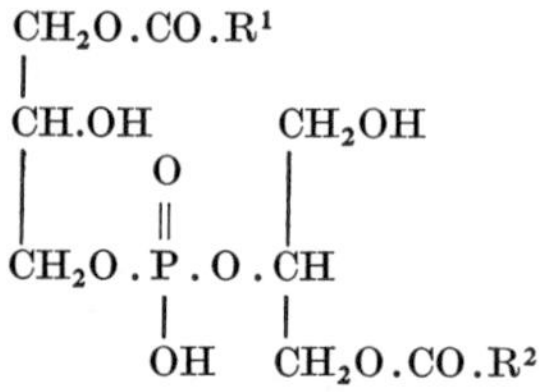

$$CH_2O.CO.R^1$$
$$|$$
$$CH.O.CO.R^2$$
$$|$$
$$O.CH_2$$
$$|$$
$$CH_2OH \qquad CH_2O.P.O.CH_2$$
$$| \qquad\qquad | \quad \parallel$$
$$CHOH \qquad CH.OH \qquad\qquad O \quad CH.OH$$
$$CH_2O.CO.R^1$$
$$| \qquad\qquad O$$
$$CH.OH \qquad CH_2OH \qquad \parallel$$
$$| \qquad O \qquad | \qquad CH_2O.P.O.CH_2 \qquad CH_2O.P.OH$$
$$\parallel \qquad\qquad | \qquad\qquad |$$
$$CH_2O.P.O.CH \qquad\qquad CH_2O \qquad\qquad CH_2O$$
$$| \qquad | \qquad\qquad | \qquad\qquad |$$
$$OH \quad CH_2O.CO.R^2 \qquad CH.O.CO.R^5 \qquad CH.O.CO.R^3$$
$$| \qquad\qquad |$$
$$VIII \qquad\qquad CH_2O.CO.R^6 \qquad CH_2O.CO.R^4$$
$$\qquad\qquad\qquad\qquad\qquad IX$$

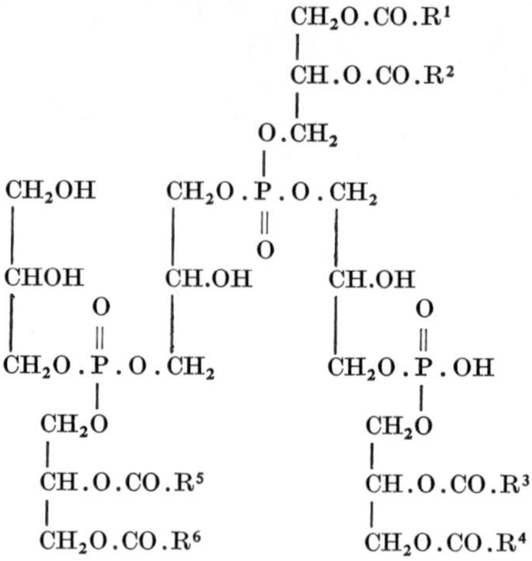

VIII and IX. Alternative structures proposed by Debuch and Rotsch (1966) for a nitrogen-free phospholipid fraction from spinach

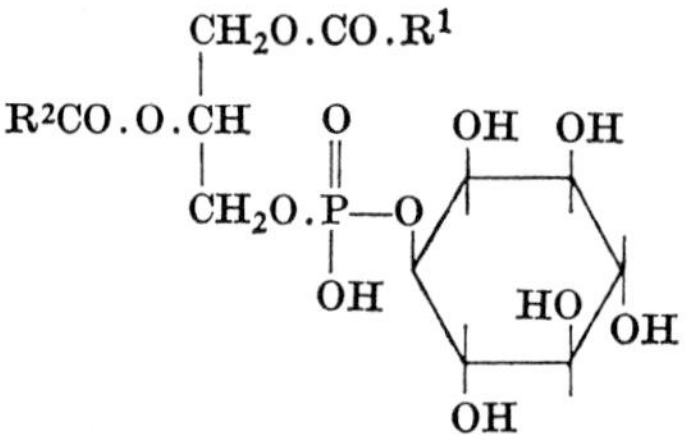

$$CH_2O.CO.R^1 \qquad\qquad\qquad\qquad CH_2O.CO.R^1$$
$$| \qquad\qquad\qquad\qquad\qquad\qquad\qquad |$$
$$R^2CO.O.CH \quad O \quad OH \quad OH \qquad R^2CO.O.CH$$
$$| \qquad \parallel \qquad\qquad\qquad\qquad\qquad\qquad | \qquad O$$
$$CH_2O.P—O \qquad\qquad\qquad\qquad\qquad\qquad \parallel$$
$$| \qquad\qquad\qquad\qquad\qquad\qquad\qquad CH_2O.P.O.base$$
$$OH \qquad HO \quad OH \qquad\qquad\qquad\qquad |$$
$$\qquad\qquad\qquad\qquad\qquad\qquad\qquad\qquad\qquad O$$
$$OH \qquad\qquad\qquad\qquad\qquad\qquad\qquad |$$
$$\qquad\qquad\qquad\qquad\qquad\qquad\qquad\qquad Sugar$$

X. Phosphatidyl-myo-inositol (monophosphoinositide)

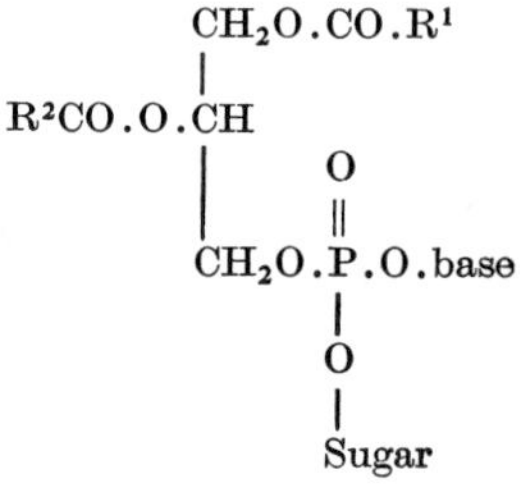

XI. General structure for triester phosphoglycolipids as proposed by Galanos and Kapoulas (1965)

Fig. 2.2 (VIII–XI) Phosphoglycerides.

Maruo (1958) and Benson and Strickland (1960) identified glyceryl-phosphorylserine in the hydrolysates of extracts from a variety of algae and higher plants, and it is likely that this class of lipid is a general, if minor, component of plant extracts.

4. Phosphatidyl glycerol

Benson and Maruo (1958) showed that a major phospholipid of many plant tissues, particularly chlorophyllous tissues, is phosphatidyl glycerol (VI, Fig. 2.2). Because this lipid has two asymmetric centres it can theoretically exist in any of four stereoisomeric forms, but Benson and Miyano (1961) suggested that the compound present in chloroplasts is L-α-phosphatidyl-D-glycerol; this configuration was subsequently confirmed by Haverkate and Van Deenen (1964) for a phosphatidyl glycerol fraction from spinach leaves.

5. Diphosphatidyl glycerol (cardiolipin)

In 1960 Benson and Strickland identified 1,3-diglycerophosphoryl glycerol among the deacylated phosphatide fractions prepared from a variety of algae, bacteria and higher plant tissues, and concluded that the original plant lipid was identical with the beef heart cardiolipin which Macfarland (1958) characterized as a 1:3 diphosphatidyl glycerol containing four fatty acid residues (VII, Fig. 2.2). The identity of plant cardiolipin with that from beef-heart was subsequently confirmed by Coulon-Morelec and Douce (1968).

Debuch and Rotsch (1966) have partially characterized a further class of plant lipid containing only glycerol, phosphate and fatty acids, which they isolated as a minor component of spinach leaf lipids. They have proposed two alternative structures (VIII, IX, Fig. 2.2) for this lipid, of which the former is considered most likely.

6. Phosphatidyl inositol

1-Phosphatidyl-myo-inositol (X, Fig. 2.2) is a major lipid of many plant tissues (Lepage et al., 1960) but, unlike animals, plants do not appear to synthesize phosphatidyl derivatives of inositol polyphosphates such as the di- and tri-phosphoinositides. On the other hand other inositol-containing lipids occur in plants, for example the various sphingolipids which are described later in this chapter (p. 53).

7. Triester phospholipids

On the basis of the countercurrent distribution behaviour of several naturally occurring phospholipid fractions and their methylation products, Collins (1959) proposed that many natural phospholipids occur in the living cell as triesters of phosphoric acid and that the analytical methods commonly applied to their isolation, such as silicic acid chromatography, result in cleavage of one of the phosphate ester bonds. This hypothesis was supported and extended by Galanos and Kapoulas (1965) who proposed that native plant phospholipids are triester glycolipids in which the more commonly accepted structures for the phospholipids are modified by the inclusion of a sugar moiety esterified with the remaining phosphate hydroxyl (XI, Fig. 2.2).

It was suggested that the phosphate-sugar bond is particularly labile and would readily rupture in the presence of mildly acidic substances. Little additional experimental evidence has been offered in support of these theories, nor has a natural triester phospholipid been isolated or characterized, so that the existence of such lipids remains in doubt. Their presence in cellular tissues has also been questioned by Davenport (1966) who pointed out that naturally occurring lipid systems give negatively charged micelles, whereas tertiary phosphate structures would be expected to give net positive micelles. It has also been claimed that such tertiary phosphate groups are not so labile as the arguments of Collins, Galanos and Kapoulas suggest.

B. Glycosyl Diglycerides

1. Mono- and di-galactosyl diglycerides

In 1961 Carter and coworkers (1961a, b) described the isolation from wheat flour of two lipids which on alkaline hydrolysis yielded mono- and digalactosyl glycerols. On the basis of structural studies on these water-soluble derivatives the authors ascribed to their respective parent lipids the structures 1,2-diacyl-3-β-D-galactopyranosyl-L-glycerol (I, Fig. 2.3) and 1,2-diacyl-3-(α-D-galactopyranosyl-1,6-β-D-galactopyranosyl]-L-glycerol (II, Fig. 2.3). Subsequently numerous publications have estab-

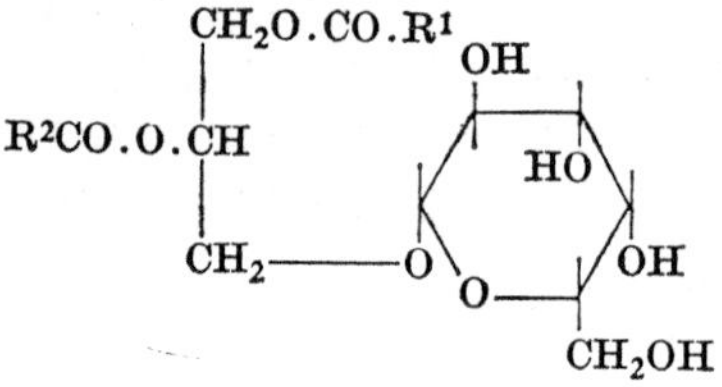

I. Monogalactosyl digylceride 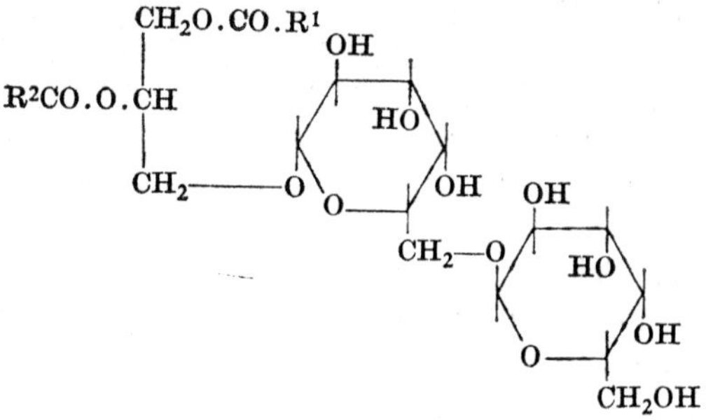   II. Digalactosyl diglyceride

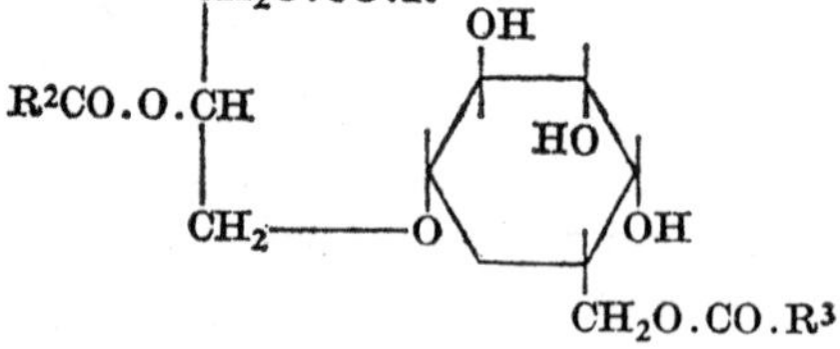

III. 6-O-Acyl monogalactosyl diglyceride 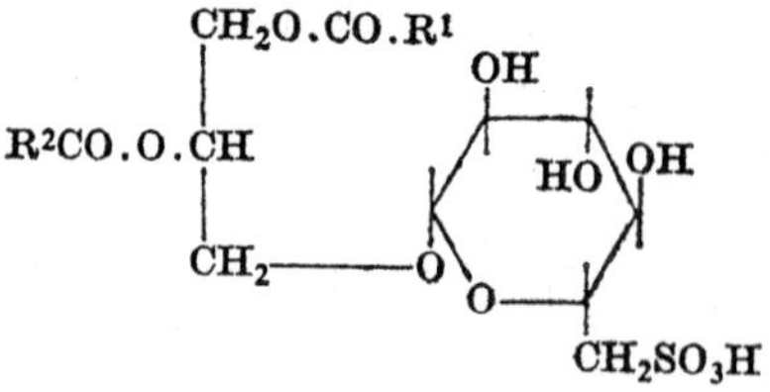   IV. Sulphoquinovosyl diglyceride (sulpholipid)

Fig. 2.3 (I–IV) Glycosyl glycerides.

lished that the mono- and di-galactosyl diglycerides are present in a wide variety of plant tissues. Udelnova and Boichenko (1967) reported that the monogalactosyl diglyceride fraction from a variety of leaves contains up to 12% of bound manganese although the site of attachment of the cation was not established.

2. Trigalactosyl diglyceride

Partial evidence that some plant tissues may contain small amounts of a trigalactosyl diglyceride has been quoted by various groups (e.g. Benson et al., 1958; Allen et al., 1966; Ongun and Mudd, 1968) and this lipid was more completely characterized by Galliard (1969) who showed that extracts from potato tubers contained a lipid in which the molar proportions of fatty acids, glycerol and galactose were 2 : 1 : 3 and in which an additional D-galactopyranosyl moiety appeared to be linked $\alpha(1 \to 6)$ to the terminal galactose unit of digalactosyl diglyceride. Trigalactosyl diglyceride was also identified in spinach chloroplasts by Webster and Chang (1969) who tentatively identified tetragalactosyl diglyceride in the same preparation.

3. O-Acyl galactosyl diglyceride

In 1967 Heinz (1967a) reported the isolation from spinach leaf homogenates of a monogalactosyl diglyceride fraction in which an acyl group was esterified to part of the sugar moiety. Myrrhe (1968) isolated a similar fraction from wheat flour and by a study of the products of oxidation of his material was able to establish that the sugar residue was esterified at the 6-position; Heinz and Tulloch (1969) subsequently confirmed that the compound from spinach has a similar structure namely 1,2-diacyl-3-(6-acyl-β-D-galactopyranosyl)-L-glycerol (III, Fig. 2.3).

Heinz (1967b) was unable to locate this compound in normal extracts from spinach leaf and concluded that the acylated galactosyl diglyceride is formed only in the disrupted cell, by acyl transfer from digalactosyl diglyceride to monogalactosyl diglyceride.

4. Sulphoquinovosyl diglyceride (sulpholipid)

Benson and coworkers (1959) noted that a variety of algae and leaves contained a unique class of sulphur-containing glycolipid which they later showed contained a sulphonic acid group (Daniel et al., 1961) and subsequently established the full identity of the lipid as 1,2-diacyl-3-(6-sulpho-α-D-quinovopyranosyl)-L-glycerol (IV, Fig. 2.3) (Benson, 1963).

C. Other Glycerides

Classes of glyceride containing long-chain alkenyl and saturated ether residues rather than acyl ester groups are of common occurrence in animal tissues but are seldom observed in plants, although Kaufmann *et al.* (1970) have detected small quantities of alk-1-enyl acyl ethanolamine- and choline-glycerophosphatides in pea seeds (II, Fig. 2.4). Monoacyl phospholipids (lyso-phospholipids, I, Fig. 2.4) are also far less abundant in plant tissues than in animals, but Wren and Merryfield (1970) have shown that lyso lecithin and lyso-phosphatidyl ethanolamine comprise the major part of a tightly bound lipid fraction in wheat-starch granules.

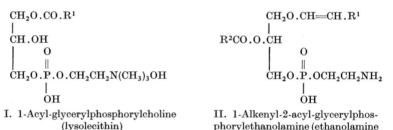

I. 1-Acyl-glycerylphosphorylcholine
(lysolecithin)

II. 1-Alkenyl-2-acyl-glycerylphos-
phorylethanolamine (ethanolamine
plasmalogen)

FIG. 2.4 (I–II) Minor phospholipids of plants.

2. DIOL LIPIDS

There is a steady accumulation of evidence (see Bergelson, 1969) which indicates that acyl esters of dihydric alcohols occur in animal and plant tissues, although their presence in the majority of lipid extracts is easily overlooked because these compounds tend to co-chromatograph with their glyceride analogues. Nevertheless in 1961 Ukita and Tanimura isolated from the seed-oil of the Far Eastern plant *Coix lachrima* a mixed fatty acid ester of *erythro*-butane-2,3-diol and which they named coixenolide (I, Fig. 2.5). Latterly, Bergelson and coworkers (1964) have reported that acyl derivatives of ethylene glycol, propanediols and butanediols occur as minor constituents of "triglyceride" fractions from corn seed.

A unique class of diol lipid which contains nitrogen has been recently described by Mikolajczak *et al.* (1969) and Seigler *et al.* (1970) who found that some 35% of *Cordia verbenacea* seed-oil comprised a compound in which two fatty acid moieties are esterified with an unsaturated five carbon dihydroxynitrile, and for which these workers proposed the

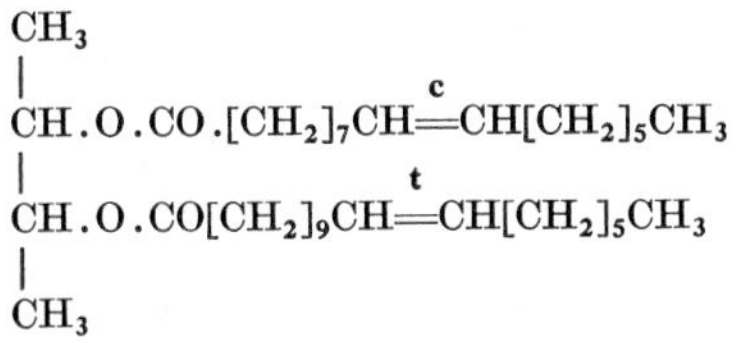

CH₃
| c
CH.O.CO.[CH₂]₇CH=CH[CH₂]₅CH₃
| t
CH.O.CO[CH₂]₉CH=CH[CH₂]₅CH₃
|
CH₃

I. Butane-2,3-diol palmitoleate *trans*-octadec-11-enoate from *Coix lachrima* oil

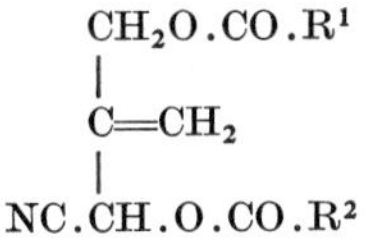

CH₂O.CO.R¹
|
C=CH₂
|
NC.CH.O.CO.R²

II. Diacyl dihydroxynitrile from *Cordia verbenacea* seed-oil

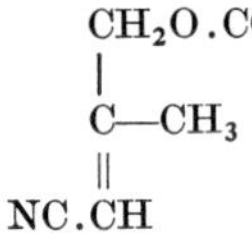

CH₂O.CO.R¹
|
C—CH₃
‖
NC.CH

III. Monoacyl cyanolipid from *Stocksia brahuica* seed-oil

FIG. 2.5. (I–III) Diol- and cyano-lipids from seed-oils.

structure depicted in II, Fig. 2.5. Incompletely characterized lipid fractions from other seed oils (Kasbekar and Bringi, 1969; Hopkins *et al.*, 1968) appear to have certain points of similarity to the lipid described by Mikolajczak *et al.* (1969).

Another class of cyanolipid has been characterized by the same group (Mikolajczak *et al.*, 1970) who isolated a fatty acid ester of an isoprenoid hydroxynitrile (3-cyano-2-methylprop-2-ene-1-ol) from the seed-oil of *Stocksia brahuica* (III, Fig. 2.5).

3. OTHER ACYL ESTERS

A. Sterol Esters and Sterol Glycoside Esters

Most plant cells appear to contain small quantities of sterol acyl esters (I, Fig. 2.6), and in 1964 Lepage reported that the sterol glycosides, which are of widespread occurrence in the tissues of higher plants, can also occur in the form of monoacyl esters in which a fatty acid is esterified to the hydroxyl group on the C-6-position of the glucose (II, Fig. 2.6).

B. Wax Esters

Wax ester fractions from plants are usually complex mixtures of esters of normal acids and primary alcohols of medium to long chain length

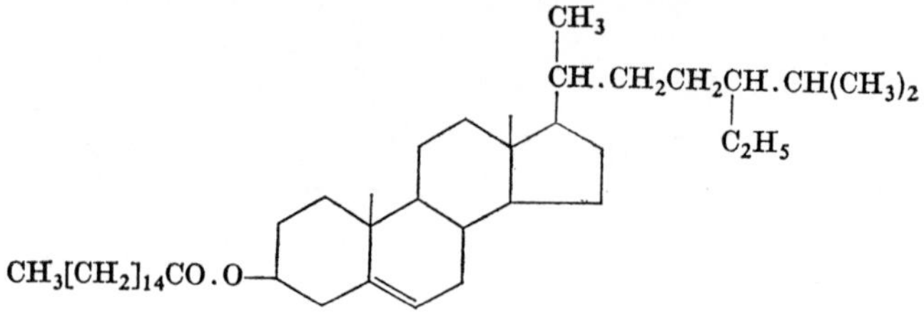

$$CH_3$$
$$CH.CH_2CH_2CH.CH(CH_3)_2$$
$$C_2H_5$$

$$CH_3[CH_2]_{14}CO.O-$$

I. β-Sitosteryl palmitate

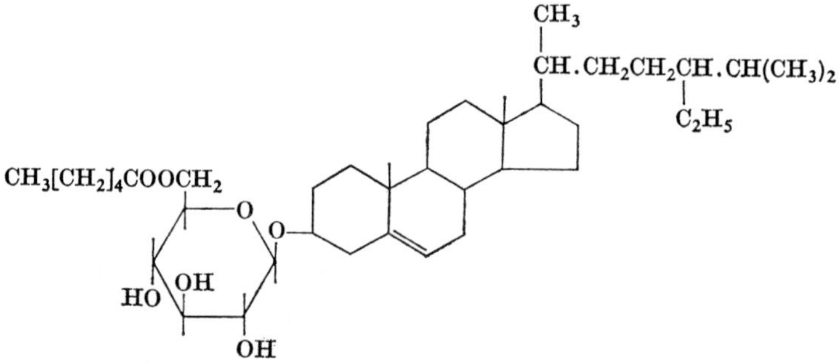

$$CH_3$$
$$CH.CH_2CH_2CH.CH(CH_3)_2$$
$$C_2H_5$$

$$CH_3[CH_2]_4COOCH_2$$

O

O—

HO   OH

OH

II. 6-O-Palmitoyl-β-sitosteryl glucoside

Fig. 2.6. (I–II) Acyl esters containing sterols.

(i.e. 10 to 30 carbon atoms). Until recently the structure of single molecular species of plant wax esters was difficult to determine because of the experimental difficulties involved in their isolation. Consequently, published data regarding the composition of natural wax esters have commonly comprised qualitative and quantitative analyses of the mixtures of alcohols and acids obtained by saponification of the wax ester fraction which gives only partial information regarding the structure of individual components. For example, Purdy and Truter (1963a, b, c) reported that the surface wax of cabbage leaves contains wax esters containing fatty acids of chain length C_{12} to C_{24} and primary alcohols of chain length C_{12} to C_{28}, while Radler and Horn (1965) have shown that wax ester fractions from grape cuticle contain a small proportion of fatty acids and alcohols of odd carbon number.

The introduction of suitable GLC apparatus and stationary phases now permits the separation of individual molecular species of wax esters, so that Hamilton and Power (1969) were able to show that the major class of ester from the surface wax of Rye grass is a C_{52} ester, probably made up from a C_{26} acid and a C_{26} alcohol (Fig. 2.7):

$$CH_3[CH_2]_{24}CH_2O.CO.CH_2[CH_2]_{23}CH_3$$

Fig. 2.7. Major wax ester from rye grass (Hamilton and Power, 1969).

C. Carotenoid Esters

There is little information regarding the occurrence and distribution of carotenoid esters in plants, although flower pigments are known to contain saturated fatty acid esters of carotenoid alcohols (Kuhn *et al.*, 1931; Booth, 1964; Alam *et al.*, 1968), and their presence in a green alga has also been reported (Kleinig and Czygan, 1969). Where a dihydroxy alcohol is involved, diesters are the fatty acid derivatives most commonly found (Fig. 2.8) although Booth (1964) has identified a lutein monoester in dandelion flowers.

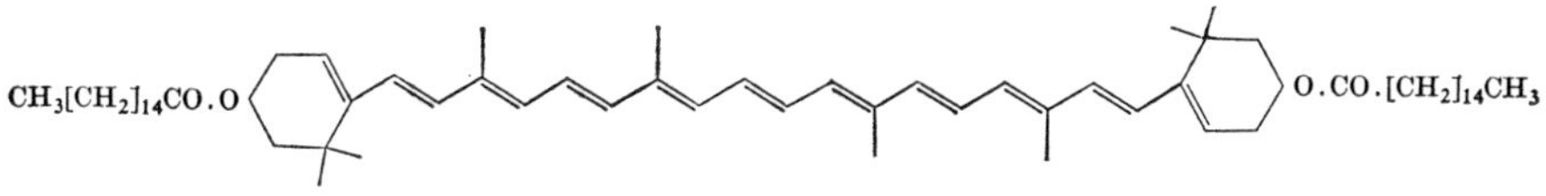

FIG. 2.8. Luteol dipalmitate.

D. Terpenoid Esters

A few published papers have referred to the presence in plants of acyl esters of terpenoid alcohols, and Dunphy and Allcock (1971) have shown that between 30 and 60% of the total monoterpenoid alcohol content of rose petals occurs in the form of acyl esters, geranyl stearate (Fig. 2.9) predominating.

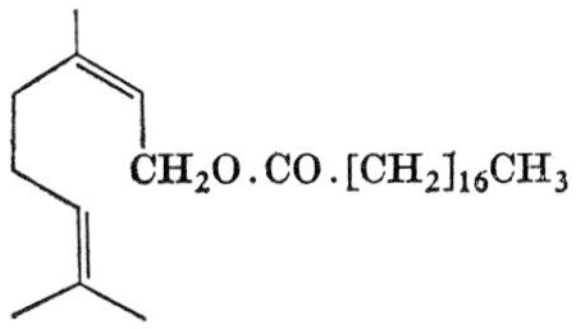

FIG. 2.9. Geranyl stearate.

4. SPHINGOLIPIDS

Sphingolipids may be defined as those lipophillic substances which contain the long-chain amino-alcohol sphingosine (*trans*-D-*erythro*-1,3-dihydroxy-2-amino-4-octadecene; I, Fig. 2.11) or other amino-alcohols structurally related to this base. Numerous classes of sphingolipid occur in animal tissues, the three most abundant classes being sphingomyelin, cerebroside and ganglioside, but of these only cerebrosides occur in plants, and then in only minor proportions.

A. Cerebrosides (Ceramide Monoglycosides)

Cerebrosides contain a nitrogenous alcohol, fatty acid and a monosaccharide moiety. Their presence in mammalian tissues was recognized for a long time before Carter and Greenwood (1952) demonstrated that the general structure of cerebrosides comprises a long chain amino alcohol with a fatty acid linked through an amide bond to the amino group and a monosaccharide moiety bound glycosidically at the C-1 position (Fig. 2.10).

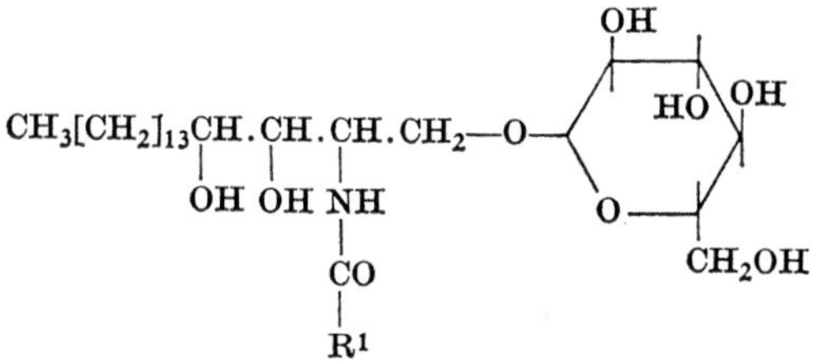

$$CH_3[CH_2]_{13}CH.CH.CH.CH_2—O$$

with OH OH NH below, CO, R^1

FIG. 2.10. Cerebroside from wheat flour.

The first unequivocal demonstration of the presence of cerebrosides in plant tissues was reported by Carter and coworkers (1961) who characterized a series of these compounds from wheat flour extracts and showed that they comprised derivatives of at least four classes of amino-alcohol. They identified these alcohols as 1,3,4-trihydroxy-2-amino-octadecane (II, Fig. 2.11), commonly referred to as phytosphingosine, its *trans*-8 unsaturated analogue (dehydrophytosphingosine; III, Fig. 2.11), dihydrosphingosine (IV, Fig. 2.11), and a mono-unsaturated dihydrosphingosine in which the location of the double bond is still uncertain (V, Fig. 2.11) (Carter and Hendrickson, 1963). Two further classes of plant long-chain base were reported by Karlsson and Hølm (1966) who found C_{19} and C_{20} analogues of phytosphingosine (i.e. 1,3,4-trihydroxy-2-amino-nonadecane and 1,3,4-trihydroxy-2-amino-eicosane in corn cerebrosides) while more recently Carter and Koob (1969) have identified a C_{16} phytosphingosine (D-ribo-1,3,4-trihydroxy-2-aminohexadecane) in flax-seed phytoglycolipid (VI, Fig. 2.11).

Apart from differences in the nature of their constituent aminoalcohols, plant cerebrosides differ from most of those of mammalian origin in containing glucose exclusively as their component sugar (Carter *et al.*, 1961; Sastry and Kates 1964a, b), whereas animal cerebrosides usually contain galactose, although glucose also appears in the cerebrosides of humans suffering from certain metabolic disorders.

The specific nature of the fatty acids which tend to associate with this class of lipid is discussed in other sections (pp. 23, 91).

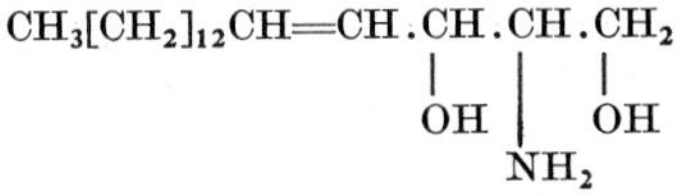

$$CH_3[CH_2]_{12}CH\!=\!CH.CH.CH.CH_2$$
$$\qquad\qquad\qquad\quad | \quad\quad | \quad | $$
$$\qquad\qquad\qquad\quad OH \quad | \quad OH$$
$$\qquad\qquad\qquad\qquad\quad NH_2$$

I. Sphingosine (D-*erythro*-1,3-dihydroxy-2-amino-*trans*-4-octadecene)

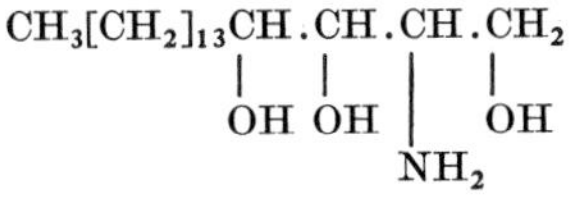

$$CH_3[CH_2]_{13}CH.CH.CH.CH_2$$
$$\qquad\qquad\quad | \quad | \quad | \quad | $$
$$\qquad\qquad\quad OH\ OH \quad | \quad OH$$
$$\qquad\qquad\qquad\qquad NH_2$$

II. Phytosphingosine (D-*ribo*-1,3,4-trihydroxy-2-amino-octadecane)

$$CH_3[CH_2]_8CH\!=\!CH[CH_2]_3CH.CH.CH.CH_2$$
$$\qquad\qquad\qquad\qquad\qquad | \quad | \quad | \quad | $$
$$\qquad\qquad\qquad\qquad\qquad OH\ OH \quad | \quad OH$$
$$\qquad\qquad\qquad\qquad\qquad\qquad\quad NH_2$$

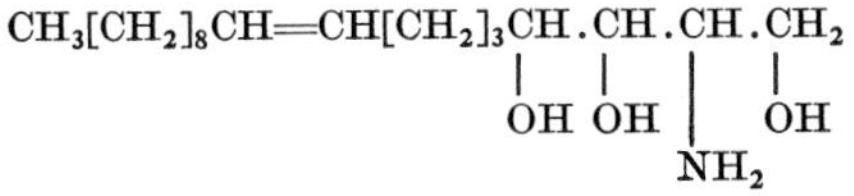

III. Dehydrophytosphingosine (D-*ribo*-1,3,4-trihydroxy-2-amino-*trans*-8-octadecane)

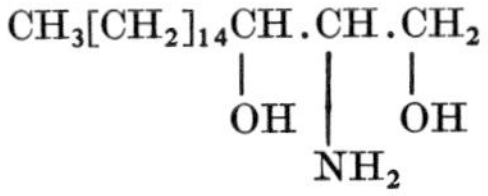

$$CH_3[CH_2]_{14}CH.CH.CH_2$$
$$\qquad\qquad\quad | \quad | \quad | $$
$$\qquad\qquad\quad OH \quad | \quad OH$$
$$\qquad\qquad\qquad\quad NH_2$$

IV. Dihydrosphingosine (D-*erythro*-1,3-dihydroxy-2-amino-octadecane)

$$CH_3[CH_2]_xCH\!=\!CH[CH_2]_{12-x}CH.CH.CH_2$$
$$\qquad\qquad\qquad\qquad\qquad | \quad | \quad | $$
$$\qquad\qquad\qquad\qquad\qquad OH \quad | \quad OH$$
$$\qquad\qquad\qquad\qquad\qquad\quad NH_2$$

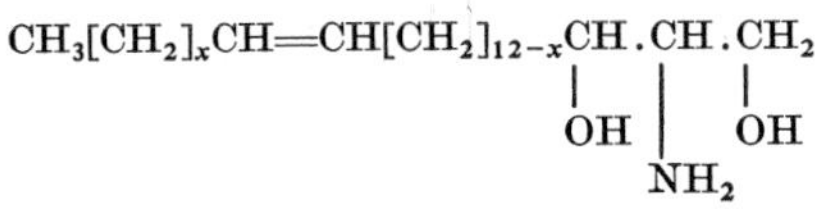

V. Partial structure for amino-alcohol from wheat-flour cerebrosides

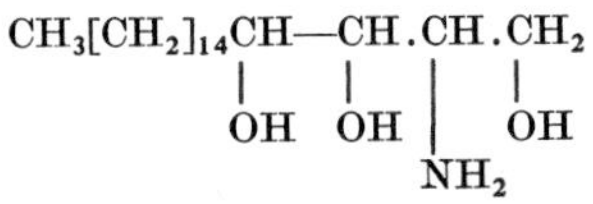

$$CH_3[CH_2]_{14}CH\!-\!CH.CH.CH_2$$
$$\qquad\quad | \quad | \quad | \quad | $$
$$\qquad\quad OH\ \ OH \quad | \quad OH$$
$$\qquad\qquad\qquad\quad NH_2$$

C_{19}-Phytosphingosine

$$CH_3[CH_2]_{15}CH.CH.CH.CH_2$$
$$\qquad\qquad\quad | \quad | \quad | \quad | $$
$$\qquad\qquad\quad OH\ OH \quad | \quad OH$$
$$\qquad\qquad\qquad\qquad NH_2$$

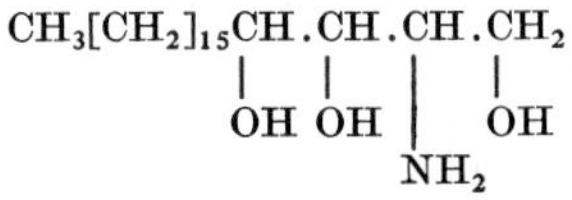

C_{20}-Phytosphingosine

$$CH_3[CH_2]_{11}CH.CH.CH.CH_2$$
$$\qquad\qquad\quad | \quad | \quad | \quad | $$
$$\qquad\qquad\quad OH\ OH \quad | \quad OH$$
$$\qquad\qquad\qquad\qquad NH_2$$

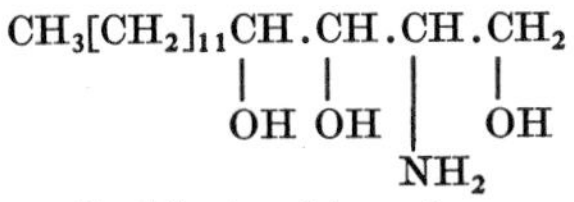

C_{16}-Phytosphingosine

VI. Other saturated amino-alcohols of plant extracts

FIG. 2.11 (I–VI) Plant sphingosines.

B. Phytoglycolipid and Related Substances

In 1958 Carter *et al.* reported the occurrence in a variety of commercial oil-seed phospholipid fractions of a series of complex phosphorus-containing glycolipids containing phytosphingosine and inositol, which they called phytoglycolipids and for which they proposed the general structure illustrated in I, Fig. 2.12. Subsequent work (Carter *et al.*, 1964a, b; 1969) was mainly concentrated on defining the structure of the complex oligosaccharide moiety which characterizes this lipid, so that this group was able to propose a detailed structure (II, Fig. 2.12) in which the only remaining feature to be determined is the point of attachment and sequence of the terminal monosaccharides.

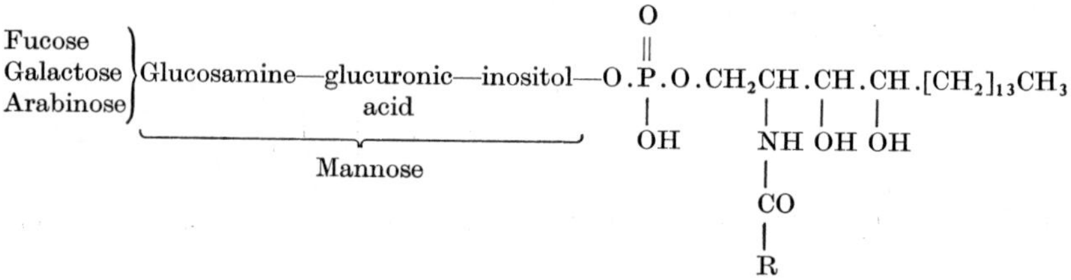

I. General structure of phytoglycolipid (Carter *et al.*, 1964b)

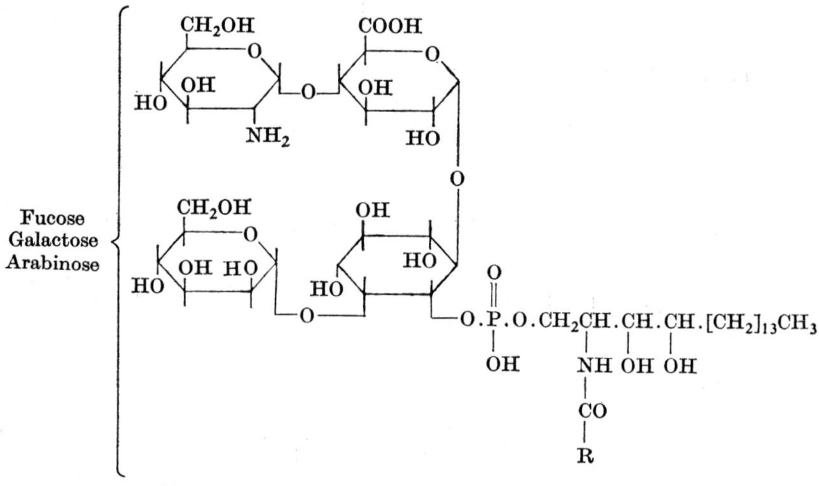

II. Structure of phytoglycolipid from *Zea mays* (Carter *et al.*, 1969)

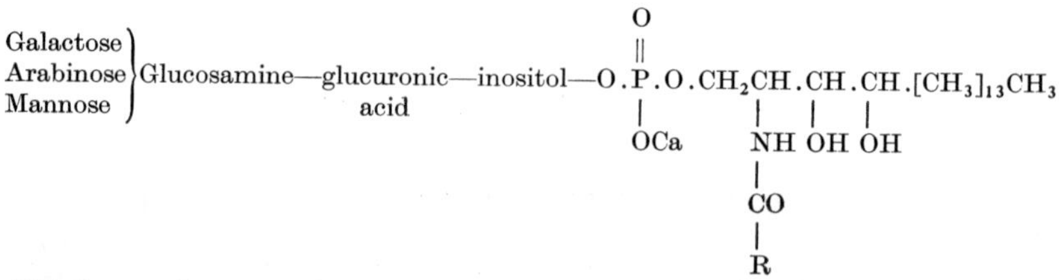

III. Proposed structure for phytoglycolipid fraction from peanut (Wagner *et al.*, 1969)

Fig. 2.12. (I–III) Proposed structures for complex sphingoglycolipids of plants.

Wagner and coworkers (1969a) isolated a similar phytoglycolipid fraction from peanut phospholipids which differed from that of flax and maize in containing no fucose. Wagner *et al.* proposed the tentative structure illustrated in III, Fig. 2.12 for their preparation, which differs from that of Carter *et al.* in the site of attachment of the mannose moiety, although the German group offered little evidence to substantiate this difference. It is probable that this peanut phytoglycolipid was a major constituent of the complex lipid isolated and partially characterized by Malkin and Poole (1953).

The procedures employed for the isolation of phytoglycolipid fractions such as those described above have usually involved mild alkaline hydrolysis of crude phosphatide fractions as a preliminary step, and this raises the question of whether the phytoglycolipid molecule is only part of an even more complex lipid occurring in the plant cell. Comparison between the countercurrent distribution properties of the purified phytoglycolipid and that of the crude lipid fraction from which this material is obtained by hydrolysis, suggests that this may be the case (Carter *et al.*, 1962).

Although almost all of the earlier work on the isolation and characterization of phytoglycolipid was carried out on phospholipid fractions from oil-seeds, Carter and Koob (1969) recently isolated a phytoglycolipid from bean leaves which differed from similar preparations from flax and maize seed, and resembled that from peanut, in containing no fucose.

A preliminary note by Wagner and coworkers (1969b) describes the partial characterization of a lipid from the green alga *Scenedesmus obliquus* which superficially resembles the phytoglycolipids in many respects, but does not contain inositol. This is the first indication that algae synthesize sphingosines or sphingolipids.

Another class of sphingoglycolipid, as yet incompletely characterized, was reported by Carter and Kisic (1969) who isolated from crude oil-seed inositol lipid fractions a substance related to phytoglycolipid but contained no aminosugar, and which they described as a ceramide phosphate-polysaccharide.

5. Cutin

Microscopic inspection of the epidermal cells of the aerial parts of plants shows that the outer walls of the cells are much thicker than those of any underlying cell. This is because the epidermal cells are coated by a continuous thick layer, usually called the cuticle, and which is composed of a variety of classes of compound some 80% of which are lipophillic.

These lipoidal substances can be fractionated into an easily solvent-extractable fraction composed of waxes (p. 71) and a non-solvent extractable fraction known as cutin which can be rendered soluble only after saponification.

Several groups of workers have shown that the major hydrolysis products of cutin are polyhydroxy acids and ordinary fatty acids and Matic (1956) proposed that the cutins are cross-linked molecules of high molecular weight formed by intermolecular interesterifications in which the hydroxyl groups of the hydroxy acids esterify with the carboxyl groups of other molecules in a fairly random fashion (Fig. 2.13).

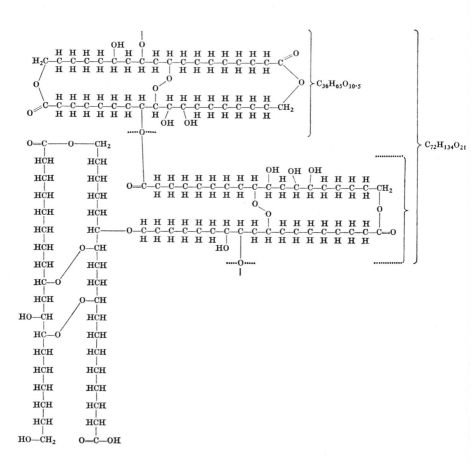

FIG. 2.13. Structure of cutin (Heinen and Brand, 1963).

CHAPTER 3

The Lipid and Fatty Acid Composition of Specific Tissues

The essential structural role of different lipids in the maintenance of specific classes of plant membrane, as well as their biochemical involvement in various aspects of plant metabolism (Chapter 10), suggests that plant cells and organelles of dissimilar structure and function will have differing lipid and fatty acid compositions. It also follows that equivalent tissues from different plants may have many biochemical features in common.

In this chapter we shall compare the acyl lipid composition of individual plant tissues and indicate the extent to which the points of similarity or difference are operative throughout the plant kingdom.

1. HIGHER PLANTS

A. Fruits

It will be appropriate to describe first some of the structural forms in which the fruits of higher plants can occur.

The outer covering of a fruit is called the pericarp; in a ripe pod this is mostly thin and dry, whereas in drupes or berries the pericarp is thick and fleshy (e.g. palm fruit, Fig. 3.1.a) and in a nut it is hard and bone-like. The pericarp often consists of two layers (endocarp and exocarp) or three layers (endocarp, mesocarp and epicarp). These layers are most easily recognized in drupes, such as olives and peaches, in which the skin is epicarp, the fleshy soluble portion is mesocarp and the hard bony stone surrounding the seed is endocarp.

In other fruits, the pericarp is much thinner than it is in a peach or olive, and in grains such as wheat, the pericarp is the thin outer coating of the seed.

A monocotyledonous seed (e.g., Fig. 3.1.b) is comprised of three principal parts, namely the embryo which develops from the fertilized egg cell and grows to form the new plant when germination occurs, the endosperm which is a food reserve for the developing plant, and finally the seed coat which protects the seed.

In many dicotyledonous seeds such as the pea, much of the reserve material is contained in the cotyledons which become the first leaves of

the developing plant, but in the castor bean (Fig. 3.1.c) most of the
reserve oil is present in the endosperm which surrounds the cotyledons
in the ungerminated seed.

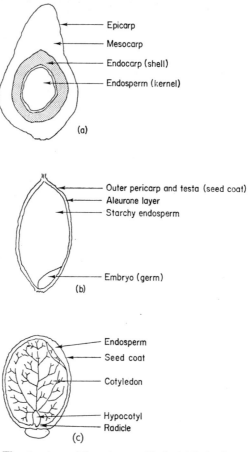

FIG. 3.1. (a, b, c) The structure of three types of fruit. (a) Palm fruit; (b) wheat grain;
and (c) castor bean.

1. Fruit coat (exocarp) lipids

The fleshy or succulent (exocarp) tissue of many fruits, such as plums,
pears, peaches and apples, contain comparatively low levels of fatty
substances. For example, Galliard (1968b) found that lipids comprised
only 0·09% of the fresh weight of the pulp of pre- and post-climacteric
apples, and 25% of these lipid fractions contained no fatty acid residues.
The remaining 75% of acyl lipid was chiefly composed of phospholipids

and glycolipids with triglyceride comprising only 5% of the total lipid extract (Table 3.1).

TABLE 3.1

Quantitative composition of lipids in pulp of pre-climacteric apples (Galliard, 1968b)

Lipid	Weight lipid mg/1000 g fresh wt	% by weight of total lipid
Phosphatidyl choline	189	21·5
Phosphatidyl ethanolamine	124	14·4
Phosphatidyl inositol	53	6·0
Phosphatidyl glycerol	27	3·1
Diphosphatidyl glycerol	5·8	0·7
Phosphatidyl serine	4·0	0·4
Phosphatidic acid	2·8	0·3
Unknown phospholipid	—	0·2[a]
Total phospholipid	405·6	46·6
Monogalactosyl diglyceride	42	4·8
Digalactosyl diglyceride	107	12·2
Total galactolipid	169	17·0
Esterified steryl glucoside	12	1·4
Steryl glucoside	51	5·8
Sterol	129	14·7
Sterol esters	18	2·0
Total steryl lipids	210	23·9
Sulpholipid	—	1·0
Glucocerebroside	34	3·9
Triglyceride	44	5·0
Chlorophyll	0·9	0·1
Combined neutral lipids	208	23·1
Combined polar lipid	672	76·9
Total lipids	880	

[a] Approximate value.

By contrast fruits also occur in which the fleshy exocarp has a high lipid content, most of which is triglyceride. For example, the fat content of the pericarp of the mature fruit of oil palm (*Elaeis guineensis*) is seldom less than 35% of the fresh tissue weight, and is often as high as 70%. Other plants bearing fruits with triglyceride-rich exocarps are the olive (*Olea europea*), avocado (*Persea americana*) and laurel (*Laurus nobilis*), and it is noteworthy that in most cases the fatty acid compositions of these exocarp oils are usually appreciably different from those of the corresponding seed. For example, the major fatty acids of the exocarp oil ("palm oil") from the palm fruit are palmitic, oleic and linoleic acids, whereas the oil from the nut of the same fruit ("palm kernel oil") is chiefly comprised of glycerides of lauric, myristic, palmitic and oleic acids (Table 3.2).

TABLE 3.2

Fatty acid compositions of the exocarp and kernel oils from the oil palm

	Fatty acid (% of total acids)								
	8:0	10:0	12:0	14:0	16:0	18:0	18:1	18:2	18:3
Palm kernel oil	2·3	2·7	42·5	16·3	16·4	2·3	15·5	2·0	—
Palm oil	—	—	—	2·0	42·3	2·5	44·1	9·1	t

2. Seed lipids

In the majority of mature seeds the quantitatively major class of lipid is triglyceride, which may constitute between 10 and 70% of the tissue dry weight (Wolff, 1966), while phospholipids and glycolipids normally represent less than 2% of the total seed lipid. One known exception to this rule occurs in the seeds of a member of the graminae, *Briza spicata*, which contains some 20% of lipid of which at least 78% is comprised of galactosyl diglycerides (Smith and Wolff, 1966). The remaining 22% of the lipid fraction may be triglyceride, although this was not established. This unusual seed lipid composition is not common in members of the *Briza* genus, in which the major class of seed lipid is usually triglyceride (I. A. Wolff, personal communication) although seeds of many other members of the graminae also have higher ratios of polar lipid to triglyceride than is found in most other classes of plant. For example, Fisher (1962) reported that in wheat, lipids form 1–2% of the endosperm, 8–15% of the germ and about 6% of the bran, with an average of 2–4% for the whole kernel. Wheat germ oil is mainly triglyceride whereas lipid extracts from endosperm and bran contain much higher proportions of phospholipids and glycolipids (Pomeranz and Chung, 1965), and because the germ represents only a small portion of the total seed weight it follows that the total seed extracts should contain a substantial proportion of polar lipids.

A second exception to the general rule that triglycerides are the major class of lipid in seeds occurs in a few species such as *Simmondsia californica* (Green *et al.*, 1936) and *Murraya koenigii* (Kartha and Singh, 1969) in which the major lipid component is wax ester and hydrocarbon.

Immature seeds contain far less lipid than the corresponding mature tissue, and this relatively small lipid fraction is chiefly composed of

TABLE 3.3

Fatty acid composition of some common seed-oils

Plant	Fatty acid (% of total acids)													
	16:0	16:1	18:0	18:1	18:2	18:3	20:0	20:1	20:2	22:0	22:1	22:2	24:0	24:1
Peanut, *Arachis hypogaea*[a]	12·9	0·2	4·5	43·1	32·5	—	0·7	—	—	3·1	—	—	1·1	—
Soybean, *Soja hispida*[b]	11·5	—	3·9	24·6	52·0	8·0	—	—	—	—	—	—	—	—
Rapeseed, (a) *Brassica napus* "Regina II"[c]	3·9	0·4	1·2	12·6	15·6	7·3	0·7	10·1	1·3	0·7	43·6	1·0	0·2	1·1
Rapeseed, (b) *Brassica napus* "Erucic acid free"[c]	4·9	0·5	2·0	47·9	25·2	15·2	1·2	1·9	—	1·1	—	—	—	—
Corn, *Zea mays* (germ)[b]	12·0	—	2·3	28·3	56·6	0·8	—	—	—	—	—	—	—	—
Wheat, *Triticum vulgare* (whole seed)[d]	22·3	—	0·8	9·6	62·9	2·6	1·3	—	—	—	—	—	—	—

[a] French (1962); [b] Craig and Murti (1959); [c] Appelqvist (1969); [d] Fisher and Broughton (1960).

phospholipid and glycolipid, together with a small proportion of triglyceride; the levels of the last class of lipid increase markedly with seed maturation (see p. 236).

Although the majority of seed oils contain only "conventional" fatty acids (Table 3.3) seeds are also the major source of the "unusual" classes of fatty acid in plants (Table 3.4; and see Chapter 1) and these acids are usually confined to this class of tissue, although a very few but notable exceptions to this rule are also known (p. 38).

TABLE 3.4

Some seed-oils containing unusual fatty acids[a]

Plant	Unusual acid[b]	Fatty acid (% of total acids)			
		Unusual acid	16:0	18:1	18:2
Helenium bigelowii[c]	*trans*-3-hexadecenoic	10	10	8	65
Petroselinum sativum[d]	petroselinic	76	2	15	6
Aleurites montana[e]	α-eleostearic	81	—	3	12
Crepis foetida[f]	crepenynic	60	5	4	28
Ricinus communis[g]	ricinoleic	90	1	3	5
Vernonia anthelmintica[h]	vernolic	72	4	6	16
Sterculia foetida[i]	sterculic	50	20	11	12
Hydnocarpus wightiana[j]	hydnocarpic[k]	49	2	6	—

[a] The distribution and structures of these and other unusual acids are discussed in Chapter 1 (p. 9); [b] Structures are illustrated in Table 1.5 (p. 10); [c] Hopkins and Chisholm (1964c); [d] Placek (1963); [e] Hilditch and Mendelowitz (1951); [f] Mikolajczak *et al.* (1964); [g] Canvin (1963); [h] Gunstone (1954); [i] Wilson *et al.* (1961); [j] Cole and Cardoso (1939); [k] Also chaulmoogric (27%) and gorlic (12%).

B. Corms, Tubers and Bulbs

Other types of plant storage organ, such as corms, tubers and bulbs, normally accumulate polysaccharides rather than triglyceride as their major source of reserve energy. Consequently the lipid content of these tissues is generally low and is mainly comprised of phospholipids and glycolipids. For example, Lepage (1968) reported that only about 0·5% of the tissue dry weight of potatoes is comprised of lipid, of which 16·5% is "neutral lipid" (mainly triglyceride), 45·5% phospholipids and 38·1% glycolipids; comparable results have been obtained by Galliard (1968a). Similarly, narcissus bulbs (Nichols and James, 1964) and turnip roots (Lepage, 1967) contain 1·2 and 1·9% lipid respectively, phospholipids and glycolipids predominating. In all such tissues examined to date, the most abundant component fatty acids were palmitic, linoleic and linolenic acids (Table 3.5).

TABLE 3.5

Fatty acid composition of some plant storage organs

	Fatty acid (% of total acids)						
Organ	16:0	16:1	17:1	18:0	18:1	18:2	18:3
Narcissus bulb[a]	17·0	1·7	—	0·7	8·9	68·9	2·9
Turnip root[b]	15·0	0·7	2·6	1·5	7·8	13·9	58·6
Potato tuber[c]	19·7	—	—	2·8	3·5	59·8	15·2

[a] Nichols and James (1964); [b] Lepage (1967); [c] Galliard (1968a).

C. Flowers

Comparatively little is known regarding the lipid composition of flower tissues, although Thompson et al. (1965, 1968) found that flower buds of the cotton plant (Gossypium sp.) contain approximately equal quantities of triglycerides and polar lipids, mono- and di-galactosyl diglycerides and lecithin comprising the major part of the latter fraction.

D. Mitochondria

Although mitochondria play an essential general role in plant metabolism and must be involved in many aspects of lipid metabolism in plants, little quantitative data is available regarding their lipid composition. Abdelkader et al. (1969) reported that the major fatty acids of potato tuber mitochondria were linoleic and palmitic acids with linolenic acid comprising most of the remainder, and these results were in general agreement with analyses previously obtained for plant mitochondria from other sources (Table 3.6). The same workers (Abdelkader et al., 1969) also noted that their mitochondrial preparation contained phospholipids, mono- and di-galactosyl diglycerides and sterol glycoside esters, as well as neutral lipids and free fatty acids, but no indication of the relative concentrations of these components was given, except that phosphatidyl choline and phosphatidyl ethanolamine were the major phospholipids present.

It is consequently difficult to assess the proportion of lipids present in whole plant cells which originate in their mitochondria, although Ongun

3

TABLE 3.6

Fatty acid composition of plant mitochondrial preparations

Source	Fatty acid (% of total acids)							
	14:0	16:0	16:1	17:0	18:0	18:1	18:2	18:3
Potato tuber[a]	—	20·0	—	—	3·5	2·0	55·0	16·0
Apple flesh[b]	—	23·0	—	—	5·2	4·0	48·0	19·0
Pear flesh[b]	—	22·0	—	—	2·2	6·0	62·3	7·5
Carrot root[c]	0·2	31·0	0·8	0·6	1·4	3·8	55·0	5·8

[a] Abdelkader *et al.*, 1969; [b] Romani *et al.*, 1965; [c] Dalgarno and Birt, 1963.

et al. (1968) showed that the chloroplasts of tobacco leaves contained at least 83% of the total cellular monogalactosyl diglyceride, 88% of the digalactosyl diglyceride, 76% of the sulpholipid and 74% of the phosphatidyl glycerol, so that the proportion of these lipids occurring in the other subcellular organelles of tobacco leaves may be quite small.

E. Leaves

1. Whole cells

The leaves of higher plants contain up to 7% of their dry weight as acyl lipids, pigments and hydrocarbons. It was originally believed that the major class of acyl lipid in leaves was triglyceride, but in 1959 Weenink presented evidence that leaves contain comparatively little triglyceride and that the major lipids of leaves are galactosyl diglycerides. Wintermans (1960) subsequently demonstrated that the major classes of lipid in leaves are galactosyl diglycerides, sulpholipid, phosphatidyl glycerol, phosphatidyl inositol, phosphatidyl choline and phosphatidyl ethanolamine. More recently, the development of improved analytical techniques permitted Roughan and Batt (1969) to carry out quantitative lipid analyses of a wide variety of photosynthetic tissues in which the relative concentrations of different classes of leaf lipid varied considerably between plant species (Table 3.7).

Phospholipids frequently found in trace quantities in leaves include phosphatidyl serine, phosphatidic acid and phytoglycolipid; other minor leaf lipids are triglyceride, diglyceride, glucocerebroside, sterol esters and (possibly) esterified sterol glycoside.

Dark-grown (yellow) leaves contain lower levels of glycolipids and phospholipids than similar tissues grown in the light (Wintermans, 1960), and Nichols (1963) presented chromatographic evidence that the

TABLE 3.7

Glycerolipid composition of photosynthetic tissues (from Roughan and Batt, 1969)

| Plant | Galactosyl diglycerides | | Sulpholipid | Lecithin | Phosphatidyl glycerol | Phosphatidyl ethanolamine | Phosphatidyl inositol | Cardiolipin |
	Mono-	Di-						
					μ Moles lipid per g fresh tissue			
Moss (unidentified)	2·68	1·50	0·48	1·26	0·40	0·40	0·15	0·05
Maidenhair tree (*Ginkgo biloba*)	4·70	2·80	0·30	1·80	0·85	0·55	0·25	0·27
White clover (*Trifolium repens*)	8·60	5·20	0·76	1·41	0·13	0·87	0·25	0·28
Tomato (*Solanum esculentum*)	5·08	2·46	0·31	1·10	0·43	0·45	0·10	0·10
Lettuce (*Lactuca sativa*)	0·68	0·68	0·03	0·31	0·10	0·21	0·06	0·13
Perennial ryegrass (*Lolium perenne*)	5·10	3·95	0·45	1·35	0·75	0·55	0·20	0·20
Maize (*Zea mays*)	3·10	2·30	0·35	0·45	0·48	0·24	0·12	0·24

inner (yellow) leaves of cabbage contain lower proportions of galactosyl diglycerides, sulpholipid and phosphatidyl glycerol, and higher proportions of lecithin and phosphatidyl ethanolamine, than the outer (green) leaves.

The fatty acid composition of the leaves of higher plants follow a generally consistent pattern in which only slight quantitative variations occur between most classes of plant. Quantitatively the major fatty acids of leaves are α-linolenic, linoleic and palmitic acids, although the leaves of some plants such as spinach and tobacco accumulate 7,10,13-hexadecatrienoic acid (Table 3.8). Oleic and palmitoleic acids are the most abundant of leaf monoenoic acids while a third monoenoic acid, *trans*-3-hexadecenoic acid is apparently ubiquitous in green leaves but usually represents only a minor proportion of the total fatty acids.

Nichols *et al.* (1967a) were unable to confirm earlier reports that certain photosynthetic tissues, namely those of Stinging nettle *Urtica dioica* (Hilditch and Meara, 1944) and the black pine, *Pinus thunbergii* (Tsujimoto, 1940) differ from those of other plants in containing no linolenic acid.

Hydroxy acids are seldom identified in leaf extracts although Sastry and Kates (1964a) have shown that runner bean leaves contain a series of these acids specifically associated with the cerebroside fraction.

Etiolated (dark-grown) leaves contain no *trans*-3-hexadecenoic acid and a smaller proportion of polyunsaturated fatty acids than the corresponding light-grown tissues, this last difference being more pronounced in some tissues than in others (see Table 3.8). Similarly, lipids deficient in linolenic acid also occur in the white parts of variegated leaves from a variety of plants (Crombie, 1958).

The fatty acids of unusual structure which characterize the seed-oils from a variety of plants are seldom found in the lipids of the corresponding leaf tissue, although two exceptions to this general rule have been established. The first occurs in plants of the family Malva, some members of which accumulate cyclopropenoid fatty acids in their leaves as well as their seeds (Shenstone and Vickery, 1961), while the γ-linolenic acid and octadecatetraenoic acids which characterize the seed-oils of the Boraginaceae are also found in the component lipids of their leaves (Jamieson and Reid, 1968, 1969).

2. Chloroplasts

Most of the acyl lipids of chloroplasts are located in the lamellae and osmiophilic grana (Bailey and Whyborn, 1963; Shibuya and Maruo, 1965) and present a much simpler composition than is found in the whole leaf cell. The major chloroplast lipids comprise three glycolipids (mono-

TABLE 3.8

Fatty acid composition of leaves

	14:0	16:0	16:1 (9c)	16:1 (3t)	16:3	18:0	18:1	18:2	18:3 (6,9,12)	18:3 (9,12,15)	18:4	20:0	22:0	24:0
Pasture grasses[a]	1·1	15·9	2·5		—	2·0	3·4	13·2	—	61·3	—	0·2	0·3	—
Stinging nettle[b]	3·0	11·7	—	2·0	—	1·2	2·2	15·3	—	66·1	—	—	—	—
Antirrhinum[c]	0·1	13·4	—	1·3	t	2·4	1·8	17·7	—	51·9	8·6	1·8	3·6	—
Spinach[c]	0·2	12·9	—	2·6	4·6	t	6·6	16·3	—	56·2	—	—	—	—
Broad bean[d]	—	11·7	6·9		—	3·2	3·4	14·3	—	56·4	—	—	4·0	—
Broad bean (etiolated)[d]	—	16·7	—		—	4·7	—	33·5	—	39·4	—	—	4·6	—
Holly (unvariegated)[e]	—	22·0	t	t	—	t	2·5	13·8	—	60·2	—	—	—	—
Water forget-me-not[f]	1·0	21·3	1·8		—	2·1	4·0	23·9	12·8	15·0	5·6	t	5·0	4·6

[a] Garton, 1960; [b] Nichols *et al.*, 1967a; [c] Debuch, 1961; [d] Crombie, 1958; [e] Nichols, 1965a; [f] Jamieson and Reid, 1968.

TABLE 3.9

Lipid composition of chloroplasts

| Source | Galactosyl diglycerides | | Sulpholipid | Lipid (% of total lipids) | | | | |
	Mono-	Di-	Tri-		Phosphatidyl glycerol	Phosphatidyl choline	Phosphatidyl ethanolamine	Phosphatidyl inositol	Phosphatidic acid
Sugar beet[a]	44·0	24·3	—	5·6	8·7	9·0	3·4	2·3	2·5
Tobacco[b]	42·0	30·6	—	10·0	10·0	4·0	2·7	0·7	—
Spinach (lamellae)[c]	38·1	25·4	3·2	11·1	14·3	4·8	—	3·2	—

[a] Wintermans, 1960; [b] Ongun et al., 1968; [c] Allen et al., 1966.

and di-galactosyl diglyceride and sulphoquinovosyl diglyceride) and one phospholipid (phosphatidyl glycerol). Whether small quantities of lecithin also occur in chloroplasts is a problem which is as yet unresolved, and is a reflection of the difficulties involved in the isolation of chloroplasts free from other cellular constituents. Analyses of a variety of chloroplast preparations (Benson and Maruo, 1958; Wintermans, 1960; Nichols, 1963; Ongun et al., 1968) have indicated the presence of small proportions of this lipid (Table 3.9) which has also been proposed as a cofactor in the conversion of oleate to linoleate in chloroplasts (Gurr et al., 1969). On the other hand Shibuya et al. (1965) could not detect lecithin in spinach lamellae preparations while Nichols (Nichols and James, 1968) was unable to obtain biochemical evidence for the presence in spinach leaves of separate lecithin pools of different metabolic function.

The fatty acids of leaf chloroplasts are notable for their high degree of unsaturation in which trienoic acids, particularly α-linolenic acid, predominate (Table 3.10).

TABLE 3.10

Fatty acid composition of chloroplasts

Source	Fatty acid (% of total acids)								
	14:0	16:0	16:1	16:3	18:0	18:1	18:2	18:3	22:0
Antirrhinum[a]	t	9·5	1·3	0·2	1·0	1·6	15·1	71·3	—
Broad bean[b]	—	7·4	9·2	—	1·2	5·2	2·6	72·0	1·2
Wheat[c]	t	17·3	1·5	—	1·5	2·3	6·0	70·0	—

[a] Debuch, 1961; [b] Crombie, 1958; [c] Wolf et al., 1966.

3. Cuticles

The cuticles of leaves and fruits consist of a network of cross-esterified hydroxy fatty acids (p. 58) embedded in "wax" (Martin, 1964), a term which should strictly be reserved for esters of long chain fatty acids and alcohols but is frequently applied to the mixture of long chain paraffins, esters, ketones, alcohols, acids and occasionally other substances which also occur in the cuticle. The wax esters which occur in these fractions are usually composed of long-chain fatty acids and primary alcohols of even carbon number having similar compositions to those of the free fatty acid and alcohols from the same wax (Table 3.11) although an exception to this general rule occurs in the leaf wax of Little Club wheat in which the free fatty acid fraction contains a much higher proportion of palmitic acid than is found in the component fatty

TABLE 3.11

Composition of some cuticular waxes (Kolattukudy, 1970)

Fraction	% in total waxes	C12	C14	C16	C18	C20	C22	C23	C24	C25	C26	C27	C28	C29	C30	C31	C32	C33	C34	C35
Cabbage leaf																				
Paraffins	36·0	—	—	—	—	—	—	—	—	—	—	—	1	93	1	3	—	—	—	—
Esters {alcohols	12·6	14	14	—	—	—	—	—	—	—	52	—	14	—	—	—	—	—	—	—
Esters {acids		14	28	—	28	—	14	—	8	—	—	—	—	—	—	—	—	—	—	—
Primary alcohols	8·7	—	—	—	—	6	12	—	6	—	36	—	6	—	—	—	—	—	—	—
Free acids	9·2	—	—	24	24	41·5	5·5	—	5·5	—	—	—	—	—	—	—	—	—	—	—
Secondary alcohols	11·1	—	—	—	—	—	—	—	—	—	—	—	—	100	—	—	—	—	—	—
Ketones	13·8	—	—	—	—	—	—	—	—	—	—	—	—	100	—	—	—	—	—	—
Ketols	0·9	—	—	—	—	—	—	—	—	—	—	—	—	100	—	—	—	—	—	—
Grape berry																				
Paraffins	1	—	t	t	t	t	t	5·7	3·5	17·2	3·8	19·5	2·6	22·1	2·4	14·8	1·1	1·9	—	0·4
Alcohols {free	40	—	—	—	—	0·1	1·3	1·2	14·2	5·7	42·6	5·3	21·3	4·4	3·3	—	t	—	t	—
Alcohols {esterified	9	—	—	—	—	0·2	1·7	1·7	11·6	6·2	44·4	6·6	20·7	1·1	1·2	—	t	—	—	—
Aldehydes	12	—	0·4	0·2	0·8	1·7	0·4	0·6	12·4	2·8	41·7	2·5	21·8	1·0	7·5	0·5	2·8	—	—	—
Acids {free	7	—	0·4	4·6	9·5	12·2	7·8	1·4	12·8	2·6	81·0	1·2	10·6	1·0	3·3	2·3	2·0	—	—	—
Acids {esterified	9	—	0·4	5·3	18·1	31·7	18·0	1·5	9·8	1·4	3·9	0·7	2·3	0·7	1·2	0·3	1·2	—	—	—
Sugar cane leaf																				
Paraffins	8·5	—	—	5·4	—	—	—	—	1·1	7·0	4·9	55·7	2·9	12·8	2·2	4·4	—	1·6	—	—
Alcohols {free	26	—	—	—	—	—	—	—	0·6	0·4	15·0	4·6	72·1	1·7	3·5	0·9	1·2	—	—	—
Alcohols {esterified		—	—	—	—	—	—	—	0·6	0·2	13·1	2·7	73·0	1·5	6·3	1·0	1·6	—	—	—
Acids	10	—	—	2·9	1·8	1·2	1·9	—	2·5	1·9	8·2	8·0	50·5	2·4	8·1	1·0	4·4	1·4	2·5	—
Aldehydes	50	—	—	0·6	—	—	—	—	0·25	0·2	7·35	1·7	66·2	2·1	12·25	1·6	4·2	—	3·45	—

acids of the corresponding wax ester fraction. Wax fatty acids have a different composition from those of the "internal" lipids of the leaf, such as the phospholipids and glycolipids. Fairly minor proportions of acids and alcohols of odd carbon number may also appear in these fractions, while in contrast the major paraffins of cuticular waxes usually have odd carbon numbers, and in many cases a single class of paraffin constitutes the major part of the wax hydrocarbons. In the cuticular wax of cabbage leaves for example, the C_{29} paraffin constitutes 93% of the wax hydrocarbons (Kolattukudy, 1965) whereas the C_{31} normal paraffin represents over 98% of the wax hydrocarbons of the leaves of *Pisum sativum* (Macey and Barber, 1970).

The major ketones and secondary alcohols usually have the same chain length as that of the major paraffin in the same wax, and the metabolic significance of these and other structural similarities between various components of cuticular waxes is discussed on p. 252.

2. MOSSES AND FERNS

The photosynthetic tissues of members of the Bryophyta and Pteridophyta have similar lipid compositions to the leaves of higher plants (Nichols, 1965a; Radunz, 1968a; Roughan and Batt, 1969), but differ considerably in their fatty acid content. Schlenk and Gellerman (1965) showed that a variety of ferns and mosses contained substantial quantities of polyunsaturated fatty acids of 20 or more carbon atoms including arachidonic acid, hitherto regarded as an acid peculiar to mammalian tissues and marine algae. Nichols (1965a) and Wolf and coworkers (1966) subsequently demonstrated that a large proportion of these higher polyunsaturated acids occur in the tissue chloroplasts (Table 3.12). Mosses and ferns also contain C_{16} and C_{18} monoenoic acids unsaturated at the 11-position, in addition to the more common 7- and 9-isomers (Schlenk and Gellerman, 1965).

C_{20} polyunsaturated acids and 11-enoic acids are also found in the leaves of *Ginkgo biloba* which is formally classified as a gymnosperm but which nevertheless possesses certain features which would justify its classification as a member of the pteridophyta.

3. ALGAE

A. Higher Algae

All algae, apart from the blue-greens (cyanophyceae) synthesize and accumulate the major varieties of acyl lipid found in the photosynthetic

TABLE 3.12

Fatty acid composition of various mosses and ferns, and *Ginkgo biloba*

	Fatty acid (% of total acids)																
	14:0	16:0	16:1	16:2	16:3	18:0	18:1	18:2	18:3	20:0	20:1	20:2	20:3	20:4	20:5	22:0	24:0
Ginkgo biloba (leaves)	1·5	19·3	1·5	0·3	4·9	1·6	4·8	8·2	48·2	—	—	1·2	1·7	3·4	—	—	3·3
MOSSES																	
Sphagnum (total)[a]	1·1	14·5	7·8	—	3·1	1·9	11·8	16·2	31·4	—	—	—	—	9·6	—	2·1	—
Sphagnum (leafy parts)[a]	0·7	20·6	1·4	—	2·2	1·6	25·6	19·5	13·2	—	—	—	3·0	10·2	—	—	—
Brachythecium and *Mnium*[b]	0·1	12·1	2·5	—	2·6	t	2·8	14·4	14·7	0·3	0·3	33·8	—	5·6	1·7	2·3	—
Hypnum cupressiforme[c]	1·2	13·5	5·1	—	1·9	1·7	7·3	19·7	23·5	—	—	—	—	11·7	7·2	4·6	—
Anomodon rostratus (chloroplasts)[d]	5·6	23·1	13·6	—	—	3·0	11·8	9·3	21·6	—	—	—	—	4·6	5·4	—	—
FERNS																	
Cyrtomium falcatum (chloroplasts)[d]	1·0	24·9	2·8	—	—	1·2	8·9	31·2	14·9	—	—	—	—	5·2	—	—	—
Onoclea sensibilis[a]	0·4	21·1	6·8	1·0	5·3	—	4·8	5·4	47·9	—	—	0·3	—	8·0	—	—	2·4
Matteucia struthiopteris[a]	2·9	17·6	3·0	—	4·1	—	3·2	6·8	54·2	—	—	—	—	8·2	—	—	—
Equisetum arvense[a]	—	28·2	4·4	1·8	9·4	—	4·0	11·5	43·0	—	—	—	0·6	2·1	—	—	—

[a] Schlenk and Gellerman, 1965; [b] Gellerman and Schlenk, 1964; [c] Nichols, 1965a; [d] Wolf *et al.*, 1966.

TABLE 3.13

Fatty acid composition of some freshwater algae

	14:0	16:0	16:1 (9 or 7)	16:2 (3)	16:3	16:4	Fatty acid (% of total acids) 18:0	18:1	18:2	18:3	18:4	20:2	20:3	20:4	20:5	U[a]
Nitella[b]	1·4	21·4	1·6	—	7·2	—	—	t	2·3	31·2	17·0	—	—	t	6·3	—	—
Scenedesmus obliquus[c]	0·9	35·4	1·6	—	t	0·2	15·0	—	7·8	6·4	29·5	1·9	0·6	—	—	—	—
Chlorella pyrenoidosa[c]	—	19·7	3·1	—	2·5	7·3	—	—	45·8	9·5	12·0	—	—	—	—	—	—
Chlorella vulgaris (dark-grown)[d]	t	25·8	10·7	—	3·6	—	—	3·6	18·1	36·3	1·3	—	—	—	—	—	—
Chlorella vulgaris (light-grown)[d]	2·0	26·0	8·4	t	6·9	2·1	—	2·2	2·2	33·5	20·1	—	—	—	—	—	—
Navicula pelliculosa (diatom)[e]	2·8	9·1	30·8	—	3·2	18·3	—	—	t	6·2	3·9	2·6	—	—	4·5	14·5	4·2

[a] U = unknown acid, possibly a C_{16} dienoic acid isomer; [b] Schlenk and Gellerman, 1965; [c] Klenk *et al.*, 1963; [d] Nichols, 1965b; [e] Kates and Volcani, 1966.

tissues of higher plants, although sterol glycosides and cerebrosides have not been detected in algae (Nichols, 1970).

The fatty acids of freshwater algae are similar to those found in leaves, although the relative proportions of the different classes of acid may differ widely. For example, photoautotrophic cultures of *Scenedesmus obliquus* and *Chlorella vulgaris* exhibit very similar fatty acid compositions to those of many leaves except that the proportions of linolenic acid are somewhat lower, but when *Chlorella* is grown on media containing substantial quantities of organic nutrients the cells accumulate markedly depressed levels of polyenoic acids (Nichols, 1965b) (Table 3.13). Such changes in levels of the polyenoic acids are undoubtedly a reflection of the reduced importance of the photosynthetic apparatus in algae growing in media rich in nutrients; such cultures produce fewer chloroplasts than those grown photoautotrophically and because the chloroplast is the major site of linolenic acid accumulation (p. 71) the concentration of this acid is largely dependent on the photosynthetic activity of the cell.

Studies with marine algae by Klenk and coworkers (1963), Wagner and Pohl (1965) and Patton *et al.* (1966), and those by Kates and Volcani (1966) on diatoms, have shown that these classes of algae differ from most freshwater algae in synthesizing polyenoic acids of 20 or more carbon atoms (Table 3.14). In cells which contain these acids, the proportion of α-linolenic acid is correspondingly low and Reid and Jamieson (1970) have shown that the proportion of 20:5 acid in the total cellular lipid increases in the order green algae ⟨ red algae ⟨ brown algae.

B. Blue-green Algae (Cyanophyceae)

Among photosynthetic organisms which perform the Hill reaction, the blue-green algae (Cyanophyceae) possess a uniquely simple acyl lipid composition. They do not synthesize phosphatidyl choline, phosphatidyl ethanolamine and phosphatidyl inositol, which are found in all other classes of algae, and their major acyl lipids are those four classes ubiquitous to the chloroplasts of leaves and algae, namely mono- and di-galactosyl diglyceride, sulphoquinovosyl diglyceride and phosphatidyl glycerol (Nichols *et al.*, 1965a). Some members of the Cyanophyceae contain lipids which are as yet uncharacterized but which have not been observed in other algae (Nichols *et al.*, 1965a; Allen *et al.*, 1966). The comparatively simple lipid compositions of blue-green algae are undoubtedly a reflection of the morphological structure of these organisms which contain few recognizable subcellular structures apart from their photosynthesizing lamellae; organized microsomes and mitochondria are not observed (Echlin and Morris, 1965).

TABLE 3.14

Fatty acid composition of some marine algae

	14:0	15:0	16:0	16:1 (9 or 7)	16:1 (3)	16:2	16:3	16:4	17:0	18:0	18:1	18:2	18:3 (9, 12, 15)	18:4	20:0	20:2	20:3	20:4	20:5	22:0	22:1	22:5	22:6	U[a]
RED ALGAE																								
Rhodomenia subfusca[b]	·9	0·7	29·1	4·9	—	0·2	0·2	t	0·5	0·3	14·9	1·4	1·1	0·8	—	—	0·5	14·0	24·1	—	—	—	0·9	—
Ceramium rubrum[b]	5·6	0·6	38·6	7·3	—	0·5	0·6	0·6	0·8	1·3	13·6	1·2	2·1	1·0	—	—	0·9	4·8	17·2	—	—	0·5	1·4	—
Placamium coccineum[b]	10·4	0·3	26·6	5·7	—	0·1	0·1	0·6	0·2	0·9	5·6	3·0	0·4	0·9	0·2	—	6·6	11·5	21·6	—	—	—	0·6	—
BROWN ALGAE																								
Fucus serratus[b]	9·7	0·7	26·4	2·2	—	0·1	0·4	0·1	0·4	0·8	18·7	9·0	6·0	6·0	0·5	0·6	0·4	10·0	8·1	—	—	—	—	—
Fucus platycarpus[b]	11·6	0·7	24·5	2·0	—	0·2	0·4	0·3	0·3	0·8	16·2	7·6	7·1	7·0	—	1·5	1·1	10·9	7·8	—	—	—	—	—
Fucus vesiculosus[b]	11·2	0·7	25·6	1·6	—	0·2	0·3	0·4	0·4	0·8	17·2	7·4	8·2	6·0	0·3	1·0	1·1	10·1	7·6	—	—	—	—	—
GREEN ALGAE																								
Codium fragile[b]	0·8	—	28·1	1·66	—	0·9	12·3	—	0·1	0·9	10·8	5·5	27·2	1·5	0·5	—	—	3·0	1·9	3·2	—	—	—	—
Enteromorpha compressa[b]	0·6	0·3	21·6	0·5	1·7	1·1	2·2	14·9	0·4	—	7·9	4·7	25·8	8·6	—	—	2·1	3·6	2·2	—	1·7	—	—	—
SALT-WATER DIATOMS																								
Nitzschia closterium[c]	5·5	0·4	25·4	24·9	—	1·8	3·7	—	—	0·6	2·6	2·5	1·7	—	—	—	—	9·4	17·2	—	—	—	—	1·7
Cyclotella cryptica[c]	4·5	1·4	15·7	34·8	—	5·0	7·8	—	—	1·7	0·9	0·7	1·6	—	—	—	—	—	18·8	—	—	—	—	7·4
Phaeodactylum tricornutum[c]	2·7	0·3	15·8	31·1	—	3·4	7·3	—	—	1·4	1·8	0·8	1·2	—	—	—	—	—	25·8	—	—	—	—	6·6

Fatty acid (% of total acids)

[a] U = unknown acid, possibly a C_{16} dienoic acid isomer; [b] Klenk *et al.*, 1963; [c] Kates and Volcani, 1966.

Within the blue-green algae there occur much larger qualitative and quantitative variations in fatty acid composition than are observed within the other algal classes, and because nearly all the acyl lipids of the blue-greens originate from the photosynthetic apparatus these differences clearly represent variations in the synthesizing capacities of the lamellae. The one generalization which can be made regarding the fatty acid compositions of this class of alga is that none of those so far examined can synthesize either arachidonic acid or *trans*-3-hexadecenoic acid. Reference to Table 3.15 shows that some blue-greens are unique among cells which perform the Hill reaction in containing no polyenoic acids (Holton *et al.*, 1964; Parker *et al.*, 1967). Others contain linoleic acid but no trienoic acids whereas a third group synthesizes α-linolenic acid and has fatty acid compositions not unlike those of green freshwater algae. One blue-green alga (*Spirulina platensis*) is apparently unique in synthesizing γ-linolenic acid rather than α-linolenic acid (Nichols and Wood, 1968).

TABLE 3.15

Fatty acid composition of some blue-green algae

	16:0	16:1	16:2	18:0	18:1	18:2	18:3 (9,12,15)	18:3 (6,9,12)
				Fatty acid (% of total acids)				
Spirulina platensis[a]	43·4	9·7	t	2·9	5·0	12·4	t	21·4
Myxosarcina chroococcoides[a]	38·2	8·6	1·2	4·0	6·8	9·2	33·3	—
Chlorogloea fritschii[a]	42·3	4·9	t	5·4	14·3	17·2	15·8	—
Anabaena cylindrica[a]	46·0	6·4	5·6	3·6	6·0	24·0	11·2	—
Anabaena flos-aquae[a]	39·5	5·5	4·3	1·0	5·2	36·5	10·7	—
Mastigocladus laminosus[a]	38·5	42·5	—	t	16·8	2·1	—	—
Anacystis nidulans[b]	47·0	38·8	—	1·4	10·0	—	—	—

[a] Nichols and Wood, 1968; [b] Holton *et al.*, 1964.

Certain protists, like *Euglena gracilis*, may develop and metabolize in a variety of ways, according to the conditions employed for their growth. Accordingly cultures of such organisms may contain extremely complex mixtures of fatty acids (Korn, 1964a) which can be divided into two groups, namely those arising from animal-like metabolism and which are particularly evident when dark- or heterotrophically-cultured cells are studied, and those more typical of photosynthetic tissues and which are present in significant quantities only in cells grown in the light on media con-

TABLE 3.16

Variations in the fatty acid composition of *Euglena gracilis* according to growth conditions

Growth conditions	13:0	14:0	15:0	16:0	16:1 (9 or 7)	16:2	16:3	16:4	17:1	18:0	18:1	18:2	18:3	18:4	20:2	20:3	20:4	20:5	20:6	22:5	22:6	Ref.
Dark grown[a]	—	11·6	—	19·4	16·0	—	—	—	—	1·8	7·2	2·7	2·2	—	3·1	1·1	12·6[e]	9·7	—	(13·0)	—	1
Light grown[a]	—	7·2	—	13·9	6·0	2·8	—	16·0	—	0·6	10·0	3·6	31·5	—	—	—	3·3	2·3	—	—	—	1
Dark grown (synthetic medium)[b]	27·6	22·9	6·5	7·5	1·6	2·0g	0·9	—	2·8	1·2	3·9	1·1	—	1·1	2·9	1·1	6·0	t	—	—	—	2
Light grown (synthetic medium)	4·6	14·2	1·4	12·7	2·3	2·1g	4·7	11·2	t	t	3·9	2·8	17·1	t	2·3	t	8·0	6·5	—	—	—	2
Dark grown (complex medium)[c]	0·7	10·0	0·7	18·1	5·9	1·9g	5·3	4·6h	1·1	—	11·2	3·3	5·2	t	12·8	2·8	15·5	t	—	—	—	2
Light grown (complex medium)	—	3·2	2·1	9·0	t	10·2g	5·8	16·5	2·6	—	4·8	9·3	17·8	t	2·3	t	6·4	4·4	—	—	—	2
Dark grown (synthetic medium)[b,f]	—	5·66	8·0	4·8	0·4	—	—	—	4·7	0·5	3·2	0·4	0·6	—	4·7	—	21·0[e]	13·4[e]	7·0[e]	4·5	7·0	3

Fatty acid (% of total acids)

[a] Containing sucrose as carbon source.
[b] Containing carbon dioxide, glutamic acid and malic acid as carbon sources. Medium pH, 3·3.
[c] Containing peptone, sodium acetate and yeast extract as carbon sources. Medium pH, 7·0.
[d] Mixture of C_{22} and C_{24} unsaturated acids.
[e] Mixture of isomers
[f] Also contained 17:0 (0·9%), 19:2 (1·1%) and 19:4 (2·2%) acids.
[g] A mixture of 16:2 and 17:0 acids.
[h] A mixture of 16:4 and 18:0 acids.
References: 1. Hulanicka *et al.*, 1964; 2. Haverkate, 1965; 3. Rosenberg, 1963.

taining little organic material. Other, more subtle, factors may also affect the fatty acid composition of *Euglena* cells. For example, Hulanicka and coworkers (1964) found that the amount of CO_2 flushed through the medium and the presence of organic carbon sources both influence the fatty acid composition of green cells, while Rosenberg (1963) and Rosenberg *et al.* (1964, 1967) showed that the concentration of energy source in dark cultures, and the phase of growth, are also of great importance (Table 3.16).

Nichols (1970) and Kenyon and Stanier (1970) have discussed the application of lipid fatty acid analyses to the study of phylogenetic relationships between photosynthetic organisms.

CHAPTER 4

Distribution of Individual Fatty Acids Between Lipid Classes

One of the more prominent features of the fatty acid composition of plant tissues, and certainly one of the most significant from the biochemical point of view, is that individual fatty acids are seldom distributed at random between the different classes of lipid present, but frequently exhibit clear-cut tendencies to associate with specific lipids. Moreover, the acyl groups in any particular lipid class are not arranged haphazardly between the 1-, 2- and 3-positions (in triglycerides) or 1- and 2-positions (in phosphatides or glycosyl glycerides) but are usually distributed between these positions in an ordered manner.

1. Seed Lipids

A. Triglycerides

1. Lipase studies

In triglycerides the three alternative sites for attachment of fatty acids theoretically permit a wide variety of isomers to be derived from even a comparatively limited number of different fatty acid classes. For example, there are 27 different ways in which three dissimilar fatty acids can be arranged between the three hydroxyl groups assuming that the 1- and 3-positions are not biochemically equivalent, as recent studies have confirmed.

In practice, however, earlier theories which proposed that the acyl groups in vegetable triglycerides obey a statistically random distribution based on the concentration of the fatty acids in the total oil were shown to be incorrect, firstly by the fact that catalytic randomization of a natural fat produced changes in its physical properties (Norris and Mattil, 1947) and more recently by studies with pancreatic lipase. The value of results from hydrolysis of natural triglycerides with this enzyme is that under the conditions generally employed for its use, acyl groups attached to primary hydroxyl groups are preferentially released (Mattson and Volpenhein, 1966) so that analysis of the component acids of the free fatty acid and monoglyceride fractions produced give values for the composition of acyl groups on the primary and secondary alcohol groups, respectively. Use of this technique in the analysis of vegetable triglycerides has indicated a marked specificity for saturated acids to accumulate at the primary positions of the glycerol molecule with correspondingly higher concentrations of unsaturated acids at the 2-position (Mattson and Volpenhein, 1961, 1963) so that in many vegetable oils 95–100% of the fatty acids in the 2-position are unsaturated C_{18} acids (oleic, linoleic and linolenic) even when the total concentration of these acids in the oil is as low as 37 or 38%.

These early results emphasized the non-random character of acyl group distribution in vegetable glycerides and has led to general accept-ance of the theory of positional distribution. Gunstone (1962) and Mattson and Volpenhein (1963) have suggested that the acids found in natural triglycerides fall into two groups, namely those which are preferentially esterified at the 1- and 3-positions (designated "saturated" by Gunstone, and "Category I" acids by Mattson and Volpenhein) and those which occur at the 2-position ("unsaturated" or "Category II" acids). Mattson and Volpenhein (1961) had earlier shown that although oleic, linoleic and linolenic acids belong to Category II, the C_{20} and C_{22} monoethenoid acids which characterize the Cruciferae seed oils belong to Category I and behave like palmitic and stearic acids. In the majority of seed oils the relative proportions of Group I and Group II acids do not permit their exact distribution between the primary and secondary glycerol hydroxyls, so that either Group I acids are found in the 2-position or, more commonly, Group II acids occur in the 1- and 3-positions. The analysis of a wide variety of natural oils has permitted extensive study of the general principles governing the arrangement of acyl groups in vegetable triglycerides and has led to the general accept-ance of a restricted randomization theory which has been formalized as the 1,3-random-2-random theory by the work of Vander Wal (1964), Coleman (1963) and others (Subbaram and Youngs, 1964; Jurriens and Kroesen, 1965).

In 1969 Evans *et al.* proposed a series of rules by which the structure of unsaturated vegetable oil triglycerides may be directly calculated from their overall fatty acid compositions. Rule 1 states that saturated fatty acids are distributed exclusively between the 1- and 3-positions, while Rule 2 requires that oleic and linolenic acid be distributed equally and randomly to all positions. Rule 3 then requires that the proportion of fatty acids on position 2 which is not attributed to oleic or linolenic acids according to Rules 1 and 2, must be taken up by linoleic acid.

Fats high in saturated acids or other category I acids, such as oils from palm kernel and members of the Cruciferae, have almost all their linoleate in the 2-position, but in other respects the principles governing the distribution of acyl groups in the Cruciferae oils appear to differ from those in most other vegetable oils.

2. *Stereospecific analysis*

In interpreting data from studies using pancreatic lipase many research groups assumed that fatty acids are distributed in the same proportion between the 1- and 3-positions, even though Schwartz and Carter (1954) and Hirschmann (1960) had already emphasized that,

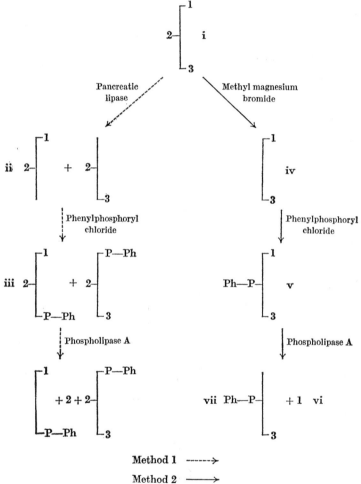

Method 1 ------>
Method 2 ------>

Fig. 4.1. Stereospecific analysis of triglycerides (Brockerhoff, 1965a, 1967).

regardless of the nature of the acyl groups involved, the 1- and 3-positions of glycerol are not interchangeable. Recently, determination of the individual fatty acid composition at the 1-, 2- and 3-positions in natural glycerides has been obtained by the technique of "stereospecific analysis" as independently developed in several laboratories. The original method devised by Brockerhoff (1965a) (Fig. 4.1) involved initial degradation of triglyceride (i) with pancreatic lipase to give a racemic α,β-diglyceride (ii) followed by the conversion of the diglyceride to a phospholipid (iii), which was then reacted with the stereospecific phospholipase A which degrades only esters of glycerol-3-phosphate. The major limitation of this method is that the fatty acids on the 3-position cannot be isolated and analysed directly; their composition has to be calculated by difference and this makes the determination of minor components in position 3 rather inaccurate and that of trace components impossible (Brockerhoff, 1965b).

In a subsequent scheme (Brockerhoff, 1967) these problems were resolved by employing specific chemical hydrolysis with methyl magnesium bromide to remove the acyl groups on the 2-position of the triglyceride, yielding a 1,3-diglyceride (Fig. 4.1.iv.). This was converted to L-2-phosphatidylphenol (v) and use was then made of the discovery by de Haas and van Deenen (1964) that phospholipase A will liberate the fatty acid (vi) from the 1-position of an L-2-phosphatidate. The fatty acid on position 3 of the original triglyceride can then be isolated by chemical hydrolysis of the lysophosphatide (vii).

A third technique for the stereospecific analysis of triglycerides has been described by Lands and coworkers (1966) who employed the diglyceride kinase of *E. coli* to phosphorylate selectively the 1,2-digly-

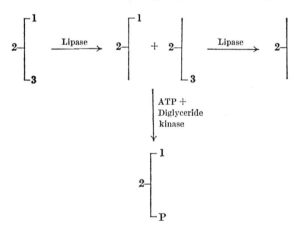

FIG. 4.2. Stereospecific analysis of triglycerides (Lands *et al.*, 1966).

cerides of the diglyceride mixture obtained by pancreatic lipase hydrolyses. Thus a comparison of the fatty acids in the intact triglyceride, the monoglyceride produced by lipase and the phosphatidate formed by the kinase reaction, indicated the acids esterified at each position in the triglyceride (Fig. 4.2).

Stereospecific analyses of a variety of vegetable oils by Brockerhoff and Yurkowski (1966) indicated that although the fatty acid compositions of the 1- and 3-positions were frequently very similar, in none of the fats examined was the distribution completely symmetrical (Table 4.1), although asymmetric distribution of C_{16} and C_{18} acids between these two positions was less pronounced than it is in many animals (e.g. Brockerhoff et al., 1965). The significance of this kind of information in relationship to the pathways by which triglycerides are synthesized by plant cells is discussed in Chapter 6.

3. Optical asymmetry in triglycerides

The question of whether optically active triglycerides occur in nature defied experimental solution for a long period. Clearly, if a single glyceride molecule contains different acyl groups on the 1- and 3-positions then the carbon atom in position 2 of the glycerol becomes asymmetric and optical activity becomes possible. The problem can therefore be regarded as comprising two basic questions, namely (1), whether individual triglyceride molecules occur in which structurally different acids are found on the 1- and 3-positions and (2) if such molecules do occur, are they present as one antipode or as a racemic mixture?

It is not always possible to answer (1) on the basis of stereospecific analyses alone since overall symmetry of fatty acid distribution between the 1- and 3-positions does not preclude asymmetry in individual triglycerides, even though overall asymmetry must preclude the complete symmetry of every individual. For example, we may consider a mixture of triglycerides which stereospecific analysis has shown to contain equal proportions of two acids, X and Y, on both 1- and 3-positions. Such a composition could be derived from a mixture containing equal quantities of two triglyceride species, one containing only X on the 1- and 3-positions (Fig. 4.3, Structure I) and the other exclusively containing Y at these positions (II). Alternatively the same overall composition would be observed in a mixture obtained by mixing a species which contains X at the 1-position and Y at the 3-position (III) with one containing Y at the 1-position and X at the 3-position (IV). In structures I and II the carbon atom at the 2-position is not asymmetric and optical activity is not possible. On the other hand both III and IV will be optically active but are optical antipodes so that if the two species are present

TABLE 4.1

Stereospecific analyses of vegetable fats (Brockerhoff and Yurkowski, 1966)

Oil or fat	Position	Fatty acid, moles %										
		16:0	16:1	18:0	18:1	18:2	18:3	20:0	20:1	22:0	22:1	24:0
Peanut	1	13·6	0·3	4·6	59·2	18·5	—	0·7	1·1[a]	1·3	—	0·7
	2	1·6	0·1	0·3	58·5	38·6	—	—	0·3[a]	0·2	—	0·5
	3	11·0	0·3	5·1	57·3	10·0	—	4·0	2·7[a]	5·7	—	2·9
Rapeseed	1	4·1	0·3	2·2	23·1	11·1	6·4	—	16·4	1·4	34·9	—
	2	0·6	0·2	—	37·3	36·1	20·3	—	2·0	—	3·6	—
	3	4·3	0·3	3·0	16·6	4·0	2·6	—	17·3	1·2	51·0	—
Soybean	1	13·8	—	5·9	22·9	48·4	9·1	—	—	—	—	—
	2	0·9	—	0·3	21·5	69·7	7·1	—	—	—	—	—
	3	13·1	—	5·6	28·0	45·2	8·4	—	—	—	—	—
Linseed	1	10·1	0·2	5·6	15·3	15·6	53·2	—	—	—	—	—
	2	1·6	0·1	0·7	16·3	21·3	59·8	—	—	—	—	—
	3	6·0	0·3	4·0	17·0	13·2	59·4	—	—	—	—	—
Maize	1	17·9	0·3	3·2	27·5	49·8	1·2	—	—	—	—	—
	2	2·3	0·1	0·2	26·5	70·3	0·7	—	—	—	—	—
	3	13·5	0·1	2·8	30·6	51·6	1·0	—	—	—	—	—
Olive	1	13·1	0·9	2·6	71·8	9·8	0·6	—	—	—	—	—
	2	1·4	0·7	—	82·9	14·0	0·8	—	—	—	—	—
	3	16·9	0·8	4·2	73·9	5·1	1·3	—	—	—	—	—
Cacao butter	1	34·0	0·6	50·4	12·3	1·3	—	1·0	—	—	—	—
	2	1·7	0·2	2·1	87·4	8·6	—	0·0	—	—	—	—
	3	36·5	0·3	52·8	8·6	0·4	—	2·3	—	—	—	—

[a] Together with 18:3.

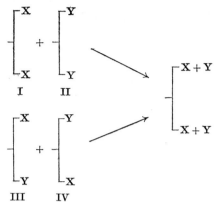

FIG. 4.3. Structures possibly contributing to a triglyceride mixture containing equal quantities of two different fatty acids (X and Y) at the 1- and 3-positions.

in equal proportions the original mixture will comprise a racemic mixture.

It is however a basic tenet of the 1,3-random-2-random hypothesis of triglyceride structure that all four structural types should be present in a natural mixture having the analytical data described, so that asymmetric triglycerides such as III and IV are invariably present. In confirmation of this the advent of superior analytical techniques has facilitated the isolation of a wide variety of asymmetric triglyceride fractions from plant tissues, the relative proportions of which in the parent

TABLE 4.2

Comparison of experimental values for molecular families with calculated values based on 1,3-random-2-random distribution (Youngs, 1961)

Fat	SSS[a]	SUS	SSU	USU	UUS	UUU
			Molecular families (mol %)			
Linseed oil						
Experimental value	0	0	0	4	22	74
Calculated value	0	1	0	2	20	76
Cocoa butter						
Experimental value	5	66	7	3	22	1
Calculated value	5	69	2	0	22	2

[a] S = saturated acids; U = unsaturated acids; SSS = trisaturated glyceride; SUS = glyceride with saturated acids on 1- and 3-positions and unsaturated acid on 2-position, etc.

oil are in accordance with calculations based on the 1,3-random-2-random theory as proposed by Richardson (Vander Wal, 1960). Thus Youngs (1961) compared experimental values for the distribution of different molecular classes in a variety of oils with those calculated by the method of Richardson, and found them to be in good agreement (Table 4.2).

The question whether the asymmetric structures found in these oils are optically active or whether they comprise racemic mixtures has been less easily tested because even optically pure enantiomeric triglycerides of the common long chain fatty acids have no measurable rotation (Sowden and Fischer, 1941), but the problem came nearer to solution through the observation of Maier and Holman (1964) that enantiomeric triglycerides containing a single "unusual" acyl group, e.g. deca-2,4-dienoic acid, exhibit measurable rotations. This phenomenon was applied by Morris (1965a, b) to a more generally applicable scheme for the detection of optical activity in natural asymmetric triglycerides (Fig. 4.4). In this scheme an asymmetric triglyceride (i) is subjected to hydrolysis by pancreatic lipase (p. 81) to give a mixture of diglycerides (ii and iii). The possibility of acyl migration and consequent racemization militates against the separation and measurement of the optical rotation of these compounds, but they may be immediately reacylated by reaction with sorbic acid chloride or trimethyl silyl chloride to yield stable reconstituted triglycerides, e.g. iv and v. These are virtually enantiomers differing only in the structure of the original acid radicals R_1 and R_3

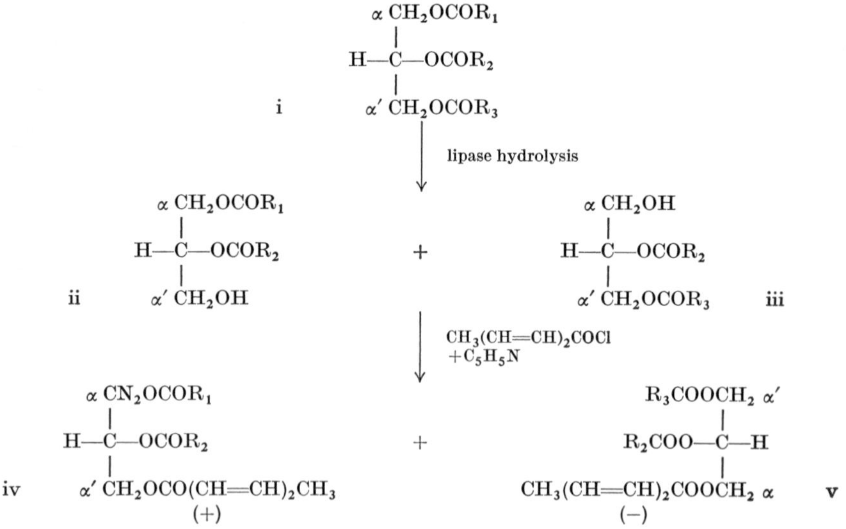

$$\begin{array}{c} \alpha\ CH_2OCOR_1 \\ | \\ H{-}C{-}OCOR_2 \\ | \\ i \qquad \alpha'\ CH_2OCOR_3 \end{array}$$

lipase hydrolysis

$$\begin{array}{c} \alpha\ CH_2OCOR_1 \\ | \\ H{-}C{-}OCOR_2 \\ | \\ ii \qquad \alpha'\ CH_2OH \end{array} \quad + \quad \begin{array}{c} \alpha\ CH_2OH \\ | \\ H{-}C{-}OCOR_2 \\ | \\ \alpha'\ CH_2OCOR_3 \qquad iii \end{array}$$

$CH_3(CH{=}CH)_2COCl$
$+C_5H_5N$

$$\begin{array}{c} \alpha\ CN_2OCOR_1 \\ | \\ H{-}C{-}OCOR_2 \\ | \\ iv \qquad \alpha'\ CH_2OCO(CH{=}CH)_2CH_3 \\ (+) \end{array} \quad + \quad \begin{array}{c} R_3COOCH_2\ \alpha' \\ | \\ R_2COO{-}C{-}H \\ | \\ CH_3(CH{=}CH)_2COOCH_2\ \alpha \qquad v \\ (-) \end{array}$$

FIG. 4.4. Scheme for the detection of optical activity in natural triglycerides (Morris, 1965a, b).

and providing these radicals differ in their degree of unsaturation the respective triglycerides can be separated from each other by argentation chromatography (p. 290). If the original triglyceride (i) was indeed optically active then the derived triglycerides will have opposite rotations of approximately equal magnitude. Morris thereby showed that asymmetric glycerides from palm oil, malabar tallow and cocoa butter were all optically active even though the specific rotations obtained generally had lower values than those expected on the basis of studies on model compounds (Table 4.3). These low values could mean that the natural glycerides were only predominantly and not exclusively of one enantiomeric form, although an alternative possibility is that the compounds studied were not absolutely pure.

TABLE 4.3

Optical activity of derivatives of natural triglycerides (Morris, 1965b)

Fat	Tri- glyceride	Diglyceride-TMSi ethers ([α])		Absolute configuration
		1,2-	2,3-	Hirschmann Cahn et al.
Palm oil	SOO[a]	OO-TMSi(+2·5°)	SO-TMSi(−2·3°)	O—[O ... S] (S)
Malabar tallow	SOO	OO-TMSi(+0·5°)	SO-TMSi(−3·9°)	O—[O ... S] (S)
Cocoa butter	SOO	OO-TMSi(+1·0°)	SO-TMSi(−3·2°)	O—[O ... S] (S)

[a] S = palmitic and/or stearic acid; O = oleic acid.

B. Phospholipids and Glycolipids

It is now well established that the phospholipid (and glycolipid) fractions from mature seeds frequently have quite different overall fatty acid compositions from those of the corresponding triglyceride fractions, and in particular seldom contain those "unusual" fatty acids which frequently typify the neutral fat.

Nevertheless few data are available regarding the distribution of individual fatty acids between the different classes of polar lipid, although Senn (1969) and Burkhardt (1970) have shown that the individual phosphatides of peanut and safflower oils possess fairly similar fatty acid compositions to those of other phosphatide fractions from the same source.

2. CORMS, TUBERS AND BULBS

The major classes of lipid in other types of energy storage tissues, such as corms, tubers and bulbs, are usually the phospholipids and glycolipids, and it seems likely that the distribution of fatty acids between the individual lipids follows a pattern similar to that established for photosynthetic tissue (see below). For example, Nichols and James (1964) found that the galactosyl diglyceride fractions from narcissus tuber contained higher levels of linolenic acid than the phospholipids from the same source, which themselves possessed high levels of linoleic acid.

3. PHOTOSYNTHETIC TISSUES

A. Distribution of Fatty Acids Between Lipid Classes

A substantial quantity of data has now accumulated regarding the distribution of fatty acids both between and within the individual lipid classes present in leaves and algae. Absolute specificities of fatty acids for single lipids seldom occur in these tissues except in the case of the *trans*-3-hexadecenoic acid (p. 10). Allen and coworkers (1964), Haverkate *et al.* (1964) and Weenink and Shorland (1964) independently demonstrated that this acid was present only in the phosphatidyl glycerol fraction from the leaves of spinach and red clover, and this absolute specificity was subsequently shown to operate in the leaves of castor (James and Nichols, 1966), holly (Nichols, 1965a) and in *Chlorella vulgaris* (Nichols, 1965b). An indication that even this specificity may not always be an absolute one has been suggested by the observations of Rotsch and Debuch (1965) that although the phosphatidyl glycerol from spinach contained the major portion of the cellular *trans*-acid, a small quantity of this acid was also located in a carefully purified phosphatidyl choline fraction. *Trans*-3-hexadecenoic acid also occurs in some seed oils, e.g. *Aster alpinus* oil (Wolff, 1966) but in these tissues it is presumably a triglyceride component.

In runner bean leaves a high degree of specificity was also observed in the distribution of 2-hydroxy fatty acids, which Sastry and Kates (1964a) showed were located almost exclusively in the cerebroside fraction (Table 4.4). This specificity may prove to be general for all leaves.

TABLE 4.4

Fatty acid composition of lipid fractions from runner bean leaves (Sastry and Kates, 1964a)

Lipid	16:0	16:1	18:0	18:1	18:2	18:3	2h-16:0	2h-18:0	2h-19:0	2h-20:0	2h-21:0	2h-22:0	2h-23:0	2h-24:0	2h-25:0	2h-26:0
							Fatty acids (% of total acids)									
Monogalactosyl diglyceride	2·3	t	t	t	2·2	95·5	—	—	—	—	—	—	—	—	—	—
Digalactosyl diglyceride	4·5	t	1·0	t	1·3	93·2	—	—	—	—	—	—	—	—	—	—
Phosphatidyl choline	26·6	t	6·0	3·7	37·7	25·7	—	—	—	—	—	—	—	—	—	—
Cerebroside	—	—	—	—	—	—	29·9	0·7	t	1·6	t	20·2	4·6	37·7	2·0	3·2

Study of a wide variety of photosynthetic tissues has emphasized that the more highly unsaturated fatty acids have a marked tendency to accumulate in the galactosyl diglyceride fractions, the monogalactosyl diglyceride normally being rather more highly unsaturated than digalactosyl diglyceride from the same cell. In the leaves of higher plants, for example, linolenic acid frequently comprises more than 90% of the component fatty acids of the monogalactosyl diglyceride fraction and more than 80% of the corresponding digalactosyl diglyceride. Phospholipid fractions from the same source usually contain less linolenic acid but more linoleic acid (Table 4.5). In tissues which accumulate substantial amounts of C_{16} or C_{20} polyenoic acids in the chloroplast, as in spinach leaves (Allen et al., 1964) Chlorella vulgaris (Nichols, 1965b) Euglena gracilis (Rosenberg et al., 1966; Rosenberg and Gouaux, 1967; Constantopoulos and Bloch, 1967) and other algae (Radunz, 1968b; Nichols and Appleby, 1969), these acids also preferentially accumulate in the galactosyl diglycerides, particularly the monogalactosyl lipid (Table 4.5).

TABLE 4.5

Fatty acid content of spinach leaf lipids (Allen et al., 1964)

Lipid	Fatty acid (% of total acids)							
	14:0	16:0	16:1 (3t)	16:3	18:0	18:1	18:2	18:3
Monogalactosyl diglyceride	—	t	t	30	—	1	1	67
Digalactosyl diglyceride	—	6	—	3	1	4	3	84
Sulphoquinovosyl diglyceride	—	27	—	—	—	6	39	28
Phosphatidyl glycerol	—	22	35	—	t	2	5	36
Phosphatidyl choline	—	20	t	t	—	11	30	40
Phosphatidyl inositol	t	41	—	—	1	6	25	27
Phosphatidyl ethanolamine	t	46	—	2	1	2	7	43

In dark-grown (etiolated) photosynthetic tissues the same specificities occur, even though the levels of different classes of fatty acid may be appreciably different from those observed in the corresponding green tissues.

Although no other general specificities have been established in regard to fatty acid distribution in green tissue, Radunz (1969) has pointed out that the sulphoquinovosyl diglyceride fraction from a wide variety of plant and algal tissues invariably contains comparatively high levels of palmitic acid.

B. Positional Distribution of Fatty Acids in Individual Phospholipids and Glycolipids

The component acids of the most ubiquitous plant phospholipids, i.e. phosphatidyl choline and phosphatidyl ethanolamine, are usually distributed between the 1- and 2-positions according to rules similar to those which govern acyl group distribution in mammalian phospholipids, namely that saturated acids tend to accumulate at the 1-position and the more highly unsaturated acids at the 2-position. For example, Sastry and Kates (1964a) and McKillican (1967) found that the 1-position in lecithin fractions from runner bean leaves and wheat contained 62·6% and 59·0% saturated fatty acids respectively, whereas the corresponding 2-positions contained only 2·7 and 3·1% of these acids.

Little information has been offered as yet regarding the structure of individual molecular species of these or similar fractions, although Haverkate and van Deenen (1965) determined the distribution of fatty acids in the phosphatidyl glycerol fraction from spinach leaves and found that the major molecular entity in this fraction is one in which linolenic acid occurs at the 1-position with *trans*-3-hexadecenoic acid occupying the 2-position. These results suggest that the distribution of fatty acids in plant phospholipids may not always conform to that observed in mammals.

Of the plant glycolipids, only monogalactosyl diglyceride fractions have been examined for positional distribution of component fatty acids. In 1969 Nichols and Moorhouse reported the fractionation of the mono-galactosyl diglycerides from heterotrophically cultured *Chlorella vulgaris* into molecular species of different fatty acid content, and demonstrated that the distribution of acyl groups in this lipid does not conform to any of the rules established for other classes of plant lipid. Subsequently, Safford and Nichols (1970) were able to show that the distribution of acids in the monogalactosyl diglyceride fractions from a variety of leaves and algae frequently depends on the chain length of the acids concerned, although the preferential accumulation of the more unsaturated acids at the 2-position as originally reported by Noda and Fujiwara (1967) also operates in those cases where the component acids are primarily of a single chain length. Where both C_{16} and C_{18} acids are present,

TABLE 4.6

Distribution of fatty acids between the 1- and 2-glycerol positions of natural monogalactosyl diglycerides
(Safford and Nichols, 1970)

Tissue	Fraction	Fatty acid (% of total acids)								
		16:0	16:1	16:2	16:3	18:1	18:2	18:3 (9,12,15)	18:3 (6,9,12)	18:4
Anabaena cylindrica	Total	24	8	10	3	3	22	30	—	—
	1-	5	6	2	1	4	35	47	—	—
	2-	47	11	19	5	2	6	9	—	—
A. flos-aquae	Total	15	6	6	13	2	6	52	—	—
	1-	6	7	2	3	5	9	69	—	—
	2-	28	5	11	26	—	2	28	—	—
Chlorella vulgaris (heterotrophic)	Total	2	12	27	4	15	22	18	—	—
	1-	3	2	3	—	24	36	32	—	—
	2-	2	20	55	9	4	5	4	—	—
C. vulgaris (photoautotrophic)	Total	3	2	18	10	—	22	46	—	—
	1-	—	3	6	2	—	29	60	—	—
	2-	5	2	30	19	—	13	30	—	—
Spinach, *Spinacea oleracea* (leaf)	Total	2	—	—	21	1	2	74	—	—
	1-	—	—	—	5	2	3	90	—	—
	2-	4	—	—	38	—	2	56	—	—
Anchusa (leaf)	Total	3	—	—	—	—	3	72	4	19
	1-	5	—	—	—	—	3	90	—	3
	2-	1	—	—	—	—	2	52	7	38

the C_{18} acids are preferentially located at the 1-position (Table 4.6). The biochemical significance of the manner in which individual acids are distributed between and within plant lipids is discussed more fully in Chapter 6.

CHAPTER 5

Biosynthesis of Plant Fatty Acids

1. Introduction

The variety of fatty acid structures found in nature and described in Chapter 1 must each be due to a number of enzymic reactions resulting in the accumulation of a particular molecular species. The wide differences of observed quantitative compositions suggest some variation of specificity in the activity or control of these enzymes, while the fact that qualitative composition is often characteristic of a particular group of related sources may be explained by the presence of particular enzymes in these sources only. In general terms, however, the molecule of a natural fatty acid has a straight chain usually composed of an even number of carbon atoms, with a few chemical features at specific positions: C_{16} and C_{18} acids predominate, with unsaturated acids characterized by monoene or methylene-interrupted polyene structures near the centre of the chain. It would be surprising if there were a multiplicity of mechanisms of biogenesis of this carbon chain, which provides the backbones of all natural fats. The biochemical relationships between individual fatty acids during anabolism is now considered; catabolic pathways and some oxidative reactions are described in Chapter 8 (p. 201).

If plant tissue is analysed at various times, e.g. during development of seeds or maturation of leaves, the variation in composition can be accurately recorded. The levels of particular components of known structure can be followed, but these data alone give only an indication of the overall effect of the enzymic reactions which are taking place. Moreover, exercises in "paper chemistry" may also be misleading if the tissue is regarded only as a container for chemical reactions. The control of a sequence of enzymically catalysed reactions, especially *in vivo*, is often more complex than mass action and direct temperature effects would suggest. The detailed mechanism of biochemical reactions in complex biological systems is therefore usually investigated with the aid of substrates labelled with isotopes.

Early hypothetical pathways of fatty acid biogenesis based on insufficient data have been disproved by experiment. The direct condensation of three hexose units to give a C_{18} acid during conversion of sugar to fat does not occur, nor do chemically plausible aldol condensations involving acetaldehyde or pyruvic acid. The ubiquity of the methylene-interrupted diene system also led to the incorrect assumption that six C_3 units, which might arise from glycolysis intermediates or malonic acid, are incorporated in toto (Hilditch and Williams, 1964).

The biochemical relationship between saturated and unsaturated fatty acids has also led to some confusion: the saturated acids could be formed first and then dehydrogenated (desaturated) to their unsaturated analogues; the converse could be true, involving a hydrogenation reaction; or the two types might be synthesized by essentially different pathways. Before labelled substrates became available, the desaturation pathway seemed to be confirmed by the fact that the iodine value of plant lipids (a measure of degree of unsaturation) increased during development; however, this was the result of accumulation of both saturated and unsaturated acids at different rates. Apparent evidence in favour of the hydrogenation pathway included the supposition that the final product of lipogenesis would be the saturated acid, which represents a slightly better energy store. This appeared to be confirmed by the well-established fact that a plant yields lipid of higher iodine value when grown in cool conditions (p. 152), under which the rate of synthesis is slower. Examination of these analytical data suggested that the plant oil from different localities had the same saturated acid content, while the unsaturated content varied; this implied different pathways of biosynthesis. It was observed that oil from the same plant had a high level of stearate, linoleate and linolenate when the oleate content was low, and vice versa. This was taken to indicate that linolenate (or a precursor) was synthesized first, and progressively hydrogenated to oleate; stearate would have to be the product of another pathway not involving oleate.

In fact, tracer experiments have now shown that the usual sequence of biological events is the formation of a saturated acid (p. 102), followed by sequential desaturation (p. 130) with the possibility of further chain elongation after desaturation (p. 123). However, hydrogenation of unsaturated acid is effected by a few rumen micro-organisms, and biosynthetic pathways of saturated and unsaturated fatty acids in certain bacteria diverge at an early stage (p. 131).

2. ACTIVATION OF CARBOXYLIC ACIDS

Most of the enzymes involved in the metabolism of acetate and its higher homologues require that the acyl group is presented in an activated form, so that the transfer to the active site of the enzyme, and the subsequent catalysed reaction, are facilitated. Coenzyme A (CoA) is the

4

ubiquitous principal acyl-transfer cofactor in biological systems: its derivatives are known to be necessary for the activity of a large number of enzymes which catalyse a variety of reactions involving acyl groups.

Originally recognized as a cofactor for the enzymic conversion of acetic acid to acetylcholine, acetoacetate or citrate, its structure was established by Lipmann, Snell and Baddiley and their coworkers (Baddiley, 1955; and Fig. 5.1). Acetyl-CoA was isolated from yeast in Lynen's laboratory; the acyl group was shown to be attached to the sulphydryl group as a thiolester, rather than at any of the other functional groups. For this reason the abbreviation CoASH is often used. Acyl-CoA thiolesters (R.CO.SCoA) are obligatory intermediates in many metabolic sequences involving fatty acids, though the actual molecular species which undergo the enzymic reactions may be other acyl esters, particularly of proteinaceous thiols such as acyl carrier protein (Fig. 5.1; p. 105) or of the enzyme itself. The formation of acyl-CoA facilitates these enzymic reactions in a number of ways. The ester is water-soluble, thus enabling the enzyme to act in an aqueous environment; moreover the substrate no longer has a free carboxyl group which could interact non-specifically with protein. In particular, the thiolester is in a higher energy state, and is more reactive towards nucleophilic attack and carbanion formation. Acyl-CoA thiolesters undergo just such reactions in biological systems (Fig. 5.2): nucleophilic attack at C-1 results in transfer of the acyl group to the nucleophile; similar attack at C-3 of 2-enoyl-CoA causes addition across the double bond. Enzymic condensation at C-2 would involve carbanion formation at this position. Acyl-CoA derivatives also take part in acyl interchange reactions.

Fatty acids undergo other types of biochemical reactions, occasionally without the direct involvement of an activated S-ester: the acyl group may react as an O-ester (p. 188), or as the free acid (pp. 213, 226). One example of this latter case is the activation of free acids by their esterification to coenzyme A.

The activation of carboxylic acids is catalysed by thiokinases (acyl-CoA synthetases), which are present in a wide variety of animal, plant and microbial systems (Green and Allman, 1968a; Mahler, 1964; Mahler and Cordes, 1966). The enzymic esterification requires ATP and magnesium (or manganese) ions; and probably proceeds via enzyme-bound acyl adenylate (Fig. 5.3), the divalent ion being concerned with the binding of the nucleoside. At least four distinct animal thiokinases, differing in their substrate specificities, are known. Acetic thiokinase (EC 6.2.1.1) accepts acetate and propionate; octanoic thiokinase (EC 6.2.1.2) covers a range C_4 to C_{12}, including substituted and unsaturated acids; palmitic thiokinase (EC 6.2.1.3) is active in the

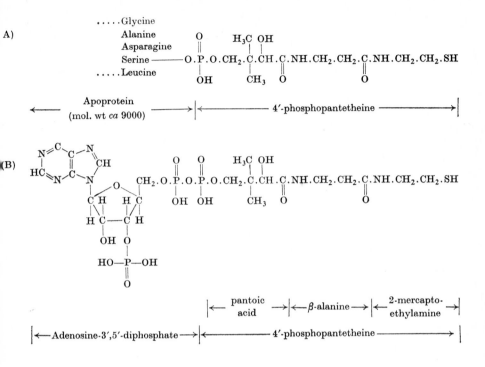

FIG. 5.1. Structures of some cofactors. A. Acyl carrier protein (ACP). B. Coenzyme A (CoA). C. Biotin (protein-bound).

Reaction	Examples	Enzymes
Nucleophilic attack at C-1	$R.COSCoA + H_2O \rightleftarrows R.CO_2H + HSCoA$ $R.COSCoA + NADH + H^+ \rightleftarrows R.CHO + HSCoA + NAD^+$ $R.COSCoA + CH_3.COSCoA \rightleftarrows R.CO.CH_2.COSCoA + HSCoA$	Hydrolase Reductase β-Ketothiolase (Fig. 8.1; p. 202)
Nucleophilic attack at C-3	$R.CH:CH.COSCoA + H_2O \rightleftarrows R.CHOH.CH_2.COSCoA$	Enoyl hydratase (Fig. 8.1; p. 202)
Condensation at C-2	$R.CH_2.COSCoA \rightleftarrows [R.\overset{\ominus}{C}H.COSCoA] \xrightarrow{CO_2} R.\underset{\underset{CO_2^{\ominus}}{\mid}}{CH}.COSCoA$ $CH_3.COSCoA + HO_2C.CO.CH_2.CO_2H \rightleftarrows CH_2.CO_2H + HSCoA$ $\qquad\qquad\qquad\qquad\qquad\qquad HO.C.CO_2H$ $\qquad\qquad\qquad\qquad\qquad\qquad\mid$ $\qquad\qquad\qquad\qquad\qquad\qquad CH_2.CO_2H$ $CH_3.COSCoA + R.COSCoA \rightleftarrows CH_2.COSCoA + HSCoA$ $\qquad\qquad\qquad\qquad\qquad\qquad\mid$ $\qquad\qquad\qquad\qquad\qquad\qquad CO.R$	Carboxylase (Fig. 5.4; p. 102) Citrate synthetase (Fig. 8.2; p. 203) β-Ketothiolase (Fig. 8.1; p. 202)
Acyl exchange	$CH_3.COSCoA + R.CH_2.CO_2H \rightleftarrows CH_3.CO_2H + R.CH_2.COSCoA$ $CH_2.COSCoA + R.CO.CH_2.CO_2H \rightleftarrows CH_2.CO_2H + R.CO.CH_2.COSCoA$ $\mid \qquad\qquad\qquad\qquad\qquad\qquad\qquad \mid$ $CH_2.CO_2H \qquad\qquad\qquad\qquad\qquad CH_2.CO_2H$	Thiophorase (p. 101) Thiophorase (p. 101)

Fig. 5.2 Enzymic reactions of acyl-CoA thiolesters. After Mahler and Cordes (1966).

C_{10} to C_{18} range. One thiokinase seems specific for certain 3-oxoalkanoic acids. The free carboxylic acids are activated at the expense of the energy inherent in the ATP supplied, which is thus made available for synthetic reactions. One fatty acid thiokinase isolated from animal sources is powered by the concurrent cleavage of GTP to GDP rather than ATP to AMP (Galzigna *et al.*, 1967; Rossi *et al.*, 1969).

The acyl-CoA synthetase from *E. coli* is an acid: CoA ligase, active on C_4 to C_{18} acids (Samuel *et al.*, 1970) but this range is apparently covered by two enzymes with different substrate specificities. Other less important activating systems which have been found in micro-organisms include the acyl kinases, which catalyse the formation of lower acyl phosphates from carboxylic acid and ATP; the phosphotransacylases, which transfer the acyl group from phosphate to CoA; and the thiophorases, which transfer CoA from acetyl-CoA to a few lower acids (C_2–C_5). The role of carnitine in the activation of fatty acid oxidation is mentioned on p. 211. The "activation" of acetyl-CoA itself by carboxylation to malonyl-CoA is described on p. 102.

In plants, acetic acid is activated by a thiokinase present in the soluble fraction of homogenates, e.g. of germinating peanut cotyledons (Rebeiz *et al.*, 1965b) and of potatoes (Huang and Stumpf, 1970). It requires CoA, ATP and Mg and is analogous to the enzyme isolated from other sources (Fig. 5.3). Long-chain fatty acid kinase is associated with microsomes and has been detected in avocado mesocarp (Barron and Stumpf, 1962) and castor bean endosperm (Yamada and Stumpf, 1965a).

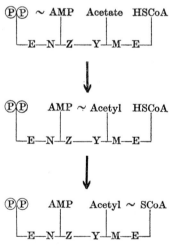

FIG. 5.3. The mechanism of thiokinases (acyl-CoA synthetases). After Green and Allman (1968a). In this model, enzyme-bound ATP is converted to bound acyl-AMP (e.g. acetyl-AMP) which generates the acyl thiolester (e.g. acetyl-CoA).

3. Biosynthesis of Saturated Fatty Acids

The fact that saturated fatty acids are totally biosynthesized from two-carbon acetate units by intact head-to-tail condensation has long been recognized (Rittenberg and Bloch, 1945; Popjak et al., 1951; Wakil, 1961; Gibson, 1963; Green and Allman, 1968b). Animals, yeasts or bacteria were used for much of the pioneering work in this area, and it is impossible to discuss fatty acid metabolism in plants without reference to data obtained with other biological systems. We shall therefore first describe the biosynthesis of saturated fatty acids in any relevant tissue before turning our attention more specifically to metabolism in higher plants (p. 124).

When the mechanism of β-oxidation of fatty acids (p. 201) was accepted, it seemed possible that the acids were synthesized by a reversal of this mechanism (Lynen, 1953). However, purified β-oxidation enzymes converted acetyl-CoA to no higher than four-carbon acyl derivatives (Stansly and Beinert, 1953), and the isolation of systems capable of producing long-chain fatty acids in the absence of the enzymes of the β-oxidation cycle led to the definition of an essentially different pathway of de novo biosynthesis. For instance, a supernatant fraction from pigeon liver was separated into two soluble subfractions, both of which were necessary to convert acetate to palmitate in the presence of ATP, CoA, isocitrate, NADP and Mn with supplementary acetate thiokinase and isocitrate dehydrogenase (Wakil et al., 1957; Porter et al., 1957). Further investigation of the purified subfractions showed that they contained no thiolase, acyl dehydrogenase or enoyl reductase (Gibson et al., 1958a), and that ATP was a necessary cofactor even if acetyl-CoA was substrate (Tichener and Gibson, 1957), and therefore an activating enzyme requiring ATP (other than acetate thiokinase) was present.

A. Acetyl-CoA Carboxylase

The demonstration by Gibson et al. (1958a, b) that the pigeon-liver system required bicarbonate (which was not incorporated into products) led to the realization that malonate was involved; one of the subfractions was indeed shown to convert acetyl-CoA into malonyl-CoA in the presence of ATP and bicarbonate (Wakil, 1958). The second subfraction contained an enzyme complex ("fatty acid synthetase") which catalysed the conversion of the malonyl-CoA to palmitate when NADPH was added. Thus malonyl-CoA was an intermediate in fatty acid biosynthesis in pigeon liver (Wakil, 1958; Wakil and Ganguly, 1959). Similar con-

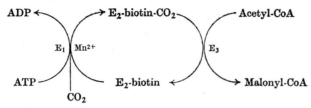

FIG. 5.4. Acetyl-CoA carboxylase. The carboxylation of acetyl-CoA in *E. coli* is catalysed by biotin carboxylase (E_1; Fig. 5.5) and a transcarboxylase (E_3; Fig. 5.6); ATP and protein-bound biotin (E_2-biotin) are also required (Alberts *et al.*, 1969d). Similar reactions occur in plants, though the enzyme is not readily separable into the subunits E_1, E_2 and E_3 (Heinstein and Stumpf, 1969).

clusions were being independently made, using preparations of rat liver (Brady, 1958) and yeast (Lynen, 1959).

The acetyl-CoA carboxylase activity depended on the presence of protein-bound biotin (Fig. 5.1) in the pigeon-liver sub-fraction (Wakil and Gibson, 1960); inhibition by avidin was reversed by addition of free biotin. The enzyme has been isolated from several sources such as liver (Numa, 1969) and yeast (Matsuhashi, 1969) and is responsible for the reactions illustrated in Fig. 5.4. Its systematic name is acetyl-CoA: carbon-dioxide ligase (EC 6.4.1.2). When isolated from *E. coli*, it dissociates into three separate protein fractions: one subunit (biotin carboxylase, E_1) can catalyse the ATP-dependent carboxylation of added biotin; another smaller subunit E_2 contains covalently bound

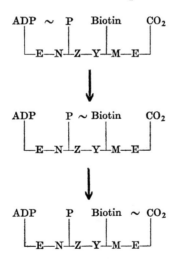

FIG. 5.5. The mechanism of carboxylation. After Green and Allman (1968b). In this model, enzyme-bound biotin phosphate is an intermediate; however, the reactions may be concerted (Lynen, 1967).

biotin which is the normal substrate for the biotin carboxylase (Alberts et al., 1969d); the third (E_3), free of biotin, catalyses the transfer of CO_2 from biotin to acetyl-CoA (Alberts and Vagelos, 1968). The carboxylation of biotin (Fig. 5.5) may take place via enzyme-bound phosphorylbiotin, or via a concerted reaction; transcarboxylation (Fig. 5.6) then occurs (Lynen, 1967; Ryder et al., 1967b). The crucial intermediate is 1'-N-carboxybiotin attached to the enzyme protein at a lysine residue through an amide bond; the product is malonyl-CoA, which is thus made available to the fatty acid synthetase. The possible control of fatty acid biosynthesis by regulation of the activity of acetyl-CoA carboxylase is discussed on p. 116.

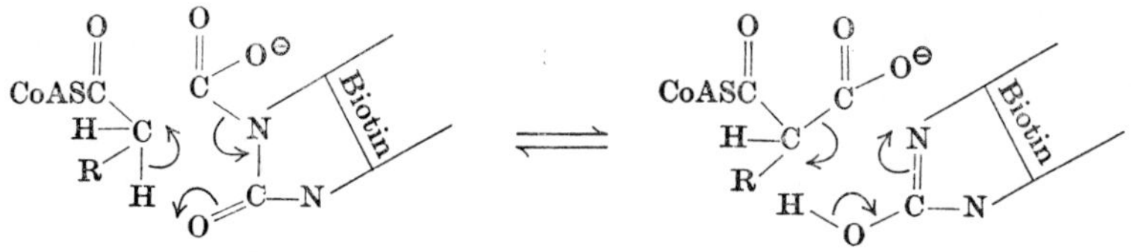

FIG. 5.6. The mechanism of transcarboxylation. After Lynen (1967).

B. Fatty Acid Synthetase

We now consider the system responsible for the biogenesis of long-chain fatty acids from acetate and malonate. When the synthetase was incubated with acetyl-CoA or malonyl-CoA, the enzyme bound each substrate with the release of CoA. When both substrates were added in the absence of NADPH, enzyme-bound acetoacetate was formed with the evolution of CO_2. Enzyme-bound acetoacetate could be reduced by NADPH to bound butyrate (Brodie et al., 1964). The evidence suggested that the synthetase catalysed a number of individual reactions, involving condensation of bound malonate and bound acetate to give bound aceto-acetate followed by reduction to bound butyrate; further similar condensations with malonate would result in fatty acid production.

The pigeon liver and yeast synthetase systems could not be resolved into their active component enzymes in order to study the detailed mechanism of fatty acid biosynthesis; in fact yeast yielded a homogeneous protein complex of molecular weight $2\cdot3 \times 10^6$ containing all the enzymes necessary (Lynen, 1969) and the avian liver contained a similar complex (Hsu et al., 1969). These complexes are discussed on p. 108. When the synthetase from E. coli was examined, it was found that standard methods yielded protein fractions containing each of the enzymes responsible and also a non-enzymic component (Majerus and

Vagelos, 1967). This was shown to be a protein acting as the acyl acceptor, and was an essential part of the synthetase system; it was named acyl carrier protein or ACP (Majerus *et al.*, 1964; 1965b; Majerus, 1967).

1. Acyl carrier protein

The ACP isolated from *E. coli* by methods described in detail by Majerus *et al.* (1969a) apparently contained 86 amino-acids and had a calculated molecular weight of 9750. A complete amino-acid sequence was determined by Vanaman *et al.* (1968, a, b), who found 77 residues and a molecular weight of 8847. The prosthetic group or substrate binding site is 4′-phosphopantetheine, which is also present in CoA (Fig. 5.1). ACP is stable to boiling at neutral pH, and to acid at low temperatures; boiling in acid or alkali causes inactivation by specific cleavage of a peptide bond or by loss of 4′-phosphopantetheine respectively.

The biosynthesis of ACP involves transfer of this functional group from reduced CoA to ACP-apoprotein, with the formation of the holo-protein and adenosine-3′,5′-diphosphate, by means of an enzyme (ACP synthetase) isolated from *E. coli* by Elovson and Vagelos (1968). *E. coli* also contains a specific ACP hydrolase (Vagelos and Larrabee, 1967; 1969) which catalyses the cleavage of ACP to 4′-phosphopantetheine and apoprotein. Investigation of the rapid turnover of the functional groups of ACP and CoA confirmed that the pantetheinate is incorporated into ACP *via* CoA and not directly (Powell *et al.*, 1969).

ACP plays the same part in *de novo* fatty acid biosynthesis (and in complex lipid biosynthesis) as CoA does in β-oxidation; however these two different forms of co-enzymically active 4′-phosphopantetheine permit the organism to distinguish between acyl precursors metabolized by synthetic and degradative pathways, though in any case the enzymes of these pathways have separate sites and normally bind their substrates tightly. An acyl carrier protein has also been isolated from plant tissues (p. 128) and its presence is inferred as an integral part of all fatty acid synthetase complexes (p. 111).

2. The enzymes of fatty acid synthetase

The individual enzymes of *E. coli* synthetase have been isolated and characterized (Vagelos *et al.*, 1969a), thus establishing the general path-way of fatty acid biosynthesis illustrated in Fig. 5.7.

(a) *Transacylases.* Synthetase enzymes (Green and Allman, 1968b) do not accept CoA esters unless a transacylase is present to convert these to ACP-substrates. Malonyl-CoA : ACP transacylase is specific for malonyl-CoA; malonyl-pantetheine can act as an efficient substitute, but

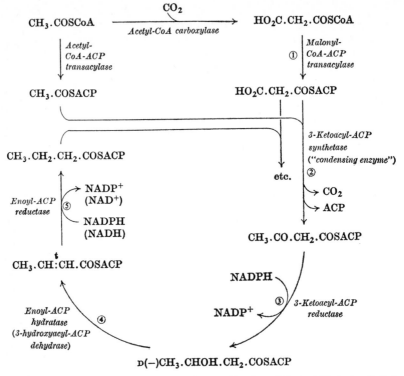

Fig. 5.7. The biosynthesis of saturated fatty acids. Acetyl-CoA yields butyryl-CoA as shown, which is subject to the same sequence of reactions to give hexanoyl-CoA, octanoyl-CoA, decanoyl-CoA, etc.; these intermediates do not normally accumulate. The terminal reaction resulting in the overall production of long-chain fatty acid (e.g. palmitate) is discussed on p. 113.

acetyl-CoA is not accepted (Alberts *et al.*, 1964; 1969b). Acetyl-CoA:ACP transacylase prefers acetyl-CoA, but acetyl-pantetheine and propionyl-CoA are also accepted, higher homologues being less active; malonyl-CoA is inactive (Alberts *et al.*, 1964; 1969a. Williamson and Wakil, 1966).

(b) *3-Ketoacyl-ACP synthetase.* The malonyl-ACP and acetyl-ACP form acetoacetyl-ACP by a condensation catalysed by 3-ketoacyl-ACP synthetase (Greenspan *et al.*, 1969; Alberts *et al.*, 1969c; Toomey and Wakil, 1966b). This "condensing enzyme" (molecular weight 66,000) shows absolute specificity for ACP-thiolesters; thiolesters of CoA or pantetheine are completely inactive. Acetyl-ACP can be replaced by propionyl-ACP, isobutyryl-ACP, butyryl-ACP, hexanoyl-ACP and octanoyl-ACP which all cause acylation of the enzyme at a cysteine residue. The acetyl-enzyme intermediate is active in transferring the acetyl group either to ACP to re-form acetyl-ACP, or to malonyl-ACP to form acetoacetyl-ACP. In the latter case the equilibrium favours the formation of the new C–C bond (condensation), since there is a concomitant cleavage of a C–S bond, and of a C–C bond to release CO_2.

(c) *3-Ketoacyl-ACP reductase*. Acetoacetyl-ACP is subject to stereospecific reduction (Alberts *et al.*, 1964; Toomey and Wakil, 1966a; Vagelos *et al.*, 1969b). The reductase is specific for NADPH; NADH is totally inactive. It accepts C_4 to C_{10} 3-ketoacyl-ACP derivatives with the formation of D(−)-3-hydroxyacyl-ACP; it will also react more slowly with CoA and pantetheine derivatives. The reverse reaction is possible, in which case the D(−)-stereoisomer is required; L(+)-3-hydroxybutyryl-ACP is not oxidized.

(d) *Enoyl-ACP hydratase*. 3-Hydroxybutyryl-ACP is converted to crotonyl-ACP by an enoyl-ACP hydratase (i.e. 3-hydroxybutyryl-ACP dehydrase; Majerus *et al.*, 1965a; 1969b. Mizugaki *et al.*, 1968a). The enzyme is surprisingly stable to boiling, and exhibits absolute specificity for 3-hydroxyacyl ACP thiolesters of the D(−) configuration, thiolesters of CoA, pantetheine and *N*-acetylcysteamine being inactive; L(+)-3-hydroxybutyryl-ACP is also not metabolised. Three such hydratases of different chain-length preferences (optima at C_4, C_8 and C_{16} respectively) cover the normal range of homologous substrates. In plant and animal systems, such multiplicity is not apparent; moreover the products are *trans*-2-enoates only (p. 133). In *E. coli* (and most other bacteria) there are also specific 3-hydroxyacyl thiolester dehydrases which convert 3-hydroxydecanoyl-ACP via the *trans*-2-enoate to the *cis*-3-enoate. This isomerization is responsible for the "anaerobic" pathway of biosynthesis of long-chain monoenoic acids in bacteria, and is described on p. 131.

(e) *Enoyl-ACP reductase*. Finally, butyryl-ACP is formed by reduction of crotonyl-ACP (Alberts *et al.*, 1964). Weeks and Wakil (1968, 1969) distinguished between two reductases isolated from *E. coli*: one was NADPH-specific, acting only on enoyl-ACP derivatives and preferring short-chain (C_4, C_6) to long-chain substrates; the other was NADH-specific, accepting ACP or CoA thiolesters and preferring longer chains. Neither contained flavin, in contrast to the corresponding reductases of yeast and *C. kluyveri* which are FMN-dependent.

(f) *Products of synthetase enzymes*. The effect of the sequential action of the enzymes of fatty acid synthetase in *E. coli* and elsewhere is thus to increase the chain length of the "starter" acetyl substrate by two carbon atoms provided by the malonyl substrate; repetition of this sequence results in the conversion of butyrate to hexanoate, and further repetitions result in the formation of long-chain acyl-ACP. The "starter" substrate is usually acetate, with the result that most natural fatty acids have straight chains of even numbers of carbon atoms. The condensing enzyme may also accept propionate, butyrate, 2-methylpropionate, 2-methylbutyrate, 3-methylbutyrate and 3-methyl-valerate; if these abnormal "starters" are available, the products have straight or branched (iso or anteiso) skeletons of odd or even carbon numbers (p. 174).

The elongation stops when the newly formed acyl chain is removed from the enzyme by transfer or hydrolysis. The chain length of the product is determined by the stage at which this terminal reaction occurs; the mechanism of this termination is discussed on p. 113. It must be fairly specific since the products of the synthetases are normally palmitate and stearate, though lower homologues can also be formed. Higher fatty acids might be biosynthesized *de novo* by synthetases with different terminal reaction specificities, or by separate elongation mechanisms discussed on p. 119.

The chemical reactions by means of which two carbon atoms are added to an acyl chain by synthetase (Fig. 5.7, p. 106) and those by which two carbon atoms are removed from an acyl chain by β-oxidation (Fig. 8.1, p. 202) are very similar. While the principles are identical, distinct details are significantly different, and these are described on p. 119.

3. *Fatty acid synthetase complex*

Fatty acid synthetase systems are evidently ubiquitous, but two distinct types are recognizable (Brindley *et al.*, 1969). Those of Type I are multi-enzyme complexes, whose components are not separable; synthetases of Type II are readily fractionated by conventional methods. The properties of individual enzymes of Type I synthetases cannot be studied, but it is generally accepted that the pathway of saturated fatty acid biosynthesis is analogous to that described for *E. coli*. The properties of known synthetase systems are given in Table 5.1.

The concept of a multi-enzyme complex was proposed by Lynen (1961; 1967) who pictured the yeast enzymes bound together and arranged round a "central" functional thiol group which holds all the intermediates in close proximity to the active sites of each enzyme in turn (Fig. 5.8). It initially accepts malonate from malonyl-CoA while a second "peripheral" thiol group accepts acetate from acetyl-CoA; condensation between these enzyme-bound substrates is followed by conversion by the established route to butyrate while the intermediates are still attached to the "central" thiol group. At this stage the saturated acyl intermediate is transferred to the "peripheral" thiol group, thus liberating the "central" group for the introduction of the next malonyl residue. The elongation of the enzyme-bound substrate can now continue in a similar efficient manner until transfer to the "peripheral" group is

TABLE 5.1

Properties of fatty acid synthetase systems[a]

Source	Molecular weight	Products	
TYPE I[b]			
Yeast	$2 \cdot 3 \times 10^6$	14:0–18:0	CoA esters
Pigeon liver	$4 \cdot 5 \times 10^5$	16:0	Free acid
Rat liver	$5 \cdot 4 \times 10^5$	16:0	Free acid
Adipose tissue	—	14:0, 16:0	Free acid
Mammary gland	—	8:0–18:0	Free acid
Euglena gracilis (etiolated)	*ca* 1×10^6	16:0	CoA ester
Mycobacterium phlei	$1 \cdot 7 \times 10^6$	14:0–26:0	Esters
TYPE II[c]			
E. coli		16:0, 18:0, 16:1	
Clostridium species		18:1, 3h-10:0	ACP esters
Pseudomonas species		3h-12:0, 3h-14:0	
Bacillus subtilis		C_{15} and C_{17} isoacids and anteisoacids	—
Avocado mesocarp		16:0, 18:0	—
Lettuce chloroplasts		16:0, 18:0	—
Spinach chloroplasts		16:0, 18:0	—
Euglena gracilis (photoauxotrophic)		18:0	ACP ester

[a] Brindley *et al.*, 1969.
[b] Synthetases of Type I are multi-enzyme complexes containing bound 4'-phosphopantetheine; they do not require an external ACP supplement.
[c] Synthetases of Type II consist of individual enzymes which require ACP.

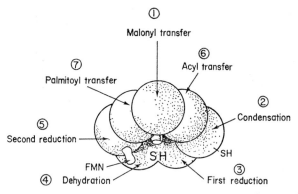

FIG. 5.8. Hypothetical structure of the multi-enzyme complex of fatty acid synthetase. After Lynen (1967). The seven enzyme sub-units shown correspond to the reactions illustrated in Fig. 5.9.

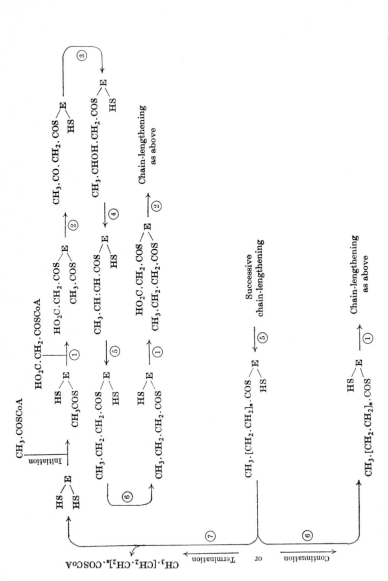

Fig. 5.9. Mechanism of the fatty acid synthetase complex; after Lynen (1967). The fatty acid synthetase is denoted by $\begin{smallmatrix} HS \\ HS \end{smallmatrix}\!\!\!\rangle E$, where the "central" binding site is represented (in bold type) by the upper SH; the "peripheral" thiol group is represented by the lower SH. The complex E contains the enzymes necessary for fatty acid synthesis (Fig. 5.7) arranged as in Fig. 5.8 to facilitate the sequential reactions (Fig. 5.10). Transfer of acetyl (or acyl) to the "peripheral" binding site initiates the process; the acyl chain so generated on the "central" site is transferred to the "peripheral" site for further chain-lengthening. Finally this transfer is superseded by dissociation of the acyl chain from the "central" site (termination reaction, p. 113) with the regeneration of the synthetase complex.

no longer possible; termination of the sequence now occurs (p. 113) by removal of the palmitate (or stearate) with regeneration of the free synthetase complex (Fig. 5.9). The intermediate reactions have been characterized with high concentrations of model substrates, and by isolation of enzyme-bound intermediates (Lynen, 1967). Synthetases from all sources appear to catalyse the same detailed reactions, with the possible exception of the terminal reaction, which yields acyl-CoA as the final product in yeast but free acids with animal and some bacterial systems (p. 113). After guanidine hydrochloride treatment, the inactive protein fragments of yeast synthetase have been separated (Willecke et al., 1969). One of the smallest polypeptide chains isolated carried the 4'-phosphopantetheine residue and evidently acted as yeast ACP in the active complex (Schweizer et al., 1970b). It could not replace E. coli ACP in a bacterial ACP-dependent reaction, its amino-acid composition was different and its molecular weight higher. However it appears that the multi-enzyme synthetase requires ACP in bound form; this corresponds with the "central" thiol group, which would then be situated at the end of the flexible pantetheine chain allowing the bound intermediates access to the synthetase enzymes in turn without release from the carrier thiol (Fig. 5.10). A similar flexible arm consisting in this case of lipoyl-lysyl residue has been postulated by Koike et al. (1963) to explain sequential reactions during oxidative decarboxylation catalysed by α-ketoacid dehydrogenase complexes. The "peripheral" group of fatty acid synthetase multi-enzyme complex was associated with a cysteine residue, and corresponds to the thiol binding site of the condensing enzyme. Another acceptor group has been revealed on which acetate and malonate can be bound, not via sulphur, but at the hydroxyl group of serine (Lynen, 1967). Schweizer et al. (1970a) have shown that the synthetase complex catalyses fatty acyl transfer between labelled and unlabelled CoA; this activity was resistant to N-ethylmaleimide and iodoacetamide inhibition, suggesting that the non-thiol (serine) binding sites are involved in acyl transfer between CoA and enzyme-bound thiol. When malonyl-CoA (or methylmalonyl-CoA) was incubated with the synthetase complex, covalent binding at two distinct sites was observed (Schweizer et al., 1970b). Analysis indicated that some malonate was present as a thiolester of the "central" 4'-phosphopantetheine, while most was attached to serine through an O-ester linkage, which was activated by some surrounding amino-acid, possibly histidine. Acetate did not compete with methylmalonate for this serine site, which corresponded to the binding site of the malonate transferase; however competition did take place for the "central" thiol site. It therefore appears that, during de novo fatty acid biosynthesis by the synthetase complex,

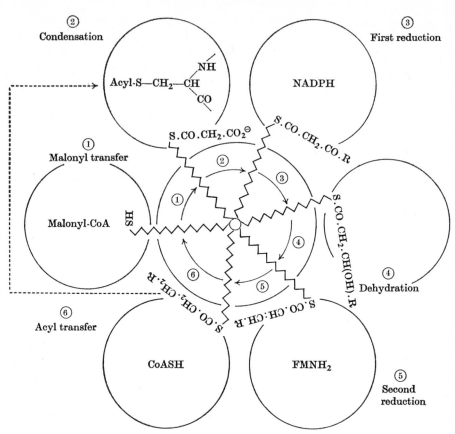

Fig. 5.10. Mechanism of the fatty acid synthetase complex. After Lynen (1967). The scheme illustrates individual events during fatty acid synthesis on the multi-enzyme complex; the initial and terminal reactions have been omitted. The six chain-lengthening reactions correspond to those in Fig. 5.9.

the "starter" acetate is accepted first by a serine site (acyl transferase) and is transferred via the "central" ACP to the "peripheral" cysteine (condensing enzyme); malonate is accepted at a serine site (malonyl transferase), transferred to the "central" ACP, and condensation occurs. Joshi *et al.* (1970) have postulated a similar sequence of events for the avian synthetase complex. The isolation, purification and assay of yeast fatty acid synthetase complex has been described by Lynen (1969).

Bound forms of 4′-phosphopantetheine are present in other multi-enzyme synthetase complexes, isolated from mycobacteria (Brindley *et al.*, 1970) or from animals (Larrabee *et al.*, 1965). *Mycobacterium phlei* also contains a free ACP (mol. wt 10,600), which may indicate some dissociation of its labile synthetase complex (mol. wt $1 \cdot 7 \times 10^6$) during fractionation (Matsumura *et al.*, 1970). Yeast synthetase probably exists naturally as a trimer; electron micrographs give an impression of a symmetrical oval particle with an equatorial ring and a central cavity (Lynen, 1967); X-ray analysis of the crystalline complex confirms the existence of this hole (Oesterhelt *et al.*, 1969). It can be dissociated into active subunits, but it is doubtful that these are single enzyme components (Sumper *et al.*, 1969b). The concept of the multi-enzyme complex may well be a representation of the mechanism of fatty acid biosynthesis in all forms of life, the bacterial complex dissociating with far greater ease than the tightly-bound enzymes from other sources.

4. Termination reactions of fatty acid synthetase

The mechanism of the termination reaction, by which the newly formed acyl chain leaves the synthetase enzyme, has not been clearly defined. This reaction governs the chain length of the product, and is evidently quite selective since purified systems normally yield palmitate (or stearate); bacterial systems produce vaccenate in addition (p. 131). Presumably the enzyme-bound product of fatty acid synthetase is acyl-ACP; the acyl chain may be removed as the ACP ester, by transfer from bound ACP to free ACP or to free CoA, or by hydrolysis to the free acid. In the case of the highly purified yeast multi-enzyme complex, this termination reaction is transfer to CoA, the final product of synthesis being acyl-CoA (Lynen, 1967; Sumper *et al.*, 1969a). On the other hand, homogeneous pigeon liver preparations yield palmitate as the free acid (Bressler and Wakil, 1962) and while the isolated *E. coli* enzymes give rise to acyl-ACP, partially purified fractions synthesize both free acids and thiolesters (Barnes and Wakil, 1968). The exact nature of the normal terminal reaction is difficult to establish in cases where it is followed by secondary reactions; the multi-enzyme complexes of yeast or pigeon liver evidently contain a transferase or a hydrolase respectively as part of the complex.

The hydrolase present in *E. coli* is not so tightly bound, and has been isolated for study (Barnes and Wakil, 1968). It was termed palmitoyl thiolesterase, and two forms have been separated, both of which hydrolyse CoA and ACP esters of fatty acids. One form (I) was inactive with octanoyl-CoA and decanoyl-CoA; the CoA thiolesters of lauric, myristic, stearic and oleic acids were hydrolysed, but the preferred substrates were palmitate, palmitoleate and vaccenate. The specificity of palmitoyl thiolesterase II is more general, accepting in addition the shorter-chain saturated esters and longer-chain 3-hydroxyacyl esters (Barnes *et al.*, 1970). This facility may have significance in the accumulation of 3-

hydroxyacids in the bacterial cell wall lipid. Palmitoyl thiolesterase is also present in purified pigeon liver synthetase, but here the substrate specificity is more definite. Lauroyl-CoA and myristoyl-CoA were not hydrolysed over a range of concentrations; however activity was observed with the CoA esters of palmitic, palmitoleic, stearic, oleic and vaccenic acids (Barnes and Wakil, 1968). This specificity could be a factor governing the pattern of fatty acids synthesized: fatty acid synthesis from acetate and malonate would occur on the multi-enzyme complex, and elongation would continue until bound palmitoyl-ACP was formed; this would then be hydrolysed to free palmitic acid, releasing the synthetase for re-use. The acyl chain might be transferred to CoA before hydrolysis, and a similar mechanism could operate in *E. coli*. Since long-chain thiolesters inhibit many enzymes (p. 118) palmitoyl thiolesterase may also regulate the activity of synthetase enzymes (and others) by controlling the level of these inhibitors.

In contrast, the mechanism in yeast must be different (Lynen, 1967). While the major product of yeast multi-enzyme synthetase is palmitate which has been transferred to CoA without hydrolysis, the rate of this acyl transfer is independent of chain length. Specificity of transfer can play no part in preventing shorter-chain intermediates from leaving the complex. However, protection of the enzyme complex from ethyl-maleimide inactivation by acyl-CoA esters was chain-length dependent, longer esters providing decreased protection (Lust and Lynen, 1968). Malonyl-CoA, which acylates the "central" thiol group of the complex (p. 108) and not the "peripheral" group, gave no protection. The environment of the "peripheral" group therefore seemed to be lipophobic in nature, repelling the longer acyl chains, which tend to remain on the "central" thiol group, thus allowing N-ethylmaleimide to attack the "peripheral" group. As the acyl chain grows, the tendency to move from the "central" group will decrease, until finally palmitate and stearate remain there, blocking the approach of a new malonyl substrate, until they are transferred to CoA (Fig. 5.9). This concept is supported by the observation by Lust and Lynen (1968) that long-chain acyl-CoA inhibits the synthetase competitively with respect to substrate malonyl CoA, but non-competitively with respect to acetyl CoA.

In general, there are two reactions open to a newly formed enzyme-bound acyl chain: it is transferred either to CoA (termination reaction) or to the "peripheral" thiol group, where chain lengthening continues at a rate limited by that of the condensation reaction (p. 117). The choice is governed by the relative velocities of the reactions: initially, elongation of shorter chains is preferred, but eventually the growing chain interacts with the enzyme protein, changing the relative velocities in favour of product formation. Schweizer *et al.* (1970a,b) further suggest that a conformational change, induced by bound palmitate and stearate, facilitates release of these products and blocks further chain-lengthening; such allosteric transitions probably play an important part in the regulation of the system.

Sumper *et al.* (1969a) have put this model on a more quantitative basis by assuming that this interaction occurs only after a C_{13} chain has been synthesized, and that thereafter each additional methylene group is associated with a change of free energy of -0.9 kcal. This leads to an equation for the product distribution which rationalizes the normal chain termination at C_{16} and C_{18}, and also could be tested experimentally under a variety of conditions. Selective denaturation of the enzyme complex by guanidine hydrochloride demonstrated that the condensation reaction was inhibited more easily than the termination reaction; its products then included shorter fatty acids at the expense of stearate, and the quantitative distribution corresponded closely to that predicted by the equation. When a deficiency of malonyl-CoA was supplied to the denatured or normal enzyme, the composition of the fatty acid mixture obtained agreed well with the calculated values. Similarly, the supply of malonyl-CoA to the synthetase could be controlled by using acetyl-CoA alone as substrate in the presence of a limited amount of acetyl-CoA carboxylase; if the activity of the carboxylase was fifty-six times less than that required for optimum synthetase activity, short-chain products predominated in amounts again approximating to those predicted.

Because the condensation reaction of fatty acid synthesis is the rate-limiting step under normal conditions, saturated acyl residues have the longest life-time as intermediates and consequently the highest probability of leaving the enzyme complex as end-products. However, the reduction reactions can be made rate-limiting if the NADPH level is sub-optimal; under these conditions, the CoA derivatives of 3-ketoacids and 2-enoic acids appear as end-products. If NADH alone is available, the first reduction step can be slowed to such an extent that 3-ketoacyl-CoA derivatives predominate. When the synthetase was coupled with an enzyme system producing NADPH five times more slowly than required for optimum fatty acid biosynthesis, some laurate, myristate, palmitate and stearate were indeed obtained; but the products also included 2-hexadecenoate as well as the three methyl ketones formed by spontaneous decarboxylation of 3-ketomyristate, 3-ketopalmitate and 3-keto-stearate. If no NADPH is available, enzyme-bound acetoacetyl-ACP cannot be reduced, and it is converted to triacetic acid lactone by a "derailment" of the normal sequence of reactions (Yalpani *et al.*, 1969).

The additional interaction between the synthetase and the normally growing carbon chain corresponding to $-0\cdot9$ kcal per methylene group is of the correct size for lipophilic (hydrophobic) interactions; increasing attraction to the transferase may therefore be responsible for the eventual termination reaction. On the other hand, lipophobic (hydrophilic) repulsion from the peripheral thiol site (condensing enzyme) may be the cause of the long-chain residue remaining on the "central" thiol until cleavage from the synthetase complex. This latter view has experimental support. The protection of the peripheral thiol from ethyl maleimide by various acyl-CoA derivatives was dependent on chain length: longer chain lengths gave decreased protection, suggesting that this thiol does have a hydrophilic environment (Lynen, 1967). Moreover, Ayling et al. (1969) have shown that yeast fatty acid synthetase interacts with decanoyl-CoA more strongly than with palmitoyl-CoA.

This concept of the termination mechanism therefore explains the variations in the products of yeast fatty acid synthetase in vitro, and probably is applicable in vivo to all such multi-enzyme complexes. It is known that the concentration of malonyl-CoA available affects the pattern of fatty acid produced in animal systems (Smith and Dils, 1966; Bartley et al., 1967; Nandedkar et al., 1969) and in plants (Yang and Stumpf, 1965a). Indeed, the biosynthesis of short-chain fatty acids by synthetase could give the false impression that another enzyme system, specific for short-chain products, was operating. The presence of methyl ketones in butter may be due to 3-ketoacids being formed during biosynthesis of fatty acids in the bovine mammary gland (Lawrence and Hawke, 1966); this might occur if the supply of NADPH to the synthetase were limited, or if the NADH:NADPH ratio were high. The methyl ketones found in fungi could be similarly formed, though here the 3-ketoacids are thought to arise by β-oxidation (Lewis and Johnson, 1966).

5. Control of the rate of fatty acid biosynthesis

The nature of the fatty acids synthesized by the enzymes of the synthetase is determined by the stage at which the termination reaction operates (p. 113). The rate of biosynthesis must also be under control, but the normal mechanism of this regulation is not known in detail. The rate-limiting step is thought to be the carboxylation of acetyl-CoA in animals (Lynen, 1967) and some yeasts (McElroy and Stewart, 1968), but this may not always be so; it has been established that in purified yeast synthetase, condensation is the critical reaction (Sumper et al., 1969a).

(a) Carboxylation. Acetyl-CoA carboxylase from some sources can be activated by the presence of citrate, isocitrate, fructose-1,6-diphosphate and other intermediates of the glycolytic and tricarboxylic acid pathways. In animal systems, this stimulating effect is connected with an aggregation of the enzyme (Vagelos et al., 1963). If citrate is present, the enzyme from avian liver consists of long filaments, 70–100 Å wide and 4000 Å long; this active form has a molecular weight of $7 \cdot 8 \times 10^6$ and is associated with some 20 moles of biotin. In the absence of citrate, the aggregates dissociate into units of 410,000, each containing one mole of biotin but being enzymically inactive (Ryder et al., 1967a). The physiological significance in vivo of this phenomenon is uncertain: intracellular citrate concentrations are in the range shown to be effective, but much is concentrated in the mitochondria, and not available to interact with extra-mitochondrial fatty acid synthetase. Moreover, the soluble acetyl-CoA carboxylase may be associated loosely with a membrane (Easter and Dils, 1968), a suggestion which would also explain its stimulation by phospholipids (Miller and Levy, 1969). The activation by citrate in vitro, the filamentous species and the possible structural role of acetyl-CoA carboxylase may be unique to animal tissues (Kleinschmidt et al., 1969).

Using a strain of yeast as an enzyme source, Rasmussen and Klein (1967) reported stimulation by a large number of compounds including citrate and fructose-1,6-diphosphate. Activation occurs without significant aggregation of the enzyme, and although a reorganization of tertiary structure involving only slight change in molecular weight or sedimentation behaviour is possible, it is apparent that the mechanism of activation is different from that of animal systems. No activation of either kind occurs with acetyl-CoA carboxylase isolated from other strains of yeast (Matsuhashi et al., 1964), from E. coli (Alberts and Vagelos, 1968) or from higher plants (Burton and Stumpf, 1966; Heinstein and Stumpf, 1969).

(b) Condensation. The fact that only unsubstituted saturated derivatives normally accumulate during saturated fatty acid biosynthesis suggests that the later condensation reactions are the slowest; it is on this basis that a rational concept of chain termination has been built (p. 114). Indeed, Sumper et al. (1969a) have shown that limiting the activity of acetyl-CoA carboxylase leads to the synthesis of short-chain fatty acids by yeast synthetase complex, which otherwise produces palmitate and stearate. This throws doubt upon the rate-determining properties of the carboxylase, at least in this system.

(c) Other factors. Pigeon liver fatty acid synthetase was inhibited by malonyl-CoA, but this inhibition was reversed by fructose 1,6-diphosphate or by phosphate; other inorganic ions had no effect (Plate et al., 1968). In E. coli all salts tested stimulated fatty acid synthetase and several of its component enzymes (Schulz *et al.*, 1969). It seemed feasible that under physiological conditions, divalent cations could exert control on the rate of synthesis in two ways: by complexing with the protein moiety of ACP substrates, thereby facilitating binding to the enzymes; and by reducing the repulsion between negatively charged groups of the substrates and their respective enzymes.

When it was demonstrated that palmitoyl-CoA inhibited synthetase activity in pigeon liver (Porter and Long, 1958) it appeared that this product might act as a negative feedback inhibitor, controlling the rate of palmitate production. However the effect has been attributed largely or completely to the non-specific detergent action of acyl-CoA (Dorsey and Porter, 1968). Nevertheless control may be exercised by long-chain acyl-CoA inhibition (Lynen, 1967) which can depress synthetase and acetyl CoA carboxylase activity, as has been demonstrated in yeast systems (Lust and Lynen, 1968). The removal of acyl derivatives by binding or by catalysing their incorporation into complex lipid would reverse this inhibition (Bartley *et al.*, 1967; Smith and Dils, 1966).

Another aspect of the regulation of fatty acid biosynthesis, besides the activity of the enzymes involved, is the supply of substrates and cofactors to them. The excretion of citrate from the mitochondria and the activity of citrate lyase may be important factors (Spencer *et al.*, 1964) since these regulate the availability of acetyl-CoA to the cytoplasmic fatty acid synthetase. Thus citrate may be critical in two distinct ways, since it can both activate and provide substrate for the acetyl-CoA carboxylase. The availability of ACP is also vital, and this depends in turn on the activity of its synthetase and a supply of CoA; the activity of ACP hydrolase would also control ACP levels effectively. The provision of reduced pyridine nucleotides is determined to some extent by rate of glycolysis, which may therefore in turn affect the rate of fatty acid synthesis during the conversion of carbohydrate to fat; however, lack of NADPH can lead to formation of β-ketoacids (p. 115).

Longer term control of fatty acid biosynthesis may also be exercised by the induction of enzyme synthesis or degradation. Thus the rate of fatty acid biosynthesis is depressed in a starving or diabetic animal; the rate is increased again by feeding or insulin treatment respectively, but the increase is prevented if inhibitors of protein synthesis are administered. This suggests that the increase is due to biosynthesis of more enzymes associated with fatty acid synthesis.

(*d*) *Rate of fatty acid biosynthesis in plants*. Some or all of these mechanisms may generally control the biosynthesis of saturated fatty acids in plants as well as in the particular animal or microbial system so far investigated. The control of unsaturated fatty acid biosynthesis in plants by regulation of desaturation is considered on p. 152. Dickson *et al.* (1969) analysed *Chlorella fusca* grown heterotrophically under different environmental concentrations of carbon dioxide: as this increased from 1% to 30%, the fatty acid content increased by nearly 50%. Most of the change was associated with the palmitic and oleic acids; one explanation is that carboxylation of acetyl-CoA is the rate-limiting step in fatty acid biosynthesis in this case. Willemot and Stumpf (1967a, b) demonstrated that, as potato slices "aged" in air, marked metabolic changes occurred within a few hours (p. 258), including an increase of severalfold in fatty acid synthesis. This increase was ascribed to *de novo* synthesis of RNA and proteins resulting from postulated derepression soon after slicing, and was not observed in the presence of inhibitors of protein synthesis. Derepression, which may also be associated with the end of dormancy in potatoes, could be triggered by slicing: a volatile repressor might escape, some tissue might become separated from the site of biosynthesis of a non-volatile repressor, or the synthesis of an inducer at the wound surface might be stimulated.

C. Fatty Acid Elongation

The chemical reactions of fatty acid biosynthesis *de novo* (Fig. 5.7, p. 106) and of the β-oxidation of fatty acids (Fig. 8.1, p. 202) are very similar: distinct differences do occur, however, and these are summarized in Table 5.2. The intermediates from one pathway are not directly available to the enzymes of the other, since they are not normally released from the appropriate system and do not accumulate. The anabolic and catabolic systems are located in different parts of the cell, and require reduced and oxidized cofactors respectively. While ACP is intimately involved with synthetases, β-oxidation requires CoA esters only. During anabolism, 3-ketoacyl-ACP is reduced by NADPH to D(−)-3-hydroxyacyl ACP (the *E. coli* enzyme will not accept CoA esters); during catabolism, L(+)-3-hydroxyacyl-CoA is oxidized by NAD$^+$ to 3-ketoacyl-CoA. The equilibria favour reduction in the former case and oxidation in the latter. The enzymic condensation of malonate with the acyl synthetase intermediate favours the anabolic reaction; conversely, thiolytic cleavage of acetyl-CoA from 3-ketoacyl-CoA is favoured during the analogous reaction of β-oxidation.

Thus the pathways of fatty acid biosynthesis from acetate and of its β-oxidation to acetate are quite distinct. Nevertheless, biosynthesis of fatty acids by elongation of existing acids can also occur (Wakil, 1961) and two mechanisms involving either malonate or acetate can be distinguished (Table 5.2).

TABLE 5.2

Comparison between the biochemistry of synthesis and degradation of long-chain fatty acids

	β-Oxidation (p. 201)	De novo biosynthesis (p. 102)	Biosynthesis by elongation (p. 121)	Biosynthesis by elongation (p. 122)
Acyl carrier	CoA	ACP	ACP?	CoA?
Coenzyme	NAD$^+$ (and NADP$^+$)	NADPH	NADPH (or NADH)	NADH (or NADPH)
C$_2$ unit	Acetate	Malonate[a]	Malonate[a]	Acetate
Configuration of 3-hydroxyacyl intermediate	L(+)	D(−)		
Nature of system	Particulate	Soluble	Particulate	Particulate
Location of system	Mitochondria	Cytoplasm	Microsomes	Mitochondria
Role of system	Cleavage of fatty acids to acetate	Elongation of acetate to palmitate[b]	Elongation of preformed C$_{10}$ to C$_{20}$ saturated and unsaturated fatty acids	Elongation of preformed C$_8$ to C$_{18}$ fatty acids

[a] Replaceable by acetate in the presence of bicarbonate and the biotin-dependent (avidin-sensitive) enzyme, acetyl carboxylase.
[b] Also stearate; vaccenate is a product in some bacteria only (p. 131).

1. *Malonate-dependent elongation*

A preformed acid may condense with malonate and extend in a manner exactly analogous to *de novo* synthesis (Nugteren, 1965; Guchhait *et al.*, 1966). This elongation system, associated with mammalian microsomes, accepts a long-chain substrate in place of acetate, and will accommodate further lengthening of the chain. Hexanoate, octanoate and stearate react slowly, as do oleate and elaidate; preferred substrates are decanoate, laurate, myristate, palmitate, hexadecenoate, linoleate, α-linolenate and γ-linolenate. Evidently the presence of double bonds enhances the reactivity of the C_{18} substrates, presumably owing to the shape of the unsaturated molecules. Since the product of *de novo* synthesis is palmitate (and possibly stearate), fatty acids longer than C_{18} are probably generally biosynthesized via this malonate-dependent ("microsomal") elongation pathway, often after desaturation. During elongation, the 3-ketoacyl, 3-hydroxyacyl and *trans*-2-enoyl intermediates occur successively (Fig. 5.11).

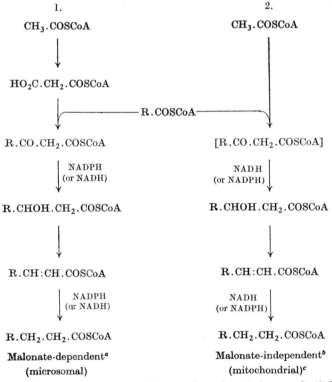

FIG. 5.11. Fatty acid elongation, (1) Malonate-dependent[a] (microsomal), (2) Malonate-independent[b] (mitochondrial)[c].
[a] Nugteren (1965); ACP thiolesters may be involved. [b] Harlan and Wakil (1963). [c] Heart mitochondrial systems catalyse these reactions but are inhibited by NADPH (Dahlen and Porter, 1968).

Rat-liver microsomes cause elongation of γ-linolenic acid $(18:3, \omega 6)$ and α-linolenic acid $(18:3, \omega 3)$ to the corresponding eicosatrienoates by means of reactions which are selectively enhanced or inhibited by the presence of other acids (Christiansen et al., 1968); such interaction between different acids may serve to regulate elongation in vivo, presumably in plant as well as animal systems. A similar elongation system has been isolated from Mycobacterium phlei, in which malonate condensed with palmitoyl-CoA or stearoyl-CoA, but not octanoyl-CoA or acetyl-CoA. Here, the chain-lengthening activity was strictly dependent on an external supply of ACP (Matsumura et al., 1970). This suggests that acyl-ACP thiolesters are the true substrates in malonate-dependent elongation as well as in de novo synthesis of fatty acids. The pathways can, however, be distinguished by their location and substrate specificity (Table 5.2).

2. Malonate-independent elongation

The second type of elongation is associated with liver mitochondria, and may occur in the absence of malonate by means of direct condensation of acetyl-CoA with an acyl-CoA (C_8 to C_{18}), the mechanism apparently being the reversal of β-oxidation (Harlan and Wakil, 1963; Dahlen and Porter, 1968). A distinguishing feature may be the presence of an NADPH-dependent enoyl-CoA reductase, which has no place in β-oxidation (Seubert et al., 1968); however either NADPH or NADH may act as cofactor (Colli et al., 1969; Mooney and Barron, 1970). ATP is required, and possibly phosphorylated derivatives of the enol form of acetyl-CoA are intermediates (Cheniae, 1964.) Carnitine may play a similar role in this malonate-independent "mitochondrial" fatty acid elongation as in β-oxidation (p. 211), since acylcarnitines are active substrates in both pathways (Warshaw and Kimura, 1970).

In mammalian mitochondria, acyl-CoA is elongated by C_2 units from acetyl-CoA but not malonyl-CoA (Colli et al., 1969; Whereat et al., 1969). The separation of the inner and outer mitochondrial membranes (Fig. 10.3, p. 268) revealed the existence of at least two systems: the inner membranes synthesize C_{14} to C_{18} acids either de novo from acetate or from a medium-chain primer, while the outer membranes are responsible for elongation of medium- and long-chain primers by one or two acetate units. While the issue is confused by some malonate-dependent de novo synthesis, the mitochondrial synthesis of stearic acid and longer saturated and unsaturated acids is primarily by elongation (Howard, 1970). Reversal of β-oxidation has also been invoked to explain butyrate production in rabbit mammary gland (Nandedkar and Kumar, 1969) and lesquerolic acid biosynthesis in plants (p. 148).

Malonate-independent elongation must depend on the tendency of acetyl-CoA to form a carbanion (p. 98), which is presumably the reacting species (Fig. 5.2). There is however no evidence for the accumulation of 3-ketoacid intermediates; unfavourable energetics of condensation of acyl-CoA with acetyl-CoA may therefore be overcome by rapid NADPH-requiring reduction, with the initial formation of the 3-hydroxyacid, which is dehydrated and further reduced (Fig. 5.11; Barron and Mooney, 1970).

3. *Biosynthesis of fatty acids by elongation*

When acetate is the substrate, elongations via malonate may be recognized by their dependence on a biotin-specific (avidin-sensitive) carboxylation. In acetate-dependent condensations, malonate is not acceptable unless a decarboxylase is present; malonyl-CoA decarboxylase is indeed quite widespread. Both elongations are distinguishable from *de novo* biosynthesis by the accumulation of label from ^{14}C-acetate at the carboxyl end of the product.

Elongation mechanisms explain the natural occurrence of some minor and unusual fatty acids (Fig. 5.12). The "major" C_{18} acids are lengthened to give rise to the "minor" $\omega 9$, $\omega 6$ and $\omega 3$ families (p. 5), which therefore owe their structures to the sequential action of synthetase desaturase and an elongation system (p. 141).

Abnormal elongations may evidently occur. If acetate were replaced by propionate, the product would be a 2-methylalkanoic acid rather than the unbranched alkanoate. In the "microsomal" system, a necessary preliminary would be the carboxylation of propionyl-CoA to methylmalonyl-CoA (p. 125); if the "mitochondrial" system operated, the reacting species would be the α-carbanion of propionyl-CoA itself.

(a) $16:0 \rightarrow 18:0$
 $18:3(9c\,12c\,15c) \rightarrow 20:3(11c\,14c\,17c)$
 $20:4(5c\,8c\,11c\,14c) \rightarrow 22:4(7c\,10c\,13c\,16c)$

i.e. $R.COSCoA + CH_3.COSCoA \rightarrow R.CH_2.CH_2.COSCoA$

(b) $R.COSCoA + CH_2.COSCoA \rightarrow R.CH_2.CH.COSCoA$
          | |
          CH_3 CH_3

(c) $R^1.COSCoA + CH_2.COSCoA \rightarrow R^1.CH_2.CH.COSCoA$
          | |
          R^2 R^2

(d) $R.COSCoA + CH_3.[CH_2]_n.CO_2H \rightarrow$
 $(R.CO.CH_2.[CH_2]_n.CO_2H) \rightarrow$
 $R.CH_2.CH_2.[CH_2]_n.CO_2H$

FIG. 5.12. Possible pathways of elongation of fatty acids.

Repetition of such elongations could lead to the 2,4,6-trimethyl and similar polymethyl fatty acids (p. 29); [1-^{14}C]propionate is indeed incorporated into the mycocerosic acids of bacteria (Gastambide-Odier *et al.*, 1963). Analogous abnormal elongations with other acyl-CoA substrates in place of propionate would lead to α-branched acids such as are found in mycobacteria (p. 28): Corynomycolic acid

$$CH_3.[CH_2]_{14}.CH(OH).CH(C_{14}H_{29}).CO_2H$$

is thus biosynthesized from two molecules of palmitate (Gastambide-Odier and Lederer, 1959).

The higher straight-chain saturated acids presumably arise by sequential chain elongation of palmitate (or stearate) to arachidate, behenate and lignocerate. In the yeast *Candida utilis*, preformed saturated acids are lengthened by a system specific for substrates of chain length between C_{20} and C_{24} (Fulco, 1967). The product is C_{26} when C_{20}, C_{22} or C_{24} are substrates, and a mixture of C_{25} and C_{27} when substrates are C_{21} or C_{23}. The formation of exceptionally long chains of carbon atoms may also take place by condensation of preformed acyl thiolester with another acyl chain longer than acetate or malonate. Kanemasa and Goldman (1965) have reported a bacterial elongation system which was insensitive to avidin; it apparently caused direct intermolecular condensation of octanoyl-CoA with acetyl-CoA or octanoyl-CoA or decanoyl-CoA. Acyl ω-carbanions are presumably intermediates in these head-to-tail reactions; while acetyl-CoA tends towards carbanion formation (p. 98), it must be postulated that longer chain terminal carbanions are produced by some form of ω-oxidation (p. 221). Such condensations may be implicated in the biogenesis of the long chains of carbon atoms found in plant hydrocarbons (p. 253).

D. *Biosynthesis of Saturated Fatty Acids in Higher Plants*

1. *Introduction*

There is every reason to suppose that the detailed mechanism of biosynthesis of *saturated* fatty acids in plants is analogous to the pathways established in *E. coli* and other systems. The structures and biosynthesis of the *unsaturated* fatty acids are however completely different in the bacterial system (p. 131). We now consider research in which plant tissue was specifically used to investigate saturated fatty acid biosynthesis.

The first of a distinguished series of papers by Stumpf and coworkers on fat metabolism in plants demonstrated that when slices of cotyledons from maturing or germinating peanuts were fed a variety of labelled substrates, fatty acids were synthesized (Newcomb and Stumpf, 1952).

Acetate was the most effective precursor. When [1-^{14}C]acetate was incubated with developing flax fruits, the individual fatty acids biosynthesized *in vitro* were labelled predominantly on the odd numbered carbon atoms (Gibble and Kurtz, 1956).

Stumpf and Barber (1957) prepared cell-free particles from avocado fruit mesocarp which catalysed the synthesis of long-chain fatty acids from acetate under anaerobic conditions. The cytoplasmic fraction had no such synthetase activity; the particulate system possessed TCA, oxidative phosphorylation and β-oxidation activity and therefore appeared to be mitochondrial; its fatty acid synthetase required ATP, CoA, and Mn as cofactors. When solubilized by Squires *et al.* (1958) it was found that carbon dioxide was also essential, though $^{14}CO_2$ was not incorporated into the acids.

2. Acetyl-CoA carboxylase

The formation of malonyl-CoA is a necessary preliminary to the operation of fatty acid synthetase. The carboxylation of acetyl-CoA in plants has been studied in detail by Stumpf and coworkers. Hatch and Stumpf (1961) purified an enzyme from wheat germ which formed malonyl-, methylmalonyl- or ethylmalonyl-CoA from acetyl-, propionyl- or butyryl-CoA respectively by carboxylation and also by transcarboxylation. The carboxylation, involving carbon dioxide, required ATP and Mg; the transcarboxylation, involving malonyl-CoA, needed no added cofactors. Both activities were associated with the same enzyme and were biotin-dependent; acetyl-CoA was the preferred substrate.

Further work on the purified system by Heinstein and Stumpf (1969) suggested that it consisted of a carboxylase and a transcarboxylase, similar to the bacterial system (p. 103). The plant system cannot be easily resolved by chromatography. Ultracentrifugal analysis revealed two protein components, one of which contained biotin; electrophoresis gave five bands, identified as the biotin-containing carboxylase and its dimer, the transcarboxylase and its tetramer, and an aggregate of enzymes. The probable mechanism of action of acetyl-CoA carboxylase obtained from all sources is illustrated in Fig. 5.4. The bacterial enzymes can be resolved, while animals apparently have a tight protein complex; the plant system has a molecular structure intermediate between these two.

Preparations of intact chloroplasts synthesize fatty acids from acetate, and evidently contain an acetyl-CoA carboxylase. Attempts by Burton and Stumpf (1966) to investigate this system were thwarted by the appearance of an inhibitor when the organelles were disrupted: this depressed the activity of wheat germ acetyl-CoA carboxylase and could account for the low activity of this enzyme in chloroplast preparations. The inhibitor was not inactivated by heat or by papain or ribonuclease and it was evidently specific for acetyl-CoA carboxylase and latent in intact chloroplasts. It might have a function in the control of fatty acid biosynthesis (p. 116) via its action on a crucial enzyme of the synthetase complex. Malonyl-CoA may also be formed from malonate by the action of a thiokinase, whose presence has been demonstrated in a wide variety of plant tissues, including germs, seeds, cotyledons, leaves and roots and in avocado mesocarp (Hatch and Stumpf, 1962). The substrate for the thiokinase would be provided by the oxidative decarboxylation of oxalacetate, a reaction which has been shown to occur in bush bean roots, where malonate accumulates and acetyl-CoA carboxylase is absent (de Vellis et $al.$, 1963; Shannon et $al.$, 1963).

3. Fatty acid synthetase

Barron et $al.$ (1961) reported that acetone-dried powders of avocado mesocarp particles yielded a source of stable soluble synthetase which converted acetate to palmitate and stearate in the presence of carbon dioxide, ATP, CoA, Mn and NADPH; it was inhibited by avidin. Acetyl-CoA plus malonyl-CoA were also similarly incorporated, but with this mixture as substrate the system did not require ATP, CoA or CO_2; nor was it sensitive to avidin. Whereas the intact particulate preparation synthesized oleate as well as palmitate and stearate, the solubilized system had lost the capacity to desaturate stearate and to elongate palmitate. The saturated fatty acids were therefore evidently synthesized in the plant particles by the same mechanism as the animal and yeast synthetases though the latter were not associated with particles. Later work (Yang and Stumpf, 1965a) showed that both the particulate and supernatant fractions from avocado possessed synthetase activity but that the former system was lacking malonate thiokinase and the latter acetyl-CoA carboxylase: the particles therefore utilized acetic acid, acetyl-CoA or malonyl-CoA, but malonic acid was inactive; the supernatant could use only malonic acid or malonyl-CoA. Examination of the soluble synthetase system by Overath and Stumpf (1964) showed that it could be resolved by ammonium sulphate precipitation into two fractions which proved to be a heat-stable ACP component and a fraction containing the synthetase enzymes.

In germinating seeds, McMahon and Stumpf (1966) found the synthetase equally divided between particulate and soluble fractions; the specific activities of the enzymes were highest in the particles. In the immature seed, the activity of the synthetase depends on the age of the tissue, but again the particulate fraction is the principal location.

Rinne (1969) observed that the developing soybean is unique in that the major site of fatty acid biosynthesis from acetate is a soluble fraction. Photosynthetic tissue also contains fatty acid synthetase: isolated leaves take up [^{14}C]acetate and convert it to long-chain fatty acid (Eberhardt and Kates, 1957; James, 1963). The major site of synthesis was found to be the chloroplast (Mudd and McManus, 1962; Stumpf and James, 1962, 1963), and the cofactors included inorganic phosphate as well as the expected ATP, CoA, Mg and CO_2; with these present, the synthesis was moreover stimulated by light, and inhibited by photophosphorylation inhibitors. Non-cyclic photophosphorylation, which produces ATP, O_2 and NADPH, was demonstrated to be an important co-reaction of fatty acid biosynthesis (Stumpf et al., 1963), though the effect of light could not be explained by cofactor synthesis alone. It has been proposed (Stumpf et al., 1967) that light stimulates lipid biosynthesis by causing the photoreduction of the inactive disulphide form of ACP (located in the chloroplast grana) to the active thiol, which is re-oxidized in the dark.

Finally a soluble synthetase obtained by breaking lettuce chloroplasts in a French pressure cell, was characterized (Brooks and Stumpf, 1966). This corresponded to the soluble E. coli and avocado systems in being resolvable into separate fractions one of which was an ACP, replaceable by ACP from the other systems. The variable activity of the acetyl-CoA carboxylase may be due to the presence of an inhibitor (Burton and Stumpf, 1966). The system present in intact chloroplasts differs from the soluble system obtained from them in that the latter prefers malonyl-CoA as substrate and requires additional cofactors (ACP, NADPH, NADH and GSH) which are presumably supplied endogenously in the particle. Moreover, the soluble synthetase is not affected by light; it synthesizes mainly stearate, while the principal products of isolated chloroplasts are palmitate and oleate (Stumpf et al., 1967). Both have lost the ability to synthesize linoleate and linolenate. In the intact organelle, Triton X stimulates incorporation of acetate, especially into oleate, though increasing the ATP concentration reversed this effect (Stumpf and Boardman, 1970). In the absence of Triton, ATP (but not other nucleosides) depressed oleate biosynthesis, suggesting that accumulation of ATP in chloroplasts may depress both biosynthesis and desaturation of fatty acids there. It has been suggested by Appelqvist et al. (1968b) that the enzymes responsible for light-induced synthesis are located in the stroma of intact chloroplasts.

Fatty acid synthesis has also been studied in spinach chloroplasts broken by osmotic shock, where it required both particulate and soluble fractions; stimulation by light could be equalled in the dark by adding enough NADPH (Mudd and McManus, 1962). The site of this synthesis is evidently particulate (Devor and Mudd, 1968).

4. Acyl carrier protein

Purified acyl carrier proteins isolated from different sources have been compared by Simoni et al. (1967). Their total amino-acid compositions are similar whether from plants (avocado or spinach) or from bacteria (*E. coli* or *Arthrobacter viscosus*). Their functions as cofactors in the synthetase systems of both *E. coli* and spinach were also compared; the plant system was not affected by the type of ACP used, but the plant ACP with the bacterial synthetase gave a mixture of saturated and unsaturated 3-hydroxy acids from C_{12} to C_{18} rather than vaccenic and stearic acids obtained from the all-bacterial system (Overath and Stumpf, 1964). This implies that plant-ACP derivatives are poor substrates for the bacterial 3-hydroxyacyl dehydrases, causing accumulation of these intermediates. A partial primary structure of spinach ACP has been determined by Matsumura and Stumpf (1968) and compared with that of ACP from two bacteria. Several amino-acids adjacent to the functional 4′-phosphopantetheine group appeared to be identical; the marked difference in biochemical activity must be related to peripheral residues, possibly resulting from differences in binding of the plant and bacterial ACPs to synthetase enzymes. The isolation of ACP from spinach leaves has been described by Simoni and Stumpf (1969), who found a molecular weight of 9700 involving 88 amino-acids.

5. Fatty acid elongation

Biosynthesis of fatty acids in plants may occur to a certain extent by chain elongation (p. 119) rather than de novo synthesis. Hawke and Stumpf (1965a) showed that, in barley seedlings, higher fatty acids (C_{20} to C_{26}) were biosynthesized from $1\text{-}^{14}C$ acetate. Distribution of label in the product suggested two stages: de novo production of palmitate followed by addition of C_2 units (probably acetate) to form saturated fatty acids from stearate to hexacosanoate. Up to 50% of the labelled octadecenoic acid was vaccenate; since bacteria were absent, this implied that the labelled palmitoleic acid present was also elongated. Palmitoleic acid arises aerobically in plants and animals from palmitic acid (p. 134); in rat liver vaccenate is synthesized from palmitate (but not stearate) via elongation of palmitoleate (Holloway and Wakil, 1964). On the other hand, some bacteria contain a synthetase capable of the anaerobic formation of palmitoleate and vaccenate de novo via cis-3-decenoate (p. 131). Vaccenic acid is a rare component of seeds; it seems to occur with an unusually high proportion of palmitoleic acid, which again suggests biogenesis by elongation (Chisholm and Hopkins, 1965).

Genetic analysis of rapeseed plants and fatty acid analysis of their seed oil indicated that 11-eicosenoic and 13-docosenoic (erucic) acids were formed sequentially from 9-octadecenoic (oleic) acid by a genetically controlled elongation system (Downey and Craig, 1964). This hypothesis was confirmed by the distribution of label found in the fatty acids synthesized from radioactive acetate in immature rapeseed pods. Yang and Stumpf (1965b) have reported that avocado mitochondria catalyse the elongation of ricinoleate (but not stearate, palmitate, myristate, 3-hydroxylaurate or 3-hydroxydecanoate) in the presence of acetyl-CoA (but not of malonyl-CoA), the product being 14-hydroxy-11-eicosenoate (lesquerolic acid, p. 25).

The alga *Euglena* synthesizes stearate from acetate *de novo*, but in this case the usual malonate pathway is not followed: the extracted enzymes do not require bicarbonate, are not inhibited by avidin, do not accept malonate and do not contain acetyl-CoA carboxylase (Cheniae and Kerr, 1965). It is possible that condensations involving a phosphorylated derivative of acetyl-CoA are involved. The possible metabolic inter-relationships of the many "minor" unsaturated fatty acids in *Euglena* have been discussed by Korn (1964a).

TABLE 5.3

The anaerobic pathway

Organisms shown to synthesize unsaturated fatty acid from acetate in the absence of oxygen[a]	
Bacteria[b]	*Clostridium butyricum*
	C. kluyveri
	C. pasteurianum
	Escherichia coli[c]
	Lactobacillus arabinosus
	Pseudomonas fluorescens
Photosynthetic bacteria[b]	*Rhodopseudomonas capsulata*
	R. gelatinosa
	R. palustris
	R. spheroides
	Rhodospirillum rubrum

[a] Erwin and Bloch (1964); Wood *et al.* (1965). The mechanism of biosynthesis is illustrated in Fig. 5.13 (p. 131) and Fig. 5.14 (p. 132).
[b] Whole cells.
[c] Active subcellular systems have been prepared.

5

4. Biosynthesis of Unsaturated Fatty Acids

The mechanism of biosynthesis of the unsaturated fatty acids was not clarified by early attempts to draw conclusions from the quantitative analysis of natural oils (p. 97). Investigations using labelled substrates by Bloch and his coworkers established that there are two distinct and mutually exclusive pathways of monoenoic acid formation, one anaerobic (p. 131) and the other aerobic (p. 134). The former is associated only with some bacteria, primitive organisms which originally lived without access to oxygen (Table 5.3); when higher forms of life superseded them as an oxygenated atmosphere became available, the aerobic pathway

TABLE 5.4

The aerobic pathway

Systems shown to synthesize unsaturated fatty acid by aerobic desaturation of the corresponding saturated acid[a]

Bacteria	*Bacillus megaterium*
	Corynebacterium diphtheriae
	Micrococcus lysodeikticus
	Mycobacterium phlei[b]
Algae	*Anabaena variabilis*
	Ochromonas malhamensis
	Porphyridium cruentum
	Poteriochromonas stipitata
	Chlorella vulgaris[c]
Euglenids	*Astasia longa*
	Euglena gracilis[b, c]
Yeasts	*Saccharomyces cerivisiae*[b]
	Torulopsis utilis
Metaphyta	Avocado (mesocarp)[c]
	Castor bean (leaf)[b, c]
	Castor bean (developing seeds)[b]
	Spinach (leaf)[b]
Metazoa	Fish (liver)[b]
	Goat (udder)[b]
	Hen (liver)[b]
	Rat (liver)[b]
	Sheep (intestinal epithelium)[b]

---

[a] Erwin and Bloch (1964); Bloch *et al.* (1967); James (1968).
[b] Active subcellular systems have been prepared.
[c] The "plant pathway" (p. 135) can be demonstrated in these systems.

evolved (Bloch, 1969). This aerobic desaturation has been demonstrated in a wide variety of systems from mammals, birds, fishes, plants, algae and bacteria (Table 5.4; James, 1968).

A. Anaerobic Pathway of Monoenoic Acid Biosynthesis

The biosynthesis of monoenoic fatty acids in anaerobic bacteria has been well established by Bloch and his coworkers (Fig. 5.13). Goldfine

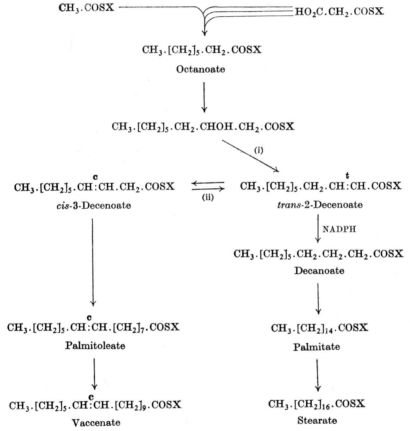

FIG. 5.13. Fatty acid synthesis in *E. coli*. After Bloch (1969); XSH represents an acyl carrier. Reaction (i) and the other stages of elongation are discussed on p. 105 (Fig. 5.7). Reaction (ii) initiates the biosynthesis of unsaturated fatty acids. Reactions (i) and (ii) are catalysed by 3-hydroxyacyl thiolester dehydratase. An analogous scheme explains the biosynthesis of 7-hexadecenoate and 9-octadecenoate from decanoate in *Clostridium butyricum* (Scheuerbrandt *et al.*, 1961).

and Bloch (1961) demonstrated that in cultures of *Clostridium butyricum* added octanoate and decanoate were extended to saturated and mono-enoic acids, laurate and myristate to saturated acids only but palmitate and stearate were not desaturated. They postulated that the double bond was introduced during the formation of the carbon chain. Scheuerbrandt *et al.* (1961) extended this study and proposed that in eubacteria the hexadecenoic and octadecenoic acids present are formed from octanoate and decanoate via dehydration of the 3-hydroxyacyl intermediates to the 3-enoates, followed by further extension. The scheme, outlined in

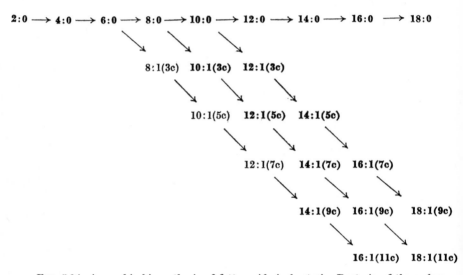

Fig. 5.14. Anaerobic biosynthesis of fatty acids in bacteria. Bacteria of the orders Eubacteriales and Pseudomonadales were considered by Scheuerbrandt and Bloch (1962). The acids in **bold type** have been isolated from bacterial lipids.

Fig. 5.14, differs from the biosynthesis of saturated fatty acids only in a single isomerization reaction (Fig. 5.13); the *cis*-3-enoate produced by the former is elongated without reduction, while the *trans*-2-enoate of the latter is subject to acyl dehydrogenase.

The critical dehydration is catalysed by 3-hydroxydecanoyl thiolester dehydrase, which yields a mixture of 2- and 3-decenoates (Kass *et al.*, 1967); methods for its purification and assay have been described by Kass (1969). It appears to be a multifunctional enzyme concerned specifically with interconversions between 3-hydroxydecanoate, *cis*-3-decenoate and *trans*-2-decenoate (Brock *et al.*, 1967), the C_8 and C_{12} substrates being twenty times less active. It is not inactivated by reagents for thiols, but an acetylenic analogue, 3-decynoyl-*N*-acetylcysteamine, is a potent and specific inhibitor, affecting all the catalysed reactions equally (Helmkamp *et al.*, 1968; Endo *et al.*, 1970); this confirmed the suggestion that a single enzyme was responsible. Its mechanism was investigated by Rando and Bloch (1968), who showed that a common intermediate (presumably enzyme-bound 2-decenoate) was involved: *cis*-3-decenoate arises only by isomerization of the *trans*-2-decenoate which is the sole direct product of dehydration of 3-hydroxydecanoate. The enzyme seems to be composed of two similar, if not identical, polypeptide chains, each of molecular weight 18,000 and each containing an active site associated with one histidine and one tyrosine residue essential for catalytic activity (Helmkamp and Bloch, 1969). Its dehydration and isomerization activity could not be separated and its function is solely to generate the *cis*-3-enoate precursor of the unsaturated fatty acids.

E. coli contains other 3-hydroxyacyl thiolester dehydrases (enoyl thiolester hydratases) which give rise exclusively to the *trans*-2-enoates (p. 107); they differ in chain length specificity (Mizugaki *et al.*, 1968a, b) and evidently contain no isomerization activity. They are responsible for diverting intermediates from the "anaerobic" unsaturated fatty acid pathway to the saturated pathway (Birge *et al.*, 1967). The fatty acid synthetase systems from *E. coli* described on p. 105 are thus responsible for the biogenesis of both saturated and monoenoic acids via these pathways, throughout which all the intermediates remain attached to ACP (Nagai and Bloch, 1967; Bloch, 1969). In plant and animal synthetase systems, the "anaerobic" pathway does not operate; they contain no 3-hydroxydecanoyl thiolester dehydrase and moreover their corresponding enoyl thiolester hydratase seems to be a single enzyme, active during saturated acid biosynthesis at all chain lengths. The multiplicity of enoyl hydratases in *E. coli* may be significant in the biosynthesis of 3-hydroxymyristate, a component of bacterial cell walls, or in the control of the anaerobic biosynthesis of saturated and monoenoic acids in the cell. Such considerations are not applicable to the plant and animal systems.

The "anaerobic" pathway explained the presence in bacterial lipids of typical 7-, 9- and 11-monoenoic acids (Scheuerbrandt and Bloch, 1962) and the apparent absence of polyenoic acids; in particular it accounted for the structure of vaccenic acid (18:1,11c) characteristic of bacteria and the only octadecenoic acid in some strains of *E. coli* (Cronan,1967). It also explained why octanoic acid was the highest homologue of the saturated fatty acid series which acted as a precursor for vaccenic acid (Fig. 5.14). The pathway is characterized not only by its lack of requirement for oxygen, but also by this maximum chain length for precursors;

these considerations clearly distinguish between the "anaerobic" pathway and aerobic desaturation, which is described below.

3-Hydroxydecanoyl thiolester dehydrase has been found only in those bacteria whose unsaturated fatty acids are synthesized by the "anaerobic" pathway (Bloch, 1969); but this pathway is not strictly confined to anaerobic bacteria, being also associated with facultative aerobes (e.g. *E. coli*) and obligate aerobes (e.g. pseudomonads), as shown in Table 5.3. Other bacteria (e.g. *Micrococcus lysodeikticus, Bacillus megaterium, Corynebacterium* species and *Mycobacterium phlei*) synthesize their unsaturated fatty acids by aerobic desaturation (Table 5.4).

B. Aerobic Pathway of Monoenoic Acid Biosynthesis

The aerobic pathway was originally described by Bloomfield and Bloch (1960), who demonstrated that the monoenoic acids in a yeast were formed primarily by direct desaturation of the corresponding saturated acids. Activation to the CoA thiolester was a necessary preliminary reaction; subsequent desaturation by cell-free particulate preparations then required only NADPH and oxygen (Fig. 5.15). The

$$CH_3.[CH_2]_{16}.COSX + O_2 + NADPH \rightarrow$$
$$CH_3.[CH_2]_7.CH\overset{c}{:}CH.[CH_2]_7.COSX + H_2O + NADP^+$$

FIG. 5.15. Aerobic pathway of oleate biosynthesis.

biosynthesis of polyunsaturated fatty acids in another yeast evidently occurred by a similar mechanism, with NADPH and oxygen as obligatory cofactors (Yuan and Bloch, 1961). In addition to these characteristic cofactors, cell-free preparations of *Mycobacterium phlei* required a ferrous salt and either FAD or FMN for desaturation of stearoyl-CoA or palmitoyl-CoA (Fulco and Bloch, 1964). This corresponds with the isolation by Nagai and Bloch (1968) from *Euglena gracilis* of a soluble desaturase system consisting of a flavin-containing NADPH oxidase, a non-haem iron-containing protein (ferredoxin) as well as the desaturase enzyme itself. Systems which require NADPH and oxygen for the direct desaturation of stearate have been isolated from many tissues (Table 5.4), but most have proved to be particle-bound and have resisted attempts to solubilize and fractionate them. Apart from the supernatant from disrupted *Euglena* cells described above, only two other soluble systems had been reported by mid-1970: a similar supernatant from homogenates of soybean cotyledons (Inkpen and Quackenbush, 1969a) and the rat-liver microsomal system solubilized by the method of Gurr *et al.* (1968).

1. Aerobic desaturation in plants

It soon became evident that while the aerobic desaturation pathway was very widespread, the behaviour of higher plants and some algae were anomalous. For a time it appeared that a distinct "plant" pathway operated. The synthesis of oleate from [^{14}C]labelled acetate in avocado mitochondria was absolutely dependent on oxygen; palmitate and stearate were rapidly incorporated into lipid, but not desaturated or elongated (Mudd and Stumpf, 1961). Similarly, oleate is formed from acetate by an oxygen-dependent route in plant leaves (James, 1963) and chloroplasts (Stumpf and James, 1963); octanoate, decanoate, laurate and myristate also served as precursors, but added palmitate and stearate were not elongated or desaturated. Similar results were obtained with barley seedlings by Hawke and Stumpf (1965b), who also tested the 3-hydroxy[3-^{3}H]derivatives of decanoate, laurate and myristate as precursors. However, these were not dehydrated and elongated by an anaerobic pathway to the appropriate monoenoic acid; oxygen was still required, and labelled oleic acid still the principal product. Since β-oxidation would eliminate the label, the hydroxyacids are presumably reduced and act as the saturated precursors. The pattern of labelling of stearate and oleate produced from [1-^{14}C]acetate was shown to be different, implying that the latter was synthesized *de novo*, while the former arose mainly from elongation of endogenous palmitate. The bulk of the stearate in the plant was obviously not being converted to oleate, though a small stearate pool (not in equilibrium with the bulk) might still be responsible. It seemed that plants were displaying a novel pathway of oleate biosynthesis in which the precursor was myristate; palmitate and stearate were apparently not precursors, but were transferred without desaturation to acyl lipids. A similar pathway of oleate biosynthesis was reported in leaves of the ice-plant *Carpobrotus chilense*, though in this case added myristate yielded 9-tetradecenoate as well as palmitoleate and oleate (Fulco, 1965). The so-called "plant pathway" (Fig. 5.16) evidently also operated in some green algae and protists (Erwin and Bloch, 1964) and apparently had the cofactor requirements of the aerobic pathway (Fig. 5.15) but substrate requirements reminiscent of the anaerobic pathway (Fig. 5.14). One possible explanation of the lack of desaturation of stearate added to plant systems might be its faster incorporation into glycerides; decanoate would be a precursor of oleate because it is not efficiently removed by this competing reaction (Mudd, 1967).

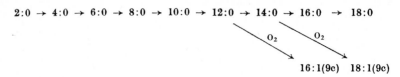

$2:0 \rightarrow 4:0 \rightarrow 6:0 \rightarrow 8:0 \rightarrow 10:0 \rightarrow 12:0 \rightarrow 14:0 \rightarrow 16:0 \rightarrow 18:0$

O_2 O_2

$16:1(9c)$ $18:1(9c)$

Fig. 5.16. The plant pathway. An apparent pathway of aerobic biosynthesis of unsaturated fatty acids in higher plants. It was distinguishable from the "anaerobic pathway" (Fig. 5.14) by its requirement for oxygen, and from the normal "aerobic pathway" (Fig. 5.15) by the failure to demonstrate direct desaturation of added palmitate or stearate (Erwin and Bloch, 1964; Harris *et al.*, 1967). Direct desaturation has now been demonstrated in green algae under special conditions (Harris *et al.*, 1965), and in sub-cellular preparations from higher plants (Nagai and Bloch, 1968).

While *Chlorella vulgaris* and *Euglena gracilis* in media containing carbohydrate failed to desaturate added palmitate or stearate, Harris *et al.* (1965) also demonstrated that if the organisms were denied access to any carbon source, direct aerobic desaturation did in fact take place. Nagai and Bloch (1968) obtained cell-free particulate fractions of etiolated *Euglena* cells which desaturated the added CoA thiolesters of palmitic and stearic acids, but not their ACP thiolesters; they also described soluble cell-free extracts of green cells which were capable of the direct desaturation of stearoyl-ACP only. They later showed that the soluble system could be fractionated to yield three components: NADPH oxidase, the desaturase and a ferredoxin (Nagai and Bloch, 1968). All three proteins were required for the conversion of stearoyl-ACP to oleate, and a similar system was isolated from spinach (Fig. 5.15). Meanwhile Harris *et al.* (1967) investigated the biosynthesis of oleate from [2-^{14}C]acetate in plant leaves. Anaerobic incubation yielded only labelled palmitate and stearate, since desaturation was inhibited by lack of oxygen. After washing and changing to aerobic conditions, analysis showed that label had evidently been transferred from endogenous stearate to oleate. Direct desaturation of exogenous stearate could not be demonstrated, even in etiolated cotyledons or leaves kept in darkness; these were expected to lack substantial carbon reserves but to require rapid synthesis of unsaturated acid on transfer to light, conditions favourable for the successful demonstration in *Chlorella* cells.

Nevertheless, it was suggested that the "plant" pathway operated by the established direct desaturation mechanism, but that added stearate or palmitate was not converted to the true enzyme substrate and was therefore not available for desaturation. Since they were evidently available for synthesis of acyl lipids, esterification to their CoA thiolesters must occur; a block therefore occurs during the process of transfer of the acyl chain from CoA to the enzyme. In some cases at least, this transfer operates via the ACP thiolester (e.g. Nagai and Bloch, 1968) and the enzymes which plants lack could therefore be the CoA-ACP palmitoyl and stearoyl transferases. These enzymes are also absent or inactive in organisms such as *Chlorella* which exhibit the "plant" pathway when growing in nutrient medium; but when the cells are suspended in an inorganic medium under illumination, the transferase enzymes are induced or activated, and direct desaturation of added stearate can take place. When growing partly heterotrophically, the organism uses acetate derived from a carbon source in the medium for the preferential *de novo* synthesis of unsaturated fatty acids. When no other nutrient is available, photosynthetic fixation of carbon dioxide becomes important. This autotrophic existence, encouraged by high illumination, requires synthesis of more unsaturated fatty acids for the photosynthetic apparatus; if *de novo* biogenesis from carbon dioxide is limited, this must be at the expense of saturated precursors, available as acyl lipids. Direct desaturation of stearate is evidently preferred, rather than catabolism and resynthesis of the unsaturated fatty acid; the appearance of the necessary transferases now enables added stearate to be also converted to oleate, and thus the aerobic pathway is demonstrable. It is of interest to note that stearate is transferred to the desaturase via stearoyl-ACP (and not stearoyl-CoA) in green *Euglena* grown on CO_2 in the light; but via stearoyl-CoA in the etiolated cells grown on sucrose in the dark, where stearoyl-ACP is a poor substrate (Nagai and Bloch, 1965). This suggests that two different transferase systems can operate in *Euglena,* possibly one of them analogous to the system inducible in *Chlorella* by a similar change to photoauxotrophic growth conditions. The existence of such a transferase has not been observed directly, but its inhibition by sterculate has been inferred (p. 156).

The appearance of the transferase activity cannot be demonstrated in higher plants, since it is not practicable to deprive the leaf cell of access to carbon sources which are used preferentially for the *de novo* synthesis of unsaturated fatty acids. The "plant" pathway seems to be only a manifestation of a specific metabolic block, which denies the exogenous saturated substrate access to the desaturase. Under these circumstances, no monoenoic and no polyenoic acids are synthesized; but if monoenoic acid is itself the substrate, no similar block exists, since direct conversion to polyenoic acid occurs (p. 142). The concept of a separate "plant" pathway of unsaturated fatty acid biosynthesis is now superfluous, since the metabolic sequence, including aerobic desaturation, is followed (Fig. 5.17). The mechanism of this desaturation is discussed on p. 146. Derivatives of CoA or ACP are active as substrates in different desaturase systems, but the effect of sterculate (p. 155) suggests that

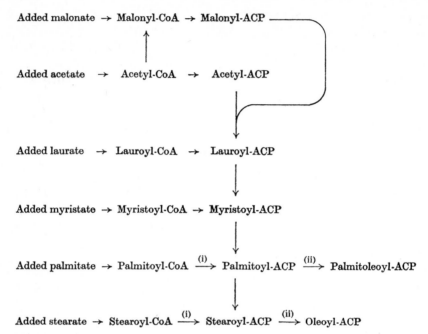

FIG. 5.17. The aerobic pathway of monounsaturated fatty acid biosynthesis in plants. After James *et al.* (1968). The "plant pathway" (Fig. 5.16) may be explained by this scheme, which involves the biosynthesis of unsaturated fatty acids from acetate via direct desaturation of endogenous saturated acid, catalysed by enzymes (ii) requiring O_2 and NADPH (Fig. 5.15). Direct desaturation of exogenous palmitate and stearate depends on the activity of the enzymes (i) responsible for transfer from CoA to ACP or to the de- saturase enzymes (ii). These transferases are not normally present (or active) in higher plants; they may be induced (or activated) in *Chlorella*. They are apparently inhibited by sterculate (p. 155).

transfer of the acyl chain from CoA (possibly via ACP) to the enzyme precedes desaturation. It may be that a complex lipid (e.g. galactosyl glyceride) acts as a substrate or acyl donor to the desaturase (p. 188).

The biosynthesis of oleic acid in non-photosynthetic tissue from higher plants has also been investigated. Oleate is synthesized from acetate and malonate in particulate fractions of avocado mesocarp (Yang and Stumpf, 1965a) and of developing castor seeds (Drennan and Canvin, 1969). Developing soybean seeds also have Δ9-desaturase activity, but here the enzymes are present mainly in the supernatant fractions from homogen- ates of the immature cotyledons (Inkpen and Quackenbush, 1969a).

2. Specificity of desaturation

The substrate specificity of desaturase systems throws some light on their mode of action. Howling *et al.* (1968) incubated the homologous series of [1-^{14}C]acids from myristic to nonadecanoic with *Chlorella vulgaris* cells, and each was desaturated to the corresponding 9-monoenoic acid; in addition myristate, pentadecanoate and palmitate yielded the 7-monoenoates. It seemed likely that two desaturases were present, each defining the double bond position with reference to the carboxyl group. The substrate is therefore probably bound to the enzyme through the carboxyl group, presumably as a thiolester. Johnson *et al.* (1969) similarly showed that hen liver preparations desaturate fatty acids of chain length from C_{12} to C_{22}: the major products were the 9-monoenoates. Two optima were observed at C_{14} and C_{17-18} suggesting two Δ-9 desaturases of different chain-length preferences; no Δ-7 desaturase was apparent. Minor amounts of the 10- and 11-monoenoates were formed, implying some lack of positional specificity.

The desaturation of stearate to oleate is highly stereospecific. Schroepfer and Bloch (1965) prepared four monotritiostearates, with label in the D-9, L-9, D-10 and L-10 positions; desaturation in growing cultures of *Corynebacterium diphtheriae* involved the loss of label from the [D-9-^{3}H] and [D-10-^{3}H]isomers, but not from the two [L-^{3}H]compounds. Isotope effects suggested that hydrogen removal from C-9 was rate-limiting and preceded hydrogen removal from C-10. Morris *et al.* (1967) incubated [*erythro*-9,10-^{2}H$_{2}$]stearate, [*threo*-9,10-^{2}H$_{2}$]stearate, [D-9-^{3}H]stearate and [L-9-^{3}H]stearate with *Chlorella vulgaris*; analysis of the oleate demonstrated that desaturation involved the loss of the D-9 hydrogen atom and of a pair of hydrogen atoms of *cis* relative configuration. The results therefore confirmed the stereospecificity of the enzymic cleavage of the D-9 and D-10 hydrogen atoms, but they also suggested that in this case the mechanism involved their simultaneous concerted removal (Morris, 1970). Bloch (1969) also incubated [*erythro*-9,10-^{2}H$_{2}$]stearate with *Corynebacterium diphtheriae* and confirmed that the relative configuration of the two hydrogen atoms removed during desaturation was *cis*. He has pointed out, however, that *trans*-eliminations are preferred in chemical and enzymic systems, and that the conformation of stearate in the enzyme-substrate complex is only inferred from that of the substrate; configurational changes during the desaturation process cannot be excluded, and the implied *cis*-elimination mechanism has not been entirely proved.

In any case it is evident that the conformation of the fatty acid chain is fixed when attached to the enzyme; this implies strong non-polar interaction between the flexible hydrocarbon chain and a lipophilic enzyme surface. These forces are apparently upset if the hydrocarbon chain is substituted: when a range of fifteen isomeric monomethyl stearic acids were compared with stearic acid itself as substrates for the

Δ9-desaturase of *Chlorella vulgaris*, the presence of the methyl groups in different positions partially or totally prevented desaturation (Brett *et al.*, 1971). The 8-, 9-, 10-, 11-, 12- and 14-methylstearates did not react to any significant extent, while substitution at the 2-, 3-, 5-, 6- or 15-positions caused a fall in desaturation rate to less than 5% of that of stearate. Only DL-4-methylstearate (20%), DL-16-methylstearate (25%), 17-methylstearate (40%) and 18-methylstearate (i.e. nonadecanoate; 8%) were acceptable, but the extra methyl group evidently still interfered with enzyme-substrate binding. The hen-liver Δ9-desaturase showed a similar preference for unsubstituted stearate, but in this case DL-2-methylstearate was desaturated at the 75% level. If only one enantiomer of each of the racemic substrates were reacting (which seems likely), then the appropriate rates of desaturation relative to stearate were double those given above. It therefore appears that the substrate (at least between C-5 and C-15) must be so closely bound to the enzyme surface that the methyl groups cannot be easily accommodated (p. 151).

C. Biosynthesis of Polyenoic Acids

1. Introduction

It has been considered that bacteria are unable to synthesize polyunsaturated fatty acids, and these have rarely been found in bacterial lipids. The anaerobic pathway (p. 131) leads to the biosynthesis in some bacteria of monoenoic acids only. The presence in *Mycobacterium phlei* of 4,8,12,16,20-hexatriacontapentaenoic and similar acids may be due to a repetitive process similar to the anaerobic pathway (Asselineau *et al.*, 1969); if so, the organism would be unique in its ability to synthesize unsaturated fatty acids by both aerobic and anaerobic pathways. However, there is as yet no direct evidence for the anaerobic biosynthesis of polyunsaturated acids.

The observed biogenesis of 5,10-hexadecadienoate in *Bacillus licheniformis* is due to aerobic desaturation (Fulco, 1969a). Here, it appears that the desaturase which converts palmitate to the 5-monoene also accepts 10-hexadecenoate (the product of another desaturase) with the formation of some 5,10-diene. This was brought about by culturing at 35°C when palmitate is desaturated to the 10-monoene only. Another desaturase was induced by lowering the temperature to 20°C, when palmitate was desaturated to both 10- and 5-monoenes; under these conditions, the diene also accumulated. Other strains of bacilli were found to have either the inducible Δ5-desaturase or the temperature-independent Δ10-desaturase but not both.

With very few exceptions (p. 76), the biosynthesis of polyunsaturated fatty acids takes place in all plants and algae by a mechanism involving sequential desaturations similar to that involved in the conversion of stearate to oleate, interspersed, if appropriate, with chain elongations (p. 119). The position of the second and subsequent double bonds is not random, but usually spaced from an existing double bond to yield a product with the familiar methylene-interrupted pattern. Since the first double bond is normally introduced in the middle of the saturated carbon chain, there is room for further desaturation on either side. In plants, fatty acids are normally desaturated between the existing double bond and the methyl end of the molecule (Fig. 5.18); oleate $(18:1,9c)$ thus gives linoleate $(18:2,9c12c)$ which in turn yields α-linolenate $(18:3,9c12c15c)$.

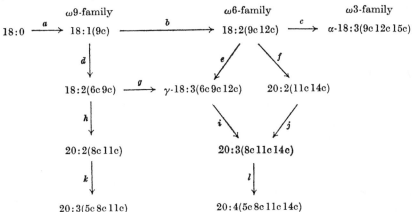

FIG. 5.18. Biosynthesis of some polyenoic acids. In higher plants, the major route is *abc* (involving α-linolenate); route *abeil* is sometimes significant, especially in lower plants and algae. Routes *dhk*, *fjl* and *eil* (involving γ-linolenate) are characteristic of animals, which can perform neither reaction *b* nor *g*. The relative importance of each route may be affected by external conditions, for example in *Euglena* (p. 78), where routes *abc* and *abfjl* may be followed (p. 157). The other "minor" plant fatty acids (e.g. Table 1.4, p. 8) are presumably formed by similar sequential elongations and desaturations from the above metabolites, possibly by more than one pathway.

The animal kingdom, however, has apparently lost the ability to desaturate in the 12-position, or indeed at any position between an existing double bond and the methyl end of the molecule; animal systems can convert stearate via oleate $(18:1,9c)$ to 6,9-octadecadienoate and not linoleate. Further metabolism causes the accumulation of 5,8,11-eicosatrienoate by chain elongation and further desaturation (Fig. 5.18). Furthermore, if linoleate is supplied to animal systems, chain elongation and desaturation similarly lead, usually via 6,9,12-octadecatrienoic (γ-linolenic) acid and 8,11,14-eicosatrienoic acid, to arachidonic acid (5,8,11,14-eicosatetraenoic acid) in which the two new double bonds arise towards the carboxyl end of the molecule (p. 157). Evidently further chain elongation provides room in the molecule for further desaturation to take place, so that highly unsaturated oils, such as those of fish, characteristically contain docosahexaenoic acids. It is therefore usual to distinguish between the α-linolenic pathway in plants and the γ-linolenic pathway in animals (Fig. 5.18) to emphasize their different capabilities for polyunsaturated fatty acid biosynthesis (Erwin and Bloch, 1964; Erwin et al., 1964). In general, plants can synthesize polyenoic acids of the ω3-family from those of the ω6-family which can arise from ω9-monoenes; animals are only able to elaborate polyenoic acids within given families. The α-linolenic pathway is followed by some blue-green algae, red algae, green algae, yeasts and fungi as well as higher plants. There is an intermediate zone: some lower plants contain fatty acids characteristic of both the α- and γ-linolenic pathways. In euglenids and phytomonads the relative importance of each pathway is affected by growth conditions; some protozoans are more animal-like since they can use the γ-linolenic pathway but only the oleate-linoleate conversion of the α-linolenic pathway. A similar situation occurs in the free-living nematode *Turbatrix aceti*, which is the only metazoan organism reported to desaturate oleate to linoleate (Rothstein and Götz, 1968).

Higher animals are restricted to the γ-linolenic pathway, and their inability to synthesize linoleic acid *de novo* leads to a deficiency disease if it is not supplied in the diet, except in mature animals who have stored enough from previous diet. Linoleic and similar acids are therefore the "essential fatty acids" of animal (including human) nutrition; one reason for this, and perhaps the only reason, is that they are precursors of the prostaglandins, a hormone-like class of fatty acids which have several essential physiological functions, but a short metabolic half-life. Such compounds have no significance in plant metabolism.

2. Biosynthesis of polyenoic acids in plants

The yeast *Torulopsis utilis* is rich in lipid with a high proportion of linoleic acid; it was therefore chosen by Yuan and Bloch (1961) for the first demonstration of the aerobic pathway of polyunsaturated fatty acid biosynthesis. Growing or resting cells converted [1-^{14}C]oleate to labelled linoleate and α-linolenate only, by an oxygen-dependent reaction the mechanism of which was therefore suggested to be analogous to the direct desaturation of stearate to oleate. Subcellular systems of the yeast, consisting of particulate and supernatant fractions, were shown by Meyer and Bloch (1963) to desaturate oleoyl-CoA.

In higher plants, leaf lipids are also rich in polyunsaturated acids; James (1962a) allowed isolated leaves to take up [1-^{14}C]oleate and demonstrated the rapid production of labelled linoleate whose only labelled product on subsequent chemical oxidation was [^{14}C]azelaic acid. Moreover, no formation of palmitate was observed, and the direct desaturation of the oleate to linoleate was inferred. The slower formation of labelled linolenate in the leaf suggested that a further desaturation of linoleate occurred (James, 1963). The site of biosynthesis of saturated and monounsaturated fatty acids within the leaf was shown to be the chloroplast (Stumpf and James, 1962; 1963. Mudd and McManus, 1962). The isolated organelles accepted acetate or malonate as precursors; but desaturation beyond the monoene stage was not observed, though α-linolenic acid is the principal acid present. Harris and James (1965) confirmed that in chopped or intact leaves, [1-^{14}C]-oleate is converted to [1-^{14}C]linoleate and [1-^{14}C]linolenate by direct sequential desaturation; their conclusions were based on a time-study of the reaction which revealed typical precursor-product relationships, and on the retention of label in the carboxyl group. Desaturation of oleate was also demonstrated in particulate fractions from safflower seeds (McMahon and Stumpf, 1964). However, all attempts to demonstrate this reaction in subcellular leaf preparations have failed. Apparently the particular enzymes responsible are lost or inactivated on disruption of leaf cells; attention was therefore turned to the green unicellular alga *Chlorella vulgaris*. While this provides a simpler photosynthetic system which can be controlled and disrupted more easily than leaves of higher plants, it cannot be assumed that all results obtained with algae necessarily apply to leaves. Nevertheless the systems have obvious similarities: one is in fatty acid composition (p. 76) and another is that both were originally classified together (Erwin and Bloch, 1964) as synthesizing monoenoic fatty acids by the "plant" pathway (p. 135). However, Harris *et al.* (1965) demonstrated that when *Chlorella* cells were suspended in buffer without nutrient carbohydrate under illumination, rapid synthesis of linoleate and linolenate occurred by the "aerobic" pathway: added [1-^{14}C]stearate was converted directly to [1-^{14}C]oleate and [1-^{14}C]linoleate. Cells grown in this way were used by Harris and James (1965) to investigate polyunsaturated fatty acid biosynthesis in subcellular preparations. Homogenates containing no viable cells were shown to retain the ability to desaturate added [1-^{14}C]oleic acid; the cofactors were oxygen and NADPH or NADH. However, on centrifugation, the activity was apparently lost. This was due to the separation of the soluble activating enzymes and the particulate desaturase system, for when [1-^{14}C]oleoyl-CoA was used as substrate, considerable desaturation to linoleate and linolenate occurred in the "chloroplast" fraction. The supernatant evidently contained a little chloroplast material as well as an activating system, since some desaturation of added oleic acid or oleoyl-CoA to linoleate occurred there.

James *et al.* (1968) have shown that sterculate partially inhibits formation of linoleate from added oleate but hardly at all from added acetate in *Chlorella* under conditions which allow the esterification of palmitate via palmitoyl-CoA to acyl lipid (p. 155). Sterculate therefore does not inhibit desaturase itself, nor the formation of CoA esters; it is therefore unlikely that the desaturase accepts oleoyl-CoA itself as substrate. The characterization of the stearoyl-ACP desaturase from *Euglena* (Nagai and Bloch, 1966) suggests that ACP esters are the true substrates or substrate donors in these desaturations (Fig. 5.19). However, Harris *et al.* (1967) have suggested that the phosphatidyl

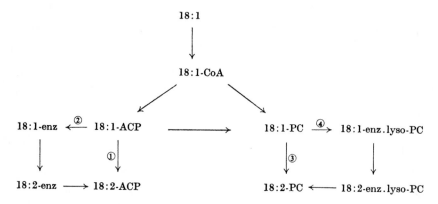

FIG. 5.19. Hypothetical pathways of desaturation of oleate to linoleate. After Gurr *et al.* (1969). Possibilities include oleoyl-CoA as substrate or substrate donor; oleoyl-ACP as substrate or substrate donor (route 2); 1-acyl-2-oleoylglycerol-3-phosphoryl-choline (18:1-PC) as substrate (route 3) or substrate donor (route 4). In *Chlorella*, route 4 appears most likely. Two oleate desaturases (enz) may exist: one accepts $\Delta 9$-monoenoates and yields $\Delta 9,12$-dienoates; the other accepts $\omega 9$-monoenoates and yields $\omega 6,9$-dienoates. Either would convert oleate to linoleate.

choline molecule is involved in the mechanism of $[1\text{-}^{14}C]$oleate desaturation in *Chlorella*, on the basis of its high specific activity with respect to newly biosynthesized linoleate, which rises to at least fifteen times as high as that of any other lipid. Nichols *et al.* (1967b) also implicated certain glycolipids and phospholipids including phosphatidyl choline. The experiments of Gurr *et al.* (1969) confirmed that phosphatidyl choline is an important intermediate in the conversion of oleate to linoleate in *Chlorella* chloroplasts. Labelled oleic acid was rapidly incorporated into this lipid, and desaturation occurred more slowly; the linoleate formed remained esterified to the lipid. Added $[1\text{-}^{14}C]$oleoyl-phosphatidyl choline was converted to $[1\text{-}^{14}C]$linoleoyl-phosphatidyl choline; oleoyl-phosphatidyl choline formed *in situ* was desaturated at a faster rate than oleoyl-CoA under conditions in which unesterified fatty acid could not be desaturated. These results indicated that either the oleate was desaturated while still attached to the lipid, or (more likely) that it was transferred directly from the lipid to the desaturase enzyme, and the linoleate produced immediately transferred back to lipid again (Fig. 5.19). Such changes in the structure of fatty acids while attached to (or closely associated with) a complex lipid are discussed on p. 188 (Fig. 6.3). Similar lipid involvement in the desaturation of linoleate to linolenate has been postulated (Roughan, 1970). In hen liver microsomes, the stearate desaturase activity is lost on washing with aqueous acetone, and then may be fully restored by adding a mixture of microsomal lipids, including phosphatidyl choline (Jones *et al.*, 1969). One explanation of these results is that some of the required lipids act as acyl acceptors or carriers during microsomal desaturation; however they may also be necessary for other unknown functions or coupled reactions associated with the organelles.

3. Specificity of desaturation

Howling *et al.* (1968) have studied the chain length specificity of the desaturases of *Chlorella vulgaris* by incubating the cells with the homologous series of labelled saturated precursors from tetradecanoate to nonadecanoate. Each of the substrates gave the corresponding 9-monoenoic acid which (with the exception of tetradecenoate) was converted to the 9,12-dienoic acid; the desaturase system therefore inserted a second double bond into the $\Delta 12$ position without regard to chain length, though the C_{18} substrate was preferred. On the other hand, of the three 7-monoenoic acids formed, only the hexadecenoate was further desaturated, and the product was the 7,10-diene. When mono-enoic acids were used as substrates, 8-heptadecenoate gave 8,11-heptadecadienoate, 9-octadecenoate gave 9,12-octadecadienoate and 10-nonadecenoate gave 10,13-nonadecadienoate. In these cases the existing double bonds were in the $\omega 9$-position and the second double bond was inserted in the $\omega 6$-position regardless of the distance between this position and the carboxyl group. The concept of location of the new double bond from the methyl end of the fatty acid molecule is rational if the desaturation indeed occurs while the acid is part of an acyl lipid (p. 151); in this case the carboxyl group is not readily available for binding and the positional specificity of the $\omega 6$-desaturase may be due to the lipophilic binding of the free hydrocarbon terminus at a suitable site on the enzyme. Two desaturases may therefore be responsible for dienoic acid biosynthesis in *Chlorella*: a $\Delta 12$-desaturase which accepts $\Delta 9$-monoenoates and a $\omega 6$-desaturase which accepts $\omega 9$-monoenoates. Oleate would be desaturated to linoleate by both these enzymes (Fig. 5.19). While the product is always a methylene-interrupted diene, the position of desaturation is not directed solely by the position of the existing double bond in the substrate, since 8-octadecenoate is not accepted by the desaturase system, and 8,11-octadecadienoate is not formed. Similarly, neither 11-octadecenoate nor 8-hexadecenoate is desaturated.

The stereochemistry of the desaturation in *Chlorella* of oleate to linoleate was shown by Morris *et al.* (1967) to be the same as that of stearate desaturation (p. 139). The conversion of the *erythro* and *threo* isomers both of $[12,13-{}^2H_2]$oleic acid to linoleic acid and of $[15,16-{}^2H_2]$-oleic acid to linolenic acid again demonstrated a *cis*-elimination of hydrogen from the 12,13 or 15,16 positions. The conversion of the D- and L-enantiomers of $[12-{}^3H]$stearate to linoleate proved that it was hydrogen in the D-12-position that was lost during the formation of the double bond, which must therefore involve the D-13-position too. By analogy, the *cis*-elimination at the 15,16-positions probably also involves hydrogen atoms both of the D-configuration, but this has not been directly proved. Double bonds are therefore formed by desaturases which are specific for the removal of *vic*-D-hydrogen atoms.

The positional specificity exhibited by the desaturases of yeast is mirrored by an absolute requirement for the specific products. Mutants of *Saccharomyces cerivisiae* which had lost the ability to desaturate, required an exogenous fatty acid with a *cis* double bond in the 9-position, and furthermore this need seemed relatively independent of chain length (Wisnieski *et al.*, 1970). Palmitoleic, oleic, linoleic or linolenic acids supported growth, and the labelled acids were not converted into other fatty acids after being incorporated into the cell. Among the acids tested on which no growth occurred were *cis*-6-, *cis*-11-, *trans*-9- and *trans*-11-octadecenoates; *cis*-5- and *cis*-11-eicosenoates; *cis*-13-docosenoate; *cis*-15-tetracosenoate; *trans*-9, *trans*-12-octadecadienoate; *cis*-9, *trans*-11, *trans*-13-octadecatrienoate; *trans*-9, *trans*-11, *trans*-13-octadecatrienoate and *cis*-5, *cis*-8, *cis*-11, *cis*-14-eicosatetraenoate (arachidonate). Other fatty acids containing epoxy, hydroxy, acetoxy, nitroxide or phenyl groups did not allow growth unless a *cis*-9-double bond was also present. Evidently the position of the double bond normally provided by the yeast desaturase is essential for membrane structure or function.

D. *Mechanism of Desaturation*

The detailed chemistry of the desaturation reaction is not clear. Any credible mechanism must take account of the requirement for oxygen, NADPH and acyl thiolester as well as the specific nature of the removal of hydrogen atoms both with regard to their position on the acyl chain and their configuration. Such a mechanism is unlikely to be similar to well-characterized dehydrogenations such as that of acyl-CoA during β-oxidation, with the formation of *trans*-2-alkenoyl-CoA (p. 203); here, the dehydrogenation can occur anaerobically, and moreover the molecule is attacked at a position activated by the thiolester group, which facilitates stretching of the C–H bonds in the 2-position. The electron transport system, involving the oxidized flavin prosthetic group, is well documented, and evidently different from that associated with desaturases, where oxygen is not replaceable by artificial electron acceptors.

The requirement for oxygen and NADPH strongly suggests the involvement of a mixed function oxygenase, in which one atom of the oxygen is reduced by the NADPH and the other provides for hydroxylation of the substrate.

One postulated mechanism of desaturation has therefore been a two-stage reaction in which hydroxylation was followed by dehydration. However, no direct experimental evidence for this process has been reported. Oxygen-containing intermediates have never been found, even in a soluble desaturase system (Nagai and Bloch, 1968); moreover appropriate exogenous hydroxy-intermediates are not dehydrated in systems containing the desaturase. Light *et al.* (1962) reported that 9- or 10-hydroxystearate (as acid or CoA ester) was not converted to oleate in growing yeast or in a particulate system containing stearate desaturase. In similar yeast systems and in rat-liver homogenates, Marsh and James (1962) found that the conversion of 9- or 10-hydroxystearate to oleate did occur, but to a much smaller extent than stearate desaturation itself; prior conversion of hydroxystearate to stearate followed by desaturation seemed likely. Elovson (1964) observed that labelled hydroxystearates were rapidly metabolized in the intact rat but specific dehydration did not occur; the radioactive oleate isolated was formed by breakdown to, and resynthesis from, acetate. When the soluble stearoyl-ACP desaturase system from *Euglena gracilis* became available (Nagai and Bloch, 1965; 1968) this was tested by Gurr and Bloch (1966) with D(–)-9-hydroxystearate (as acid, CoA thiolester and ACP thiolester), *cis*-9,10-epoxystearoyl-ACP, 9-acetoxystearoyl-ACP and 9-oxostearic acid. No evidence for oleate formation could be found.

The involvement of hydroxy-intermediates in the biosynthesis of polyunsaturated fatty acids has been similarly suspected; for instance, ricinoleic (D-12-hydroxyoleic) acid might be expected to undergo specific dehydration to give linoleic acid. The metabolism of ricinoleic acid is discussed on p. 169; it is indeed formed from oleic acid in castor seeds by a reaction requiring oxygen and reduced pyridine nucleotide cofactor. No direct evidence for the enzymic dehydration of ricinoleate to linoleate has been forthcoming, though the reverse reaction has been observed in ergot (p. 170). When DL-12-hydroxy[12-³H]oleic acid was supplied to yeast known to convert oleate to linoleate, the hydroxyacid was incorporated, but not converted to nonhydroxylated products (Yuan and Bloch, 1961). Hydroxyacids do not accumulate during the desaturation of stearate *in vitro*. However, Yang and Stumpf (1965b) found that, in a particulate preparation from avocado, labelled acetate gave rise to 14-hydroxy-11-eicosenoate (p. 129) as well as other fatty acids, but no linoleate. The hydroxyacid was biosynthesized by elongation of ricinoleic acid, though neither is a constituent of avocado lipid. It was suggested that ricinoleate was a precursor of the linoleate found in avocado, but that when the dehydrase responsible was inactivated during the preparation of the subcellular fraction, elongation occurred instead.

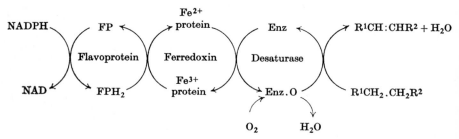

FIG. 5.20. Electron-transport chain for oxygen activation in the desaturase reaction. After Bloch (1969).

If an oxygenated intermediate is formed during the desaturation reaction, it must be irreversibly bound to the enzyme and have a transient existence; moreover since the enzymic activities responsible for the hydroxylation and dehydration did not separate when a soluble system was fractionated (Nagai and Bloch, 1968) it seems likely that they are associated with the same protein. The isolation of a flavin-containing NADPH oxidase and ferredoxin from this system indicates that an electron transfer chain illustrated in Fig. 5.20 operates; this is similar to those associated with characterized oxygenases. Nevertheless, it is still not known whether the desaturase catalyses the production of a covalent carbon–oxygen bond during the overall reaction, or whether some other intermediate is formed. The isotope effects observed during desaturation of labelled substrates (p. 139) lead to conflicting conclusions: if the desaturation is a concerted process (Morris, 1970), no hydroxy-intermediate can be involved; on the other hand if loss of hydrogen is sequential (Bloch, 1969) such an intermediate is possible.

The precise chemical form in which the substrate is accepted by a desaturase is not clear. The possibilities for the case of oleate desaturation are illustrated in Fig. 5.19. Certainly an acyl thiolester is involved; the CoA derivatives are desaturated in systems from yeast (Bloomfield and Bloch, 1960) and M. phlei (Fulco and Bloch, 1964) while the plant systems from spinach and Euglena accept ACP derivatives (Nagai and Bloch, 1968). It may be that the acyl substrates are transferred from CoA to free or enzyme-bound ACP before they are desaturated; this is analogous to transfer of acetate and malonate to fatty acid synthetase before condensation can occur (p. 105). If so, an apparent exception would then be the particulate system from Euglena which desaturated palmitoyl-CoA and stearoyl-CoA, but not palmitoyl-ACP or stearoyl-ACP. Nevertheless, this could be due to the non-acceptability by the particles of exogenous acyl-ACP.

The requirement by the desaturase for some form of thiolester suggests that this form is activated in a way which assists subsequent enzymic reaction. This suggestion is chemically plausible for all other cases where thiolesters are involved and the reaction occurs near the carboxyl group (Fig. 5.2). The aid to C–H stretching at the 2-position which occurs by electron withdrawal is not transmitted down the acyl chain; direct activation of (say) the 9,10-positions during stearate desaturation can only occur by specific curling of the molecule such that these positions interact directly with the thiolester group. This suggestion has been made by Richards and Hendrickson (1964), who point out that this pseudo-annular conformation could explain the positional specificity of desaturation. They also speculate that the mechanism of desaturation of the stearate in this form might involve the conversion of stearoyl-CoA to a perthiolester by oxidative attack at a sulphur atom rather than at carbon. The products would then be oleate and either CoA or an oxidized derivative of CoA, which would require NADPH for regeneration; no carbon–oxygen bond formation need be invoked (Fig. 5.21I). There is no experimental evidence supporting this interesting hypothesis which could equally well involve perthiolesters of ACP or of the enzyme. On the other hand, it has been pointed out by Nagai and Bloch (1967) that the activation of the substrate as a thiolester may have no such mechanistic significance, reflecting only its mode of origin in the cell. The saturated thiolesters released by fatty acid synthetase after de novo synthesis are conveniently desaturated in the same form, so that the products are immediately available for esterification to form the acyl lipids (Bloch, 1969). However, the effect of sterculate on desaturation of exogenous or preformed saturated fatty acids (p. 155) suggests that transfer of substrate from CoA to the enzyme precedes desaturation.

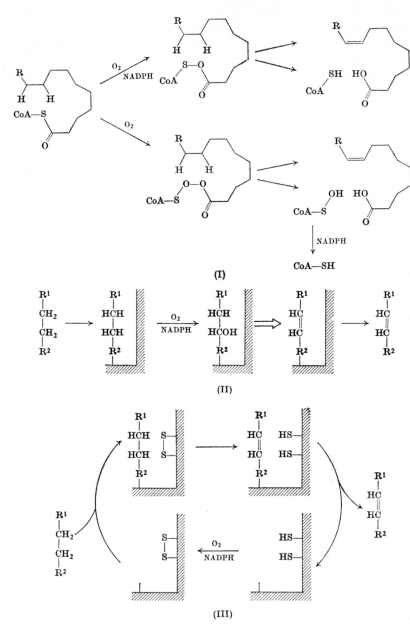

FIG. 5.21. (I–III). Speculative schemes for the mechanism of desaturation. I. Via perthiolesters of CoA. After Richards and Hendrickson (1964). II. Via unstable enzyme-bound hydroxy-intermediate. III. Direct desaturation by activated enzyme. The hatched area denotes the enzyme surface.

In some cases (if not all) a fatty acid is incorporated into a complex lipid before desaturation rather than afterwards (p. 188). Either the desaturase accepts the lipid itself as substrate, or it is coupled with a deacylase which is responsible for the transfer of the acyl residue between lipid and desaturase (p. 144); thiolesters act as substrates since they are precursors of the appropriate lipids.

In those desaturations which seem to take place at a fixed distance from the carboxyl group of the substrate, it seems possible that a specific enzyme–substrate bond is formed: transfer of the fatty acid from CoA, ACP or lipid to the enzyme as an ester would explain the specific anchoring of the carboxyl group. A powerful nonpolar interaction between the hydrocarbon chain and the enzyme surface (or a cleft in it) must be responsible for the specific orientation of the substrate, which is necessary to explain the positional specificity and stereospecificity of the ensuing reaction. Such a close juxtaposition of enzyme and hydrocarbon chain is indeed implied by the substrate specificity of Δ9-desaturase, which excludes methyl-substituted substrates (p. 140). The steric constraints are so precise that it is likely that the active site is on the same enzyme surface which binds the substrate; and that the primary reaction occurs without the intervention of another molecule for which there can be little room. If the position of desaturation at C-9 is governed by the length of a polypeptide chain linking the binding site to the active site, it is possible that loss of one amino-acid residue from this part of the enzyme structure would shorten it sufficiently to cause Δ7-desaturation; the existence of Δ5- and Δ3-desaturases could be explained by postulating further omissions from their amino-acid sequences. The biosynthesis of trans-3-hexadecenoic acid by the action of a Δ3-desaturase is discussed on p. 157. The Δ9-desaturase is ubiquitous; the Δ12-desaturase, which accepts Δ9-monoenoic substrates (p. 145), may operate in a similar manner, but it must have a different structure which is synthesized by plants and not by animals.

In those desaturations which seem to take place at a fixed distance from the methyl terminus of the substrate, one may postulate that it is this end of the molecule which is enzyme-bound. There is evidence for an ω6-desaturase responsible for linoleate biosynthesis in *Chlorella* (p. 145), where the substrate is oleoylphosphatidyl choline (p. 188). It is possible that transfer of substrate from CoA or ACP to lipid precludes its binding to the desaturase at the carboxyl group; specific desaturation can still take place after lipophilic interaction between the hydrocarbon end of the substrate and the desaturase.

The detailed mechanism of desaturation can only be elucidated after thorough investigation of a purified system. Two different speculative schemes are illustrated. In Fig. 5.21II a specific enzyme–substrate complex is formed, and a mixed function hydroxylase gives rise to a bound hydroxyacid, which is rapidly dehydrated. In Fig. 5.21III, the desaturation of the bound substrate is a concerted reaction linked with reduction of the enzyme, and possibly with a conformational change in the enzyme protein to provide energy; the cofactors are then required to re-oxidize the enzyme to its original active state.

E. Control of Desaturation

1. Introduction

It has long been known that the composition of the oil obtained from some plants varies according to the temperatures at which they grow. At lower temperatures, a higher proportion of fatty acids present are unsaturated, measured by a higher iodine value or by chromatographic analysis. In their classic consideration of the chemical constitution of the natural fats, Hilditch and Williams (1964) noted that these variations are generally confined to oleic, linoleic and (if present) linolenic acids, but that climatic temperature has little effect on the proportion of saturated acids. Warmer conditions appear to result in accumulation of oleate at the expense of linoleate (and linolenate) while stearate is not affected. Hilditch and Williams considered that this provided evidence that the biosyntheses of saturated and unsaturated fatty acids followed entirely different routes; moreover, assuming that the sole effect of a rise in temperature was increase in reaction velocities, it appeared that the reactions so affected were the conversion of linolenate *via* linoleate to oleate. At higher temperatures the reaction would tend to go to completion, i.e. more oleate would be produced. This hypothesis appeared reasonable, because the more saturated acids are more stable and represent a greater store of potential energy, thus representing the ultimate goal of lipid storage in a seed. However, this argument neglects the fact that more sophisticated control mechanisms are possible in a seed than in a chemical reaction vessel, and that the physical properties of the lipid biosynthesized may be of critical importance to the viability of the plant. Since unsaturated lipids generally have lower freezing ranges than their saturated analogues, variation of temperature causes precisely those changes in lipid composition which tend to keep its mobility constant, thus protecting the plant from its environment. Plants can live and grow under quite severe extremes of temperature, and the properties of their storage lipids and those associated with membrane function are doubtless vital.

This effect on fatty acid composition seems fairly general in all organisms which assume the temperature of their environment and therefore must remain viable over a wide range of temperatures. Those poikilotherms which have been studied include bacteria, yeast, algae and fish as well as plants. It has been observed by Canvin (1965b) that the fatty acid composition of the seed oils of rape, sunflower and flax depended on temperature when grown under laboratory conditions; safflower and castor oils were not affected.

2. Effect of temperature

(a) *Oxygen availability.* Harris and James (1969a, b) compared the biosynthesis of radioactive fatty acids in samples of plant tissue grown under the same conditions and then incubated aerobically with [2-^{14}C]-acetate at different temperatures. The rate of formation of unsaturated acids relative to saturated acids was termed "desaturation" and was found to increase with decrease of temperature in chopped narcissus bulb, and in the seeds of castor, sunflower and flax. The effect was most marked in castor and least in flax, but could not be demonstrated in leaf discs (castor or spinach) nor in intact green algal cells (*Chlorella vulgaris*). It was discovered that in the bulb tissue, "desaturation" was limited by the availability of oxygen; since oxygen is more soluble in water at lower temperatures, this suggested that the controlling factor was oxygen concentration rather than temperature. The two parameters were separated by comparing the results of incubating the bulb tissue at different temperatures under oxygen tensions necessary to keep the availability of oxygen constant; "desaturation" now decreased with decrease of incubation temperature, which would be due to the expected effect of temperature on the reactions involved. Conversely, if the incubations were carried out at constant temperature but various applied oxygen tensions, "desaturation" increased with increasing oxygenation in the bulb, castor seed and sunflower seed. The absence of this effect in *C. vulgaris*, leaves and flax seed was ascribed to the fact that sufficient endogenous oxygen was produced by photosynthesis inside these tissues to override variations outside. This suggestion was confirmed when *C. vulgaris* was incubated in the absence of light; some inhibition of "desaturation" was observed, which was reversed by increasing the oxygen tension. Further confirmation was provided by the failure of anaerobic conditions to prevent unsaturated fatty acid biosynthesis in *C. vulgaris* and in flax seed; endogenous oxygen must therefore be available for desaturation, and indeed both tissues contained chloroplasts.

The effect of temperature on the fatty acids of *Chlorella sorokiniana* appears anomalous (Patterson, 1970). As the culture temperature increases (up to 22°C) shorter-chain acids of greater degree of unsaturation are produced; above 22°C, increase of culture temperature results in shorter and more saturated acyl chains.

(b) *Induction of desaturase activity*. The regulation of desaturation through the concentration of oxygen available to the desaturases is therefore a major mechanism of immediate control of unsaturated fatty acid biosynthesis in plant tissue. Longer-term changes *in vivo* may also be due to selective synthesis or activation of desaturases as the temperature falls, and probably both forms of control (oxygen-dependence and enzyme induction) are exercised. Induction of desaturase activity can also occur during plant development. McMahon and Stumpf (1966) demonstrated that [^{14}C]acetate was incorporated into palmitate, stearate and oleate in particulate fractions of germinating safflower seedlings, but into palmitate, palmitoleate, oleate and linoleate in similar fractions from developing seeds. Plant fatty acid composition is presumably also under genetic control. The relative amounts of saturated and unsaturated fatty acids which accumulate are then determined by a combination of these effects.

The activity of desaturases may be determined after growth of an organism under different conditions, by estimating desaturation *in vitro* under standard conditions with an adequate oxygen supply. The effect of variation of oxygen availability may be examined separately, using the same system under different conditions of temperature or dissolved oxygen concentration. The first approach was used by Meyer and Bloch (1963) who demonstrated that the greater proportion of unsaturated fatty acids in yeast grown at sub-optimal temperatures was due (at least in part) to a stimulation of desaturase activity, estimated under standard conditions. The alternative approach, taken by Rinne (1969) demonstrated that an enzyme system from developing soybean cotyledons converted [1-^{14}C]acetate to fatty acids *in vitro*, and that the temperature of incubation determined the distribution of label: cooler temperatures favoured the unsaturated acids. Here, availability of oxygen was doubtless the critical factor. The separate effects of oxygen and temperature on yeast fatty acids have been investigated by Brown and Rose (1969); again, decreased dissolved-oxygen tension led to diminished unsaturated fatty acid accumulation.

Thus may be explained the wide variation in composition of the seed oil of a plant grown in different conditions or climates (p. 152). Changes also occur in root lipids as winter approaches (Gerloff *et al.*, 1966), and this ability to increase unsaturated fatty acid biosynthesis may have a direct bearing on the hardiness of a plant.

Other poikilotherms display an analogous effect. The membrane phospholipids of *E. coli* are more unsaturated if the cells are grown at lower temperatures (Haest *et al.*, 1969). The liver and adipose tissue of frogs acclimatized to 7°C contain more unsaturated (and less saturated) fatty acids than corresponding tissues from frogs kept at 25°C (Baranska and Wlodawer, 1969). The desaturase activity, estimated at 37°C, is increased in the tissues of cold-adapted frogs. Fish kept at low temperatures also accumulate more unsaturated fatty acids due to increased biosynthesis of polyunsaturated relative to monounsaturated and saturated acids (Knipprath and Mead, 1966; 1968).

The desaturation of palmitate to 5-hexadecenoate in certain bacilli appears to be under strict temperature control (Fulco, 1969b, 1970). The Δ5-desaturase is not present at 30–35°C, but induction takes place at 20°C; moreover, the induced enzyme is active at 20°C but not at 30°C. On the other hand, the rate of desaturation to palmitoleate in other bacilli is not affected by temperature, and presumably in these cases neither enzyme induction nor oxygen-dependent control is operative.

3. Inhibition of desaturation by sterculate

Cyclopropenoids might be expected to affect biological systems because of the reactivity of the strained ring towards thiols, which are irreversibly blocked. When hens are fed a diet containing sterculate [8-(2-octylcyclopropenyl)octanoate], their eggs show decreased oleate and increased stearate levels; their livers show marked reduction in ability to desaturate stearate to oleate *in vitro* (Johnson *et al.*, 1967c). The inhibition of mammalian desaturase by cyclopropene fatty acids was studied by Raju and Reiser (1967) who compared the ratios of specific radioactivities of stearic to oleic acids in adipose tissue and liver triglycerides of rats given [1-^{14}C]-stearate dissolved in corn oil (control) or in oil from *Sterculia foetida*, which contains sterculate (p. 29). Desaturation of stearate to oleate was greatly inhibited, though conversion of injected [1-^{14}C]acetate to oleate was not much affected by a diet containing sterculate (Raju and Reiser, 1969).

Since sterculic acid is biosynthesized in some plants (p. 175), its effect on plant desaturase systems is of interest. Using *Chlorella vulgaris*, James *et al.* (1968) confirmed that sterculate was a potent inhibitor of oleate formation from added stearate, but that little inhibition occurred when acetate was the substrate. Desaturation of added palmitate to 9- and 7-hexadecenoate was also inhibited, but that of added oleate to linoleate was less sensitive. The formation of acyl lipids and of *trans*-3-hexadecenoate (p. 157; Fig. 5.22) was not affected. In leaf preparations,

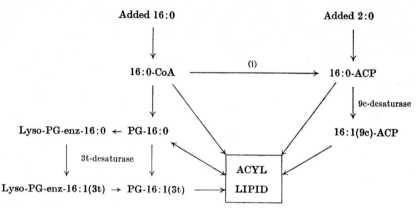

Fig. 5.22. Biosynthesis of *trans*-3-hexadecenoic acid. After Bartels (1969) and James (1968). In this scheme, 1-acyl-2-palmitoyl-glycerol-3-phosphoryl-1'-glycerol (PG-16:0) is the substrate or substrate donor for the 3t-desaturase; palmitoyl-ACP is the substrate (or substrate donor) for the 9c-desaturase. The transferase (i) is apparently inhibited by sterculate (p. 155).

stearate generated internally by anaerobic incubation from acetate was still converted to oleate on aerobic incubation with sterculate.

Evidently the algal and leaf desaturases themselves were not being inhibited by sterculate. The acyl-CoA synthetases (thiokinases) were also still active, since they were presumably necessary in the transfer of exogenous fatty acid to lipid. It therefore seems that sterculate inhibits the transfer of added saturated fatty acid from CoA to ACP, to another carrier or to the desaturase enzyme itself (Fig. 5.17).

The conversion of added oleate to linoleate in a seed system was not affected by added sterculate (Haigh et al., 1968). It appears that there exists a transferase system specific for oleate which is resistant to interference by sterculate, and is associated with a $\Delta 12$- or $\omega 6$-desaturase. A separate system must be responsible for stearate (and palmitate) transfer to the $\Delta 9$- and $\Delta 7$-desaturases, and this is sterculate-sensitive. Such transferase systems have been invoked to explain the anomalous "plant" pathway of aerobic biosynthesis of monoenoic acids (p. 137). Plants evidently contain the oleate transferase, and can convert added oleate to linoleate; their lack of the stearate transferase prevents formation of oleate from added stearate when the "plant" pathway operates.

5. Biosynthesis of Unusual Fatty Acids

The pathways of biosynthesis of saturated (p. 102), unsaturated (p. 130) and polyunsaturated (p. 140) fatty acids discussed in previous sections explain the natural occurrence of all the "major" and "minor" fatty acids. (p. 2; e.g. Table 1.4, p. 8).

For instance, arachidonic acid (20:4,5c8c11c14c) is a "minor" acid of the ω6 family, biosynthesized *de novo* in plants via linoleate, by further desaturations and chain elongation. It is a typical product of the γ-linolenic pathway and is characteristic of vertebrates which cannot synthesize their own linoleate. In rats (Mead and Howton, 1957) and some algae (Nichols and Appleby, 1969) the following reactions occur:

$$18:2(9c\,12c) \rightarrow 18:3(6c\,9c\,12c) \rightarrow 20:3(8c\,11c\,14c) \rightarrow 20:4(5c\,8c\,11c\,14c)$$

Other sequences are possible, and the following pathway operates in a soil amoeba (Korn, 1964b) and a euglenid (Hulanicka *et al.*, 1964):

$$18:2(9c\,12c) \rightarrow 20:2(11c\,14c) \rightarrow 20:3(8c\,11c\,14c) \rightarrow 20:4(5c\,8c\,11c\,14c)$$

It may be that, as in liver microsomes, both pathways exist side by side, and their relative importance is governed by selective inhibition of the various reactions by specific fatty acids present (Stumpf, 1969).

Examination of the structures of the "unusual" fatty acids (p. 9) indicates that their biogenesis is the result of significant departures from the established routes involving known fatty acid synthetases, desaturases and elongation systems. In this section we consider possible metabolic pathways leading to the accumulation of unsaturated fatty acids with unusual double bond positions, of acetylenic acids, of oxyacids and of branched-chain acids. These pathways range from those which have been established by radiochemical experiments to those which are no more than speculations based on analytical data.

A. Non-conjugated Ethylenic Acids

While the aerobic pathway of biosynthesis of *cis*-9-monoenoic acids (with some *cis*-7-isomers) has been intensively studied (p. 134), the biogenesis of double bonds in other less usual positions has attracted less attention.

1. Trans-3-hexadecenoic acid

Photosynthetic tissues of higher plants and algae contain *trans*-3-hexadecenoate, whose structure (established by Debuch, 1961) sets it apart from the other acids present. Moreover, it occurs in these tissues almost entirely at the 2-position of phosphatidyl glycerol (Haverkate and van Deenen, 1965). *Chlorella vulgaris* cells grown under conditions favouring photosynthesis (i.e. inorganic medium in light) similarly contain *trans*-3-hexadecenoate entirely in the phosphatidyl glycerol fraction; if organic nutrient is available, the acid is almost completely absent from the cells whether grown in light or dark (Nichols, 1965b). The biosynthesis of the acid was investigated in the greening cells obtained by transferring them from tryptone-glucose nutrient medium in the dark to phosphate buffer in the light; palmitate was established as the direct precursor (Nichols *et al.*, 1965b). The *trans*-3-hexadecenoate formed by desaturation of added labelled palmitate was found exclusively in the isolated phosphatidyl glycerol. In the dark, no desaturation at the 3-position was observed, though some labelled 9-hexadecenoate was found in other lipids. The need for light may only reflect the elaboration of chloroplast structures in the greening cell, with which the *trans*-3-hexadecenoate is associated. The desaturation was inhibited by the absence of exogenous oxygen, indicating that the reaction was oxygen-dependent like the established aerobic pathway of other monoenoic acid biosynthesis. Endogenous oxygen doubtless accounted for the desaturation observed. Bartels *et al.* (1967) observed that added *trans*-3-hexadecenoate was reduced to palmitate in *Chlorella* cells or chopped lettuce leaves; the residual unreduced substrate was found in all the lipids of the system. It was therefore unlikely that the high specificity of the endogenous *trans*-3-hexadecenoate for the 2-position of phosphatidyl glycerol could be due to a specific transacylation at this position of the intact lipid or of a precursor; it was therefore suggested that phosphatidyl glycerol or lyso-phosphatidyl glycerol was a required substrate or cofactor for the desaturation. The fact that added palmitate is 99% esterified to the acyl lipids of *Chlorella vulgaris* before desaturation starts, is in keeping with this suggestion; moreover the *trans*-3-hexadecenoate arises in the position on the lipid previously occupied by palmitate rather than any other acid. Direct evidence could not be obtained, however, since the appropriate labelled lipid did not penetrate the *Chlorella* cell and subcellular preparations had lost the ability to synthesize *trans*-3-hexadecenoic acid (Bartels, 1969). Nevertheless it seems that 1-acyl-2-palmitoyl-glycerol-3-phosphoryl-1'-glycerol is the substrate for the desaturase or that this lipid acts as a substrate-donor while coupled to the enzyme, the lyso-lipid then accepting the product (Fig. 5.22, p. 156, also p. 188) (James, 1968; Bartels, 1969).

The fact that the biosynthesis of *trans*-3-hexadecenoic acid is not inhibited by sterculate is in keeping with this concept, since it appears that sterculate inhibits the transfer of palmitate from CoA to ACP, to another carrier or to the Δ9c-desaturase (p. 156); transfer to lipids such as phosphatidyl glycerol is not affected (Fig. 5.22). The production of substrate for the Δ3t-desaturase is therefore apparently resistant to inhibition by sterculate, as is the enzymic action of both Δ3t- and Δ9c-desaturases.

2. Other unsaturated acids

Desaturation in the 5-position has been demonstrated (Fulco *et al.*, 1964; Fulco, 1969a, 1970): cells of some bacilli converted stearate to *cis*-5-octadecenoate, and palmitate to *cis*-5-hexadecenoate. Oxygen was an absolute requirement, and iron was a cofactor; the mechanism appeared to be similar to that of the aerobic desaturation in the 9- and 7-positions. Other bacilli converted palmitate to a mixture of 8-, 9- and 10-hexadecenoates; sometimes the unusual 8- and 10-isomers predominated. Stern *et al.* (1969) have reported that in *Leptospira canicola*, a bacterium requiring exogenous fatty acid for growth, direct desaturation occurs of added palmitate or stearate to the corresponding *cis*-11- as well as the *cis*-9-monoenoates.

Hen liver preparations apparently desaturated fatty acids to a minor extent in the 10- and 11-positions (Johnson *et al.*, 1969); the desaturation was little affected by sterculate, which inhibited the major desaturation at the 9-position. Green cells of *Euglena gracilis* contain an unusual variety of monoenoic acids, including 5- and 7-tetradecenoates, 7-, 9-, and 11-hexadecenoates, and 9- and 11-octadecenoates. Similar isomers are found in bacteria which synthesize them by the anaerobic pathway described earlier, and it therefore seemed that *Euglena* employed a similar pathway in addition to the aerobic desaturation mechanism (Nagai and Bloch, 1965; Bloch *et al.*, 1967). Extracts of photoauxotropic *Euglena* were indeed found to convert the ACP derivatives of octanoate, decanoate and dodecanoate to long-chain saturated and unsaturated acids, but only in the presence of oxygen. It appeared that the second *Euglena* pathway, which could not be demonstrated in *Chlorella* or spinach, involved an aerobic desaturation of octanoate, decanoate and dodecanoate to the corresponding *cis*-3-monoenes, followed by chain elongation (Fig. 5.23). Attempts to characterize this pathway experimentally have not been successful (Nagai and Bloch, 1967). Raju and Reiser (1969) and Donaldson (1967) have also postulated a similar alternative pathway of oleate biosynthesis in rats and chickens respectively.

With present data, it seems likely that double bonds arise at unusual positions in plant fatty acids by the action of a normal or abnormal desaturase possibly followed by further modification of the molecule. The ubiquity of the Δ9-desaturase and the coexistence of a Δ7-desaturase suggest that the abnormal loss of an amino-acid residue from the enzyme could change its specificity; further losses could presumably result in

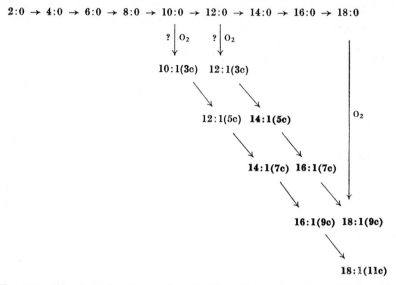

FIG. 5.23. Hypothetical pathway of aerobic biosynthesis of unsaturated fatty acids in green *Euglena gracilis*. After Nagai and Bloch (1965) and Bloch *et al.* (1967). The acids in **bold type** are present in the cells. The biosynthesis of *cis*-3-monoenoic acids by aerobic desaturation has not been confirmed (Nagai and Bloch, 1967). Their biosynthesis by an anaerobic pathway is illustrated in Fig. 5.14 (p. 132).

Δ5- and Δ3-desaturases (p. 151). The production of *trans*-double bonds by these enzymes remains unexplained. If such desaturases were not absolutely specific, a little desaturation at intermediate positions could presumably occur.

Among possible subsequent modifications to the desaturated molecule are chain elongation (p. 119), α-oxidation (p. 213) and β-oxidation (p. 201). For instance, vaccenic acid arises by chain elongation of palmitoleic acid (p. 128), which itself can be a product of a Δ9-desaturase. The biogenesis of *cis*-8,*cis*-11,*cis*-14-heptadecatrienoate (p. 13) evidently involves α-oxidation of linolenate. The presence of *trans*-2,*cis*-9,*cis*-12-octadecatrienoate in pollen (p. 11) implies the involvement of the acyl-CoA dehydrogenase associated with the β-oxidation of linoleate, though such intermediates do not normally accumulate.

B. *Conjugated Ethylenic Acids*

Since there is no direct evidence about the metabolic interrelationships of the conjugated ethylenic fatty acids (p. 14), the subject invites speculation. It seems likely that each of the *cis*-9, *cis*-12, *trans*-9 and *trans*-12 families (Table 1.7, p. 14) represents a metabolic sequence of some kind. Gunstone (1965) considered that linoleic acid might be a precursor, which could give rise to hypothetical oxy-derivatives at the 8-, 11- or 14-positions; these (e.g. 11h-18:2, 9c 12c) could dehydrate and/ or rearrange to give the known conjugated acids. Since the postulated intermediates were unknown, a new theory (Gunstone, 1966) was based on epoxy-derivatives, involving hypothetical enzymic reactions which could convert a 9,10-epoxide to a 9-yne, a 10-en-9-ol, a 12-en-9-ol or a 9,12-diene. Epoxidation of linoleic acid at either double bond would thus give rise to coronaric or vernolic acids, which could be the precursors of dimorphecolic or coriolic acids respectively. These hydroxydienes could dehydrate to give the known conjugated trienes (Fig. 5.24); moreover, the hypothesis also provided a mechanism for the biosynthesis of the major fatty acids and the prostaglandins. However, no evidence has been forthcoming to establish epoxyacids as precursors of any of these three classes of unsaturated fatty acid; indeed the biogenesis of the last two does not now appear to involve them.

Morris and Marshall (1966b) have suggested that it is unnecessary to invoke hydroxylinoleic or epoxy acids: their hypothesis is based on a lipoxygenase-like reaction (p. 226) with linoleic acid or its known

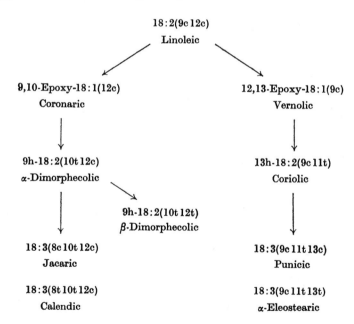

FIG. 5.24. Hypothetical pathways of conjugated fatty acid biosynthesis. After Gunstone (1966).

6

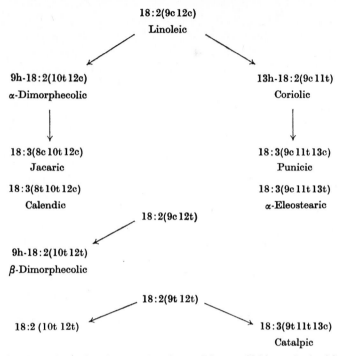

FIG. 5.25. Hypothetical pathways of conjugated fatty acid biosynthesis. After Morris and Marshall (1966b).

cis,trans- or *trans-trans*-isomers as substrates (Fig. 5.25). These could give rise to the hydroxydienes directly, which could be precursors of the trienes: or, more likely, an intermediate in the lipoxygenase-type reaction might stabilize in a number of ways, each leading to the accumulation of a diene, dienol or triene: in Fig. 5.26, the labile intermediate is pictured as a free radical. Thus linoleic acid (18:2, 9c12c) gives either coriolic acid (13h-18:2, 9c11t) and the *cis*-9-family; or α-dimorphecolic acid (9h-18:2, 10t12c) and the *cis*-12-family. The *cis,trans* isomer of linoleic acid (18:2, 9c12t) yields the *trans*-12-family, while *trans,trans*-linoleate gives rise to either *trans*-family. Unlike Gunstone's schemes, the *trans* double bonds are introduced into the molecule before the oxygen function, thereby rationalizing the biogenesis of *trans*-10,*trans*-12-octadecadienoate. Significantly, postulated intermediates are known to occur in the same tissue: this may be supporting evidence, though known intermediates of other metabolic pathways (e.g. fatty acid synthetase) do not normally accumulate and conversely some falsely postulated intermediates do so (e.g. ricinoleate during linoleate formation in castor

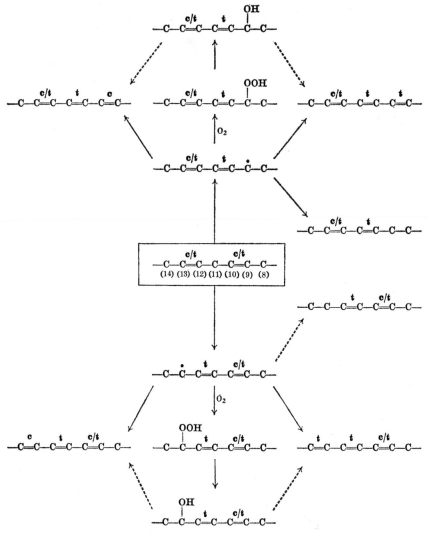

FIG. 5.26. Possible mechanism of biosynthesis of conjugated fatty acids.

seeds). Nevertheless, the classification of the conjugated acids into the four series is made more significant by the dearth of similar natural ethylenic acids which do not fit this hypothetical metabolic scheme. The apparent mutual exclusiveness of α-eleostearic and punicic acids (p. 15) is explicable, since each is formed from a common precursor. Moreover, the scheme may be logically extended to include the crepenynic family of acetylenic acids (p. 19) with the postulated biosynthesis of helenynolic acid (9h-18:2,10t12a) from crepenynic acid (18:2,9c12a). The unusual conjugated 2,4-dienoic acids (p. 17) could be formed by the same mechanism: for example, linoleate (18:2,9c12c) gives α-dimorphecolate (9h-18:2,10t12c) followed by oxidative cleavage at the 8,9-bond to yield *trans*-2,*cis*-4-decadienoate (Heinz and Jennings, 1966). The natural occurrence of this acid esterified to 8-hydroxy-5,6-octadienoate leads to the speculation that the C_{18} ester is formed *in toto* from a C_{18} acid such as 5,6,9,12-octadecatrienoate via 9-oxygenation (Fig. 5.26) and an internal oxidative rearrangement of the Baeyer-Villiger type.

C. *Acetylenic Acids*

Little is known about the detailed mechanism of biogenesis of the acetylenic bond. The carbon skeletons of several fungal acetylenes were found to be derived from acetate with the involvement of malonate; Bu'Lock and Smalley (1962) suggested a biosynthetic pathway related to the saturated fatty acids. The hypothesis invoked the facile elimination reaction in the chemistry of enol esters (including phosphates) of acylmalonic acids, which might therefore lead either to polyacetylenes, or saturated fatty acids (Fig. 5.27). It now seems less likely that this reaction actually occurs in biological systems, though a similar mechanism has been postulated for the cyclization reaction during the biosynthesis of aromatic polyketides (Bu'Lock, 1967). The biogenesis of a fatty acid chain in which some of the 3-ketoacyl intermediates remain un-

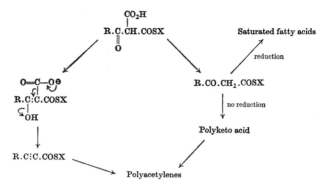

FIG. 5.27. Hypothetical biogenesis of acetylenic bonds. After Bu'Lock and Smalley (1962) and Bu'Lock (1967).

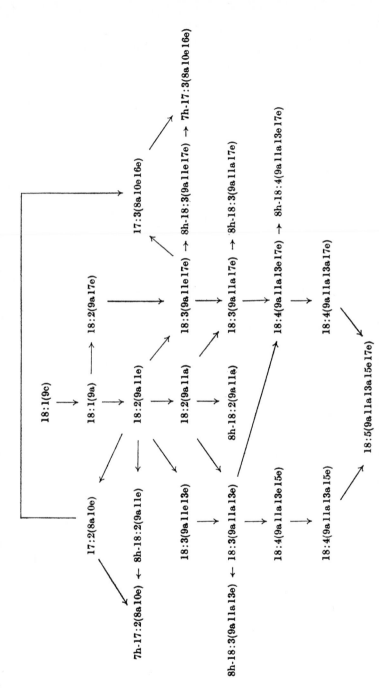

FIG. 5.28. Possible pathways of biogenesis of the stearolic family of acetylenic fatty acids. After Hopkins *et al.* (1968b), Morris and Marshall (1966a), Bu'Lock and Smith (1963). The scheme involves successive desaturations, hydroxylations and α-oxidations. Common names of some of these acids are given in Table 1.8 (p. 18).

reduced (Fig. 5.27) has also been considered; the growing labile polyketo-acyl chain might be protected by chelation to metal. The polyketoacids so produced are hypothetical precursors of polyketides, and are at the same oxidation level. Both these suggested pathways are reminiscent of the anaerobic pathway of monoenoic acid biosynthesis (p. 131), in which a structural feature is introduced into the saturated acyl chain as it grows. Whether these "anaerobic" pathways exist or not, they are not responsible for those acetylenic fatty acids whose biogenesis is analogous to the "aerobic" pathway of ethylenic fatty acid biosynthesis (p. 134), where elaboration of the whole carbon chain precedes insertion of the unsaturated bond.

In a system from *Santalum acuminitum*, Bu'Lock and Smith (1963) demonstrated that labelled acetate was incorporated into ximenynic acid (18:2, 9a 11c) and related poly-ynoic acids of the stearolic family; but the relative rates of incorporation suggested that non-acetylenic acids were synthesized first, and then converted sequentially to the range of acetylenic acids (Fig. 5.28). This dehydrogenation pathway was supported (Bu'Lock and Smith, 1967) by conversion in the fungus *Tricholoma grammopodium* of [^{14}C]oleic acid to [^{14}C]crepenynic acid (18:2, 9c 12a). However, no direct evidence was reported that linoleate was an intermediate, which would be required to substantiate the theory that ethylenic bonds are converted *in vivo* to acetylenic bonds.

Haigh *et al.* (1968) observed that [1-^{14}C]oleic acid was also converted to [^{14}C]crepenynate in the seeds of *Crepis rubra*; analysis showed that the product was labelled specifically in the carboxyl group, and must have been derived directly from oleate by desaturation, with no break-down to acetate and re-synthesis. The conversion required oxygen and was stimulated by traces of copper. Sterculic acid, a known inhibitor of the desaturation of stearate to oleate (p. 155) also inhibited the conver-sion of oleate to crepenynate, but had little effect on the oleate-linoleate reaction. No crepenynate was obtained from *cis,cis*-[^{14}C]linoleate, *cis*-9,*trans*-12-[^{14}C]linoleate or *cis*-12,13-epoxy[^{14}C]oleate (vernolate), all of which occur together with crepenynate in *C. rubra* seeds. Evidently desaturation of a preformed long-chain fatty acid gives rise to the acetylenic acid; the mechanism does not involve free linoleate or vernolate, though such intermediates may be enzyme-bound. Perhaps unknown labile intermediates (X, Fig. 5.29) of desaturation of oleate to linoleate are involved. Another possibility is the biogenesis of the acetyl-enic bond by rearrangement of an allene formed by a double desaturation at neighbouring positions.

The composition of seed oils from different *Crepis* species (Earle *et al.*, 1966) also suggests that crepenynate, vernolate and linoleate are bio-

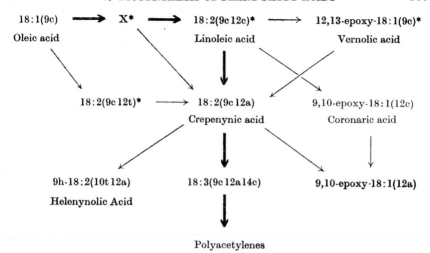

FIG. 5.29. Possible pathways of biogenesis of the crepenynic family of acetylenic fatty acids. After Haigh *et al.* (1968) and Bu'Lock and Smith (1967). Postulated enzyme-bound intermediates are denoted by an asterisk. Oleic and linoleic acids are the precursors of two plant polyacetylenes (Bohlmann and Schulz, 1968a). The biosynthesis of epoxyacids is discussed on p. 172.

chemically related; moreover a distinct inverse relationship was found between linoleic acid and the sum of vernolic and crepenynic acids in the oils (p. 19).

Bohlmann and Schulz (1968a,b) have reported that when [³H]-substrates were fed to *Chrysanthemum floculosum* or *Coreopsis lanceolate*, linoleate yielded two labelled conjugated polyacetylenic derivatives to practically the same extent as oleate; neither vernolate nor 12,13-dihydroxyoleate were so incorporated. This suggests that crepenynate arises from linoleate, and then further desaturation and/or oxidation leads to the crepenynic family of acetylenes (Fig. 5.29) with the retention of the characteristic $\omega6$ triple bond.

Helenynolic acid (9h-18:2,10t 12a) is exceptional in being the only known C_{18} hydroxyacid of the crepenynic family (p. 20). Its biosynthesis (Fig. 5.29) may involve the oxidative rearrangement of crepenynic acid (18:2, 9c 12a), similar to that suggested during the postulated conversion of linoleic acid (18:2, 9c 12c) to α-dimorphecolic acid (9h-18:2, 10t 12c), discussed on p. 162.

The mechanism of biosynthesis of the acetylenic hydroxyacids of the stearolic family (p. 18) is also uncertain, but the specificity of the position of the substituent is significant. Presumably hydroxylation occurs at the $\Delta8$-position of the unsubstituted C_{18} acid; this may be similar to the formation of ricinoleic acid (p. 169). The 8-methylene group adjacent to a triple bond or a conjugated sequence of triple and double bonds will be quite acidic and therefore rather susceptible to nucleophilic attack. The structures of the acetylenic C_{17} fatty acids (Fig. 5.28) strongly suggest that the C_{18} acids are subject to α-oxidation (p. 213). The acetylenic C_{17} hydroxyacids are also probably formed by α-oxidation of their C_{18} homologues; if hydroxylation of the C_{17} acids (as well as of the C_{18} acids) occurs, the enzyme may be a ω10-hydroxylase rather than a Δ8(7)-hydroxylase.

D. Substituted Acids

Oxygen may be incorporated into a metabolite by direct oxygenation at a saturated carbon atom, or by hydration at an unsaturated bond. In the first case, molecular oxygen provides the substituent, as in the biogenesis of 2-hydroxyacids and ω-hydroxyacids. The second route is anaerobic and involves oxygen from water, such as in the formation of 3-hydroxyacyl from 2-enoyl intermediates during β-oxidation of the fatty acids. The biosynthesis of ricinoleic acid (p. 169) involves each route in different tissues. We here consider the biogenesis of some of the natural oxyacids described in Chapter 1.

1. 2-Hydroxyacids

The accumulation of D-2-hydroxyacids (p. 23) in micro-organisms, plants and animals involves an α-hydroxylation mechanism similar to that followed during α-oxidation (p. 213). When [1-^{14}C]palmitate is incubated with leaf systems, D-2-hydroxy-[1-^{14}C]palmitate accumulates (Hitchcock et al., 1968b); it is accompanied by a trace of L-2-hydroxy-[1-^{14}C]palmitate, but this enantiomer is more quickly degraded. The mechanism involves direct replacement of hydrogen by hydroxyl with retention of configuration (Morris and Hitchcock, 1968) and is not the result of dehydrogenation followed by hydration. The 2-hydroxyderivatives of ordinary fatty acids present in plant cerebrosides (Sastry and Kates, 1964b) and in phytoglycolipid (Carter and Koob, 1969) may be similarly formed directly from the unsubstituted acid; it is not known whether hydroxylation of intact lipid-bound substrate can take place.

Other less usual plant 2-hydroxyacids could be formed in the same way. Ustilic acid B (2,15,16-trihydroxypalmitate) may be produced in the smut *Ustilago zeae* from the 15,16-dihydroxypalmitate which is also present (Lemieux, 1953). The similar biogenesis of 2-hydroxysterculate (Morris and Hall, 1967) has been discussed on p. 175; here α-oxidation is strongly implicated. The coexistence in a seed oil of 2-hydroxylinolenic acid and linolenic acid suggests an α-hydroxylation mechanism (Smith and Wolff, 1969). Similar arguments apply in bacterial systems, to lactobacillic acid and its 2-hydroxyderivative in a thiobacillus (Knoche and Shively, 1969), to 13-methylmyristate and its 2-hydroxyderivative in *Streptomyces sioyaensis* (Kawanami et al., 1969). The 2-hydroxyacids of *Corynebacterium simplex* and *Arthrobacterium simplex* are formed by α-hydroxylation (Yano et al., 1969, 1970a).

2. 3-Hydroxyacids

In subcellular fractions of the slime mould *Dictyostelium discoideum* and in guinea-pig liver mitochondria, long-chain fatty acids are converted to their D(−)-3-hydroxyderivatives (Davidoff and Korn, 1964), probably via the 2-enoates. Oxidative biogenesis by β-oxidation would lead to L(+)-3-hydroxyacid (Fig. 8.2, p. 203). Presumably D(−)-3-hydroxyacids accumulate during an anabolic sequence, or by partial reversal of fatty acid biosynthesis, which involves this enantiomer. In those lower plants where accumulation occurs, cleavage of the intermediate 3-hydroxyacyl residue from fatty acid synthetase is the most likely mechanism of biogenesis, since it has been shown that a bacterial palmitoyl thiolesterase will allow this (p. 113).

3. Ricinoleic acid

The biosynthesis of ricinoleic acid (D-12-hydroxyoleic acid) has been investigated in the castor plant (*Ricinus communis*) where it is a major component of the seed oil (castor oil). However when mature seeds were supplied with [^{14}C]acetate, no label appeared in the ricinoleic acid, though all other saturated and unsaturated fatty acids were labelled (Coppens, 1956). An investigation of immature seeds showed that biosynthesis of ricinoleic acid began 12 days after manual pollination (Canvin, 1963) or 3 to 5 weeks after flowering (James et al., 1965). Before this stage, other fatty acids are synthesized, but the ricinoleate-producing system is absent or latent; thereafter it becomes very active, until, about 6 weeks after flowering, a large mass of triglyceride has been laid down, in which ricinoleic acid accounts for up to 90% of the acids. The capacity to synthesize ricinoleate then declines, and turnover in the mature seed is evidently slow. By selecting the seed at its critical stage of development, biosynthesis of ricinoleate *in vitro* was demonstrable.

In the isolated embryo (James, 1962b; James et al., 1965), it was shown that ricinoleic acid was synthesized from radioactive acetate, octanoate, decanoate, dodecanoate, tetradecanoate and oleate, but not from palmitate, stearate or linoleate. The retention of label in the carboxyl group of ricinoleate biosynthesized from oleate showed that direct conversion had taken place rather than breakdown and resynthesis. Linoleate was evidently not an intermediate in this conversion. The fact that palmitate and stearate are ineffective as precursors of oleate in higher plants has been discussed (p. 135). While the embryo was mainly responsible for ricinoleate biosynthesis, the endosperm was also active; [^{14}C]acetate was incorporated into fatty acids, including oleate, linoleate and ricinoleate. The variation of their radioactivities with time indicated that oleate was the precursor of ricinoleate, while linoleate was not a free intermediate (Canvin, 1965a). Yamada and Stumpf (1964) demonstrated that ricinoleate was similarly synthesized in fractions obtained from homogenates of immature castor seed. Labelled oleoyl-CoA gave rise to ricinoleate with retention of activity in the carboxyl group; this system was soluble, but later work by Galliard and Stumpf (1966) showed that the enzymes were associated with a microsomal fraction. They exhibited marked substrate specificity, since stearate, linoleate, vaccenate and elaidate were not converted to hydroxyacids. Oxygen and NADH were obligatory cofactors; oxygen could not be replaced by other electron acceptors, and NADH could only be replaced by NAD$^+$ or by higher concentrations of NADPH when a supernatant fraction was present. These results ended some previous uncertainty about cofactor requirements and illustrated the similarity between the hydroxylation of oleate to ricinoleate and the desaturation of oleate to linoleate (p. 142) which requires oxygen and NADPH. Nevertheless the fact that linoleic acid and linoleoyl-CoA are not converted to ricinoleic acid in the castor system implies that ricinoleate is biosynthesized directly from oleate, and not by hydration of linoleate (Fig. 5.30).

This pathway has been confirmed by Morris (1967) who, by the use of tritium-labelled substrates, excluded the possibility that ricinoleate was formed by hydration of a double bond (or double bond precursor) which was enzyme-bound and therefore not exchangeable with a pool of exogenous linoleate. Incubation of racemic [*erythro*-12,13-^{3}H$_2$]oleic acid with immature castor seed endosperm slices revealed that only one hydrogen atom was lost from the 12-position of oleate per molecule of ricinoleate formed; none was lost from the 13-position. The mechanism must therefore involve direct hydroxyl substitution at the 12-position, and the use of [D-12-^{3}H]oleate and [L-12-^{3}H]oleate as substrates demonstrated that this hydroxylation proceeds with overall retention of configuration at the 12-position (Fig. 5.30).

Ricinoleic acid also occurs in ergot oil, extracted from the sclerotia of the fungus *Claviceps purpurea*. In mycelial cultures or (more particularly) in immature sclerotia isolated from infected rye plants, labelled oleate was desaturated under aerobic conditions, but no ricinoleate was formed in the presence or absence of oxygen (Morris et al., 1966a). However, in this tissue, [1-^{14}C]linoleate was efficiently converted to ricinoleate with

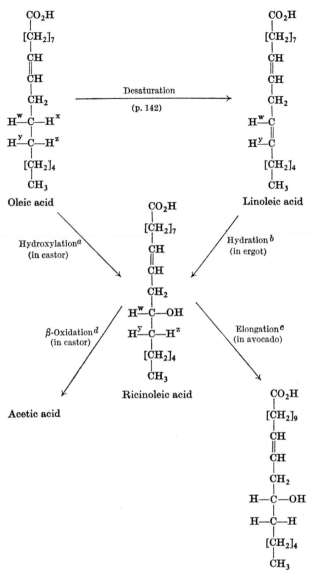

FIG. 5.30. Metabolism of ricinoleic acid. [a] Morris (1967); [b] Morris et al. (1966a); [c] Yang and Stumpf (1965b); [d] Yamada and Stumpf (1965b).

retention of label in the carboxyl group; in air, competing oxidative reactions led to loss of some precursor, but anaerobiosis enhanced the yield. The mechanism is therefore the anaerobic specific hydration of linoleate to 12-hydroxyoleate (Fig. 5.30). An analogous hydration of oleate to 10-hydroxystearate has been observed in soluble extracts of a pseudomonad (see below).

The biosynthesis of ricinoleic acid therefore takes place by two different routes in two major natural sources of this acid. This interesting difference may be associated with the different fates of the ricinoleate in the two tissues. In castor oil, where it accounts for up to 90% of the fatty acids, it occurs only in the normal triglyceride; it is absent from the seed phospholipid. Ergot oil, whose fatty acids contain up to 44% ricinoleate, contains no free hydroxyl groups; the hydroxyacid is esterified to glycerol at the carboxyl group and to unsubstituted fatty acids at the hydroxyl group. These "estolides" were characterized by Morris and Hall (1966) who raised the possibility that their biosynthesis did not involve the formation of a free hydroxyl group, but that linoleate, either free or esterified to glycerol, reacted directly with a fatty acid by an addition across the double bond in which the proton was attached to the 13-position and the acyloxy cation to the 12-position.

The structure of ricinoleic acid, and its biosynthesis from oleate or linoleate, suggests that it might be an intermediate in the conversion of oleate to linoleate; there is however, no evidence that a hydroxy-intermediate is associated with the desaturation reaction, though it is possible that a transient bound form of ricinoleate is formed (p. 148). The degradation of ricinoleic acid in castor tissue has been investigated by Yamada and Stumpf (1965b); the biosynthesis of its homologue lesquerolic acid (14h-20:1,11c) is considered on p. 129.

4. Other oxyacids

The biosynthesis of 10-hydroxystearate in soluble extracts of a pseudomonad (Niehaus and Schroepfer, 1965) has been shown to involve hydration of oleic acid by *trans*-addition of the elements of water across the double bond (Davis *et al.*, 1969; Morris, 1970). Oleate yields D-10-hydroxystearate and palmitoleate yields D-10-hydroxypalmitate (Niehaus *et al.*, 1970a); linoleate yields D-10-hydroxy-*cis*-12-octadecenoate (Schroepfer *et al.*, 1970). On the other hand, the biogenesis of ω- and $(\omega - 1)$-hydroxyacids in yeasts occurs by direct hydroxylation, the mechanism of which is discussed on p. 222. Presumably similar reactions explain the occurrence of similar hydroxyacids in higher plants.

Biosynthesis of *cis*-9,10-epoxystearate from acetate occurs in *Puccinia graminis* when sporulation is proceeding in rust-infected wheat tissue (Knoche, 1968). Labelled stearic or oleic acids are also incorporated into the epoxyacid without undergoing β-oxidation. The rate of conversion of oleate is greater than that of stearate, thus indicating that oleic acid is an immediate precursor of 9,10-epoxystearic acid. Similarly *cis*-12,13-

epoxyoleic (vernolic) acid is biosynthesized from linoleic acid in the seeds of *Xeranthemum annuum* and of *Euphorbia lagascae* (Morris, 1970). Vernolic acid (p. 27) is converted to 12,13-dihydroxyoleic acid in the crushed seeds of *Vernonia anthelmintica* (Scott *et al.*, 1963) and of *Euphorbia lagascae* (Scott and Krewson, 1966). The stereochemistry of this enzymic hydration is attack by hydroxyl at the 12-position of (+)D-12,D-13-epoxyoleate with inversion at that position, and the formation of (+)*threo*-L-12,D-13-dihydroxyoleate (Morris and Crouchman, 1969). In the spores of plant rusts, *cis*-9,10-epoxystearate is similarly hydrated to (+)*threo*-9,10-dihydroxystearate (Tulloch,1963; Hartmann and Frear, 1963; Jackson and Frear, 1967); here the L-9,L-10-epoxide yields the L-9,D-10-diol (Morris, 1970). The site of attack was confirmed as the site of inversion in both hydrations by using ^{18}O-enriched water: the label was found exclusively at C-12 of 12,13-dihydroxyoleate produced by the seeds and at C-10 of the 9,10-dihydroxystearate from the spores. Analogous stereospecific hydrations of epoxyacids occur in soluble extracts of a pseudomonad (Niehaus *et al.*, 1970b).

vic-Dihydroxyacids also arise in plants in the form of acetoxyhydroxy-derivatives of triglycerides, which could arise from the corresponding monoenes via the epoxides by acetolysis after the elaboration of the triglyceride structure (Mikolajczak *et al.*, 1968b). The natural occurrence of 9,10,18-trihydroxyoctadecanoate and 9,10,18-trihydroxy-*cis*-12-octadecenoate (Mikolajczak and Smith, 1967) suggests a similar biosynthetic pathway from oleate and linoleate respectively, involving epoxidation, hydrolysis or acetolysis, and ω-oxidation. While ω-oxidation may account for the rare accumulation of dicarboxylic acids, traumatic acid (1-decene-1,10-dicarboxylic acid, p. 28) apparently arises from oxidative breakdown of polyenoic acids initiated by lipoxygenase (p. 276). It has been suggested that the polyhydroxyacids of cutin are formed from the peroxides produced by lipoxygenase (p. 191).

The oxiran ring is biosynthesized from an ethylenic bond and can be hydrated to a *vic*-diol in plant systems. Similarly, the furan ring (p. 21) may be formed from a 1,3-diyne: 9,11-octadecadiynoate could be the precursor of the 9,12-epoxide of 9,11-octadecadienoate, and in fact both these compounds are found together in nature (Elix and Sargent, 1968). It has been suggested that a suitable poly-ynoic acid gives rise to wyerone (p. 21) in bean shoots by the same mechanism (Jones, 1966).

E. Branched-chain Acids

The biogenesis of a fatty acid whose carbon skeleton is branched or alicyclic can take place in one of two ways. Either a normal chain is synthesized first, and alkylation occurs subsequently, or an abnormal metabolite is incorporated into the chain during biosynthesis. The iso- and anteisoacids (p. 108) provide examples of the latter case; here 3-methylbutyrate or 2-methylbutyrate (derived from leucine or isoleucine respectively) act as "starter" in place of acetate in bacterial synthetase systems (Lennarz, 1961; Kaneda, 1963b). The products are isoacids and anteisoacids of odd carbon number; however the presence of 2-hydroxy-13-methylmyristate in *Streptomyces sioyaensis* (Kawanami *et al.*, 1969) and the existence of bacterial α-oxidation (Yano *et al.*, 1969, 1970) suggest that primary products could be degraded to any member of the homologous series of $(\omega - 1)$- and $(\omega - 2)$-methylalkanoic acids.

Methyl branches arise near the carboxyl group by incorporation of propionate in place of acetate in the latter stages of synthetase action. Propionate can be carboxylated to yield methylmalonate which takes part in the condensing reaction, with the formation of a 2-methyl-substituted acyl chain; if this process is repeated, the 2,4,6-trimethyl-alkanoic and similar acids can be produced (Gastambide-Odier *et al.*, 1963). Similar abnormalities can account for longer carbon chains at the 2-position (p. 124).

While the biogenesis of branches at either end of the fatty acid molecule occurs during synthesis of the carbon chain, any substitution at the middle of the chain occurs after its completion. Such methylations occur at a double bond, by transfer of carbon from the methyl group of methionine. In *Mycobacterium phlei*, oleic acid yields 10-methylstearate by such a reductive methylation (Lennarz *et al.*, 1962) whose mechanism involves loss of one of the three hydrogen atoms of the transferred methyl group (Jauréguiberry *et al.*, 1965) to yield 10-methylenestearate as intermediate (Jauréguiberry *et al.*, 1966). The substrate is the oleoyl chain of a phospholipid; this is converted to the 10-methylenestearoyl derivative which is subsequently reduced to lipid-bound tuberculo-stearate (Akamatsu and Law, 1970).

Such a mechanism may be initially the same as that involved in the biosynthesis of bacterial cyclopropane fatty acid, which also requires methionine and a preformed unsaturated acid. In this case, *S*-adenosyl-methionine reacts at the double bond of vaccenate attached to phospha-tidyl ethanolamine (p. 274), with the production of lipid-bound lacto-bacillate as in *Clostridium butyricum* (Chung and Law, 1964; Thomas and Law, 1966). Similar reactions could conceivably occur in plants, though accumulation of these acids is rare. Isoacids and anteisoacids have been reported in plants, but the major sources of these and other acyclic branched acids are bacterial and animal (p. 28).

The biosynthesis of cyclopropane and cyclopropene fatty acids in plant tissues has been investigated by Johnson *et al.* (1967a). Incubations of radioactive precursors with sliced tissue from four Malvaceae and three Sterculiaceae plants demonstrated that [Me-^{14}C]methionine was the most specific precursor of the alicyclic acids, and that dihydro-sterculate was selectively labelled. This implied that oleic acid is alkylated by methionine (or more probably by *S*-adenosyl methionine) to dihydro-sterculate; time studies indicated that this was desaturated to sterculate. This pathway was also deduced by Hooper and Law (1965). The postulated biosynthesis by methylation of stearolic acid (Smith and Bu'Lock, 1964) was not supported. The formation of cyclopropene by desaturation of cyclopropane acids is suggested by their coexistence in seed oils (Wilson *et al.*, 1961). It is possible that the substrate for methyl-ation in plants is lipid-bound, by analogy with the bacterial systems.

The biogenesis of malvalic acid is less clear. The results of Smith and Bu'Lock (1964) showed that both sterculic and malvalic acids were biosynthesized from [1-^{14}C]acetate, with labelling patterns suggesting that the shorter chain had been formed by α-oxidation (p. 213) of the preformed C_{18} chain, either before or after formation of the cyclopropene ring; the methylene bridge was unlabelled. However, the differences in rate of incorporation of the methyl carbon of methionine indicated to Johnson *et al.* (1967) that dihydromalvalic acid was the immediate precursor of malvalic acid; hence the latter is formed via α-oxidation of oleate or dihydrosterculate rather than of sterculate itself. The natural occurrence of 2-hydroxysterculic acid (p. 24) supports the possibility of an α-oxidation mechanism; though the direct hydroxylation of sterculate seems more likely (Morris and Hall, 1967) the desaturation of 2-hydroxydihydrosterculate (formed by α-hydroxylation of dihydro-sterculate) is possible.

CHAPTER 6

The Biosynthesis of Acyl Lipids

1. GLYCERIDE BIOSYNTHESIS

A. *The Initial Acyl Acceptor—Biosynthesis of Phosphatidic Acid*

Nearly all the fatty acyl groups produced by fatty acid synthesizing systems (Chapter 5) are eventually incorporated into acyl lipids via fatty acid transferases, which transport the acids from their site of synthesis to the site of acyl lipid synthesis. Most of the available evidence indicates that L-α-glycerophosphate is the initial acceptor of acyl groups at the primary site of glyceride synthesis, and that all classes of glyceride (phosphoglycerides, glycosyl glycerides, neutral glycerides) are derived from the diacyl-L(α)-glycerophosphoric acid (phosphatidic acid) thus formed.

Following the observations of Kennedy who established this pathway for the biosynthesis of glycerides in *E. coli*, Barron and Stumpf (1962) found that microsomal preparations from the mesocarp of the avocado pear can convert glycerol into diglyceride by a pathway involving glycerophosphate and phosphatidic acid as intermediates. This work, and that of Cheniae (1965) and Sastry and Kates (1966) with homogenate and microsomal preparations from spinach leaves, indicated that in plants L-α-1-glycerophosphate is converted into phosphatidic acid by stepwise acylation with CoA fatty esters:

(i)   Fatty acid + CoA $\xrightarrow{\text{ATP}+\text{Mg}^{2+}}$ Fatty acyl-CoA

(ii) Glycerol + ATP $\xrightarrow{\text{glycerol kinase}}$ L-α-glycerophosphate + ADP

(iii) Fatty acyl-CoA + L-α-glycerophosphate

$$\Big\downarrow \begin{array}{l}\text{Glycerophosphate}\\ \text{acyl transferase}\end{array}$$

Monoacyl glycerophosphate (lysophosphatidic acid)

(iv) Monoacyl glycerophosphate

$$\Big\downarrow \text{Fatty acyl-CoA}$$

Diacyl glycerophosphate (phosphatidic acid)

This stepwise mechanism is supported by the results of Sastry and Kates (1966) who detected labelled lysophosphatidic acid in their incubation mixtures, although it is conceivable that this compound may have resulted from the action of endogenous phospholipase (Chapter 7) on labelled phosphatidic acid.

In addition, Renkonen and Bloch (1969) found that extracts from photoautotrophically cultured *Euglena gracilis* can catalyse the transfer of acyl groups from either ACP or CoA thiolesters during the synthesis of monogalactosyl diglyceride by reactions stimulated by L-α-glycerophosphate. Similar extracts catalysed transfer of acyl groups from CoA esters into phospholipids but not those from ACP esters. The fact that ACP esters could not provide acyl residues for phospholipid synthesis indicates that the system was incapable of transferring acyl groups from ACP to CoA, so that the galactosyl diglyceride synthesis from ACP esters must have involved acylation of glycerophosphate by direct transfer of fatty acid from the protein. This suggests that in chloroplasts (the major site of galactosyl diglyceride synthesis) ACP thiolesters can also donate fatty acid residues for the conversion of glycerophosphate to phosphatidic acid in reactions analogous to those involving CoA illustrated above. It is even conceivable that the systems studied by Cheniae and Kerr, and Sastry and Kates, may also have involved the transfer of acyl groups from ACP to glycerophosphate, the acyl groups from the CoA ester being first transferred to ACP before becoming attached to glycerophosphate.

There has been comparatively little experimental evidence in support of the suggestion of Bradbeer and Stumpf (1960) that plants may also synthesize phosphatidic acid by phosphorylation of 1,2-diglyceride (diglyceridephosphokinase pathway) although Mazliak (1967) has suggested that such a mechanism may operate in apple parenchyma.

Another pathway to the *de novo* synthesis of phosphatidate is that established by Hajra and Agranoff (1967, 1968a, 1968b) for guinea pig liver microsomes and mitochondria (Fig. 6.1) and involves the reaction of dihydroxyacetone phosphate (c) with acyl CoA to give acyl dihydroxyacetone phosphate (d). This lipid is reduced by NADPH to give lysophosphatidate (e) which can then react with further CoA ester to give phosphatidic acid (f), and Hajra and Agranoff have pointed out that the possible selective incorporation of saturated fatty acids in the acylation of dihydroxyacetone phosphate may in part explain differences in the positional distribution of fatty acids in glycerides and phosphoglycerides.

To date there have been no reports of attempts to identify this pathway in plant tissues, but the presence of dihydroxyacetone phosphate in chloroplasts and other particles, where it may be derived either from

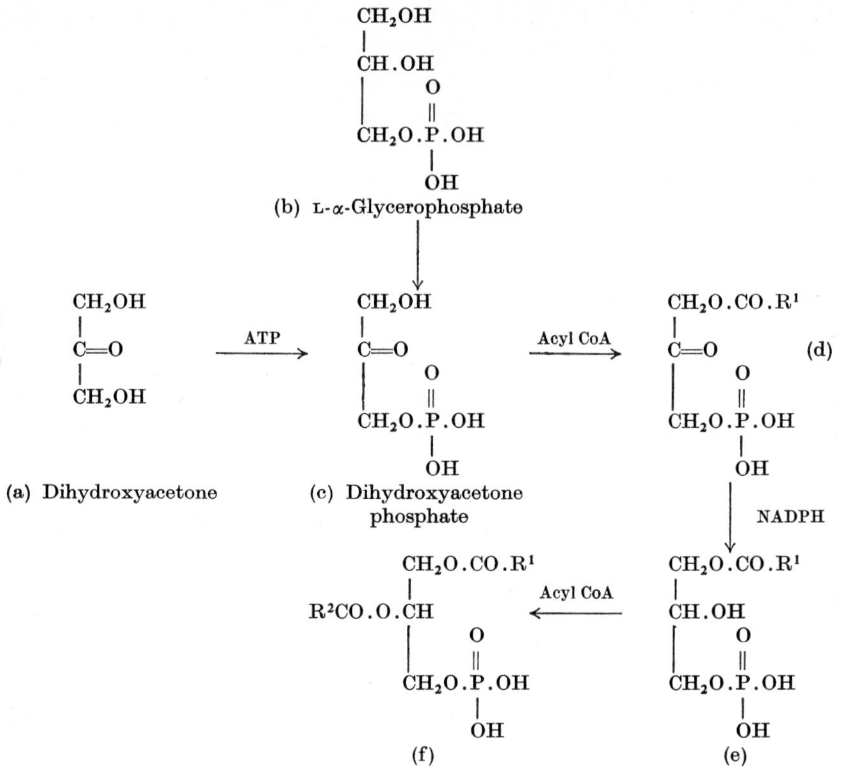

CH₂OH
|
CH.OH
|　　　O
|　　　‖
CH₂O.P.OH
|
OH
(b) L-α-Glycerophosphate

CH₂OH　　　　　　　CH₂OH　　　　　　　　　CH₂O.CO.R¹
|　　　　　　　　　　|　　　　　　　　　　　|
C=O　　——ATP——→　C=O　　——Acyl CoA——→　C=O　　　　(d)
|　　　　　　　　　　|　　　O　　　　　　　　|　　　O
CH₂OH　　　　　　　|　　　‖　　　　　　　　|　　　‖
　　　　　　　　　　CH₂O.P.OH　　　　　　　CH₂O.P.OH
　　　　　　　　　　|　　　　　　　　　　　|
　　　　　　　　　　OH　　　　　　　　　　　OH
(a) Dihydroxyacetone　(c) Dihydroxyacetone　　　　　　　　| NADPH
　　　　　　　　　　　　phosphate

CH₂O.CO.R¹　　　　　　　　　　　CH₂O.CO.R¹
|　　　　　　　Acyl CoA　　　　　|
R²CO.O.CH　　←——————　　　　　CH.OH
|　　　O　　　　　　　　　　　　|　　　O
|　　　‖　　　　　　　　　　　　|　　　‖
CH₂O.P.OH　　　　　　　　　　　CH₂O.P.OH
|　　　　　　　　　　　　　　　|
OH　　　　　　　　　　　　　　　OH
(f)　　　　　　　　　　　　　　(e)

FIG. 6.1. Synthesis of phosphatidate via dihydroxyacetone phosphate (Hajra and Agranoff, 1967, 1968a, 1968b)

dihydroxyacetone (1) by reaction with ATP or from glycerophosphate (2) through glycerophosphate dehydrogenase activity, suggest that phosphatidate in plants may not be formed by acylation of glycerophosphate exclusively.

B. Di- and Tri-glycerides

Several groups of workers (e.g. Barron and Stumpf, 1962; Cheniae, 1965; Sastry and Kates, 1966; Mazliak, 1967) have shown that phosphatidic acid is the precursor of plant di- and tri-glycerides, and that the metabolic sequence involved is probably:

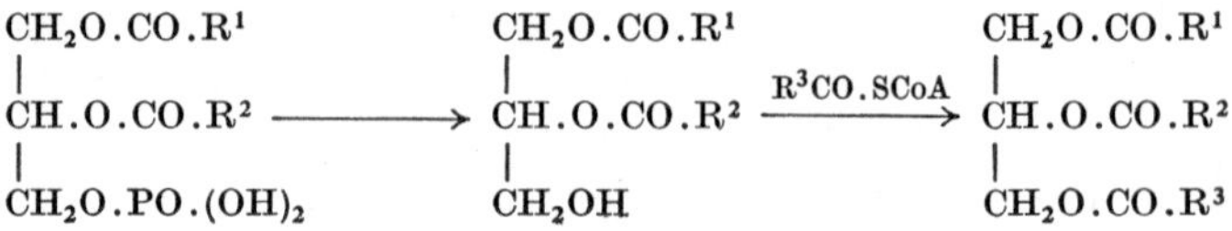

CH₂O.CO.R¹　　　　　　　　CH₂O.CO.R¹　　　　　　　　　CH₂O.CO.R¹
|　　　　　　　　　　　　　|　　　　　　　　　　　　　|
CH.O.CO.R²　——————→　CH.O.CO.R²　——R³CO.SCoA——→　CH.O.CO.R²
|　　　　　　　　　　　　　|　　　　　　　　　　　　　|
CH₂O.PO.(OH)₂　　　　　　　CH₂OH　　　　　　　　　　　CH₂O.CO.R³

Thus the three fatty acyl residues of a triglyceride are incorporated sequentially rather than concomitantly, and in each case a different enzyme may be involved. The first two residues are inserted by glyceryl-phosphoryl acyl transferases which initiate acylation of the 1- and 2-positions respectively (see p. 176), while the third residue is inserted by means of a diglyceride acyl transferase. Recent evidence that different transacylases operate most efficiently for specific fatty acid structures consequently implies that the structure of natural triglycerides will not be random with respect to the acyl residues incorporated and will be controlled by the specificities of these enzymes. It is therefore not surprising that few plant triglycerides conform to the laws of random distribution of fatty acids, and that the 1- and 3-positions frequently have different fatty acid compositions (Chapter 4).

On the basis of studies in which relatively large amounts of mono-glycerides were found in immature soybeans, Hirayama and Hujii (1965) proposed that plant triglycerides may also be synthesized *via* mono-glyceride intermediates. Although Privett and coworkers (see Roehm and Privett, 1970) were unable to detect monoglycerides in similar tissues, this does not necessarily eliminate the possibility of monoglyceride involvement in triglyceride biosynthesis, because intermediates having high turnover rates are frequently undetectable unless radioisotopically labelled substrates are used.

C. Phosphoglycerides

On the basis of biochemical evidence obtained from all classes of living tissues it is clear that there are alternative biosynthetic pathways for most classes of phospholipid, and it would be misleading to suggest that there exists a single "plant pathway" to each class. Even within a single type of cell or subcellular particle, alternative pathways to a single phosphoglyceride are possible, and evidence for the activity of one bio-synthetic mechanism does not necessarily exclude the concomitant (or alternative) operation of another.

1. Phosphatidyl ethanolamine

In mammalian tissue and bacteria, phosphatidyl ethanolamine is synthesized by reaction between CDP-ethanolamine and a 1,2-digly-ceride viz.,

1,2-Diglyceride + CDP ethanolamine
↓
Phosphatidyl ethanolamine + CMP

and Sastry and Kates (1965) have suggested that this is more likely to be the major pathway occurring in algae than the transphosphatidylation mechanism involving phosphatidyl glycerol or phosphatidyl inositol proposed by Ferrari and Benson (1961).

Vandor and Richardson (1968) have reported a synthesis of phosphatidyl ethanolamine from free ethanolamine in pea seedling microsomes which did not require cytidine phosphates but was catalysed by phospholipase D (see p. 197). The mechanism involved is apparently an exchange reaction involving endogenous phospholipids such as lecithin, i.e.

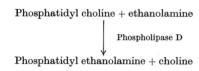

Phosphatidyl choline + ethanolamine

↓ Phospholipase D

Phosphatidyl ethanolamine + choline

but the system lacks the specificities peculiar to the transphosphatidylation mechanism characterized by Dawson and Hemington (see below). In addition Vandor and Richardson obtained evidence which suggested that phosphatidyl ethanolamine was also produced in their system by decarboxylation of phosphatidyl serine, a mechanism believed to operate in tomato roots (Willemot and Boll, 1967) and which was originally defined by Borkenhagen et al. (1961) for rat liver.

2. Phosphatidyl serine

Few studies concerning the synthesis of phosphatidyl serine in plants have been reported, although Vandor and Richardson (1968) have shown that pea seedling microsomes can incorporate free serine into phosphatidyl serine by a mechanism not requiring CTP, and involving base exchange with endogenous phospholipid:

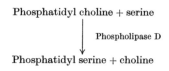

Phosphatidyl choline + serine

↓ Phospholipase D

Phosphatidyl serine + choline

3. Phosphatidyl choline (lecithin)

The most commonly described pathway leading to the formation of phosphatidyl choline in mammals is that involving the interaction of 1,2-diglycerides with CDP-choline:

1,2-Diglyceride + CDP-choline → Phosphatidyl choline + CMP

Alternatively some mammalian tissues (Arvidson, 1968) and bacteria (Kaneshiro and Law, 1964) synthesize lecithin by methylation of endogenous phosphatidyl ethanolamine, S-adenosyl methionine acting as the methyl donor:

Phosphatidyl ethanolamine

$\downarrow$ S-Adenosyl methionine

Phosphatidyl choline

As regards the plant kingdom, Kates and Volcani (1966) found that many marine and freshwater diatoms contain small quantities of phosphatidyl N-methyl ethanolamine and suggested that lecithin synthesis in these organisms occurs by stepwise methylation of phosphatidyl ethanolamine, a process which they considered to be also consistent with the kinetic data obtained from ^{32}P-labelling studies with the green alga *Chlorella vulgaris* (Sastry and Kates, 1965). Similarly, Tipton and Swords (1966) found that both dark- and light-grown cells of *Euglena gracilis* synthesize lecithin by N-methylation of phosphatidyl ethanolamine rather than by the CDP-choline-diglyceride pathway.

Conversely, the N-methylation mechanism is apparently inconsistent with the results of Nichols *et al.* (1967b) who found that the fatty acids of the lecithin fraction of *Chlorella vulgaris* were labelled much more rapidly than those of phosphatidyl ethanolamine when this alga was incubated with [^{14}C]acetate, and a recent paper by Morré *et al.* (1970) has established the capacity of onion stem tissue to effect lecithin synthesis from diglyceride and CDP-choline.

Several different groups have also proposed a variety of transphosphatidylation mechanisms for the formation of lecithin in plants. The first was made by Ferrari and Benson (1961) on the basis of the rates of labelling of lipids when algae were incubated with $^{14}CO_2$, from which the authors suggested that either phosphatidyl glycerol or phosphatidyl inositol could provide the phosphatidyl group for lecithin, e.g.

Phosphatidyl glycerol + choline

$\downarrow$

Phosphatidyl choline + glycerol

The existence of this type of pathway for lecithin biosynthesis in algae has also been proposed by Sastry and Kates (1965), but direct experimental confirmation is still lacking.

The synthesis of phosphatidyl choline in pea seedling ribosomes by a single enzyme which catalyses the incorporation of choline, serine and ethanolamine by base exchange with endogenous phospholipid has also been reported (see Section 1, above). This system is not activated by ATP, but is catalysed by phospholipase D, and appears to resemble the single enzyme system suggested for rat liver homogenates; it differs from the phospholipase-D catalysed transphosphatidylation mechanism reported by other groups (Dawson and Hemington, 1967; Dawson, 1967; Yang *et al.*, 1967) in that the exchange reactions can take place with ethanolamine-, serine- and choline-containing phosphatides whereas in the latter system, only lecithin can provide the phosphatidyl group.

4. *Phosphatidyl glycerol*

In bacteria and mammals a basic difference exists between the mechanisms leading to the conversion of diglycerides to phosphoglycerides containing ethanolamine or choline, and to those containing glycerol, inositol or serine.

We have already described the biosynthesis of the former lipids which are commonly produced by the enzymically catalysed reaction between 1,2-diglycerides and the relevant CDP-amine. Lipids of the second group are also synthesized via cytidine phosphate-linked intermediates, but in these cases it is the diglyceride acceptor which must be nucleotide-bound. Thus Chang and Kennedy (1967) demonstrated that the phosphatidyl glycerol of *E. coli* is synthesized by reactions (i → iii) and that of diphosphatidyl glycerol (cardiolipin) by reaction (iv):

(i) Phosphatidic acid + CTP → CDP-diglyceride + (P-P)$_i$

(ii) CDP-diglyceride + α-glycerophosphate
↓
Phosphatidyl glycerophosphate + CMP

(iii) Phosphatidyl glycerophosphate → Phosphatidyl glycerol + P$_i$

(iv) Phosphatidyl glycerol + CDP-diglyceride
↓
Diphosphatidyl glycerol + CMP

The synthesis of CDP-diglyceride in plant tissues was subsequently established by two groups of workers studying lipid metabolism in cauliflower mitochondria (Sumida and Mudd, 1968, 1970; Douce, 1968), so that the utilization of this intermediate in the synthesis of certain complex glycerides in plants seems likely.

Earlier, Benson and Miyano (1961) and Haverkate and van Deenen (1964, 1965) had demonstrated that plant phosphatidyl glycerol has a similar stereochemical configuration to that from animals and bacteria and the latter authors suggested that synthesis of the lipid might proceed by the same pathway in all cases, namely that illustrated above; Douce and Dupont (1969) subsequently confirmed that the phosphatidyl glycerol of cauliflower mitochondria is synthesized by this pathway.

An alternative biosynthetic route to phosphatidyl glycerol in plants has been suggested by three groups (Douce *et al.*, 1966; Dawson, 1967; Yang *et al.*, 1967) who all noted that the phospholipase D of plant tissues can catalyse the transfer of a phosphatidyl unit from lecithin to various alcohols, including glycerol, with the formation of the equivalent phosphatide. The synthesis of phosphatidyl glycerol by this route could therefore be represented by

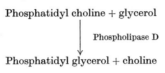

Phosphatidyl choline + glycerol

Phospholipase D

Phosphatidyl glycerol + choline

It has not been established whether this mechanism is physiologically important, or whether syntheses of this kind occur at all *in vivo* (p. 198).

5. Phosphatidyl inositol

Ferrari and Benson (1961) suggested that, in algae, phosphatidyl glycerol might provide the phosphatidyl group in the biosynthesis of phosphatidyl inositol. Such a pathway was not altogether precluded by the studies of Sastry and Kates (1966) on cell-free preparations from spinach leaves, although these authors reported evidence that in their system phosphatidyl inositol was more probably synthesized from phosphatidic acid in a manner similar to that which occurs in animal tissues, namely

CDP-diglyceride + inositol → Phosphatidyl inositol + CMP

This pathway was also proposed by Sumida and Mudd (1968, 1970) on the basis of their studies of CDP-diglyceride metabolism in cauliflower.

D. Glycosyl Glycerides

1. Galactosyl diglycerides

The earliest studies of the biosynthetic routes leading to galactosyl diglyceride formation in plants were carried out by Ferrari and Benson (1961) who observed a rapid incorporation of label into monogalactosyl diglyceride, and a slower entry into digalactosyl diglyceride, when *Chlorella pyrenoidosa* was grown photoautotrophically on $^{14}CO_2$. On the basis of these results the authors concluded that digalactosyl diglyceride was formed by galactosidation of the monogalactosyl lipid, and proposed the following mechanism for the biosynthesis of these lipids in plants:

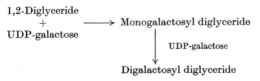

1,2-Diglyceride
+   ⟶ Monogalactosyl diglyceride
UDP-galactose

UDP-galactose

Digalactosyl diglyceride

Partial confirmation of this pathway was obtained by Neufeld and Hall (1964) who demonstrated that spinach chloroplasts catalyse the transfer of galactose from UDP-galactose to an uncharacterized endogenous acceptor with the apparent formation of mono-, di-, tri- and tetra-galactosyl diglycerides, and the scheme proposed by Ferrari and Benson received final experimental confirmation in the studies of Ongun and Mudd (1968). Using an acetone powder prepared from spinach chloroplasts these authors obtained a 95% incorporation of label into lipid from UDP-[^{14}C]galactose when a 1,2-diglyceride (diolein) was employed as acceptor, and they also found that although the diolein served as an efficient acceptor for monogalactosyl diglyceride synthesis, it was not directly implicated in synthesis of digalactosyl diglyceride. For the second galactosylation, monogalactosyl diglyceride was the preferred acceptor and the evidence clearly indicated that two separate enzymes are involved in galactolipid synthesis in spinach, the enzyme responsible for the synthesis of monogalactosyl diglyceride being more tightly bound to the chloroplast membranes than that involved in the formation of digalactosyl diglyceride. Ongun and Mudd also found that the galactosyl diglyceride synthesizing systems are not confined to chlorophyllous tissue, although they and others (Chang and Kulkarni, 1970) observed that these systems are normally slightly stimulated by light.

Mudd *et al.* (1969) later showed that biosynthesis of monogalactosyl diglyceride in the presence of spinach leaf acetone powder proceeds most efficiently with more highly unsaturated diglyceride acceptors, an observation which may be of great importance when considering the significance of the high degree of unsaturation usually found in naturally occurring galactosyl diglyceride fractions (see Section E). The same workers found similar galactosyl diglyceride-synthesizing activity in that fraction of a spinach leaf homogenate not sedimented by centrifugation for 60 min at 100,000 g. The pH optimum for galactolipid synthesis from UDP-galactose in their chloroplast preparation was 7·2, the proportion of monogalactosyl diglyceride decreasing and that of digalactosyl diglyceride increasing as the pH was lowered. The galactosyl glyceride synthesizing system was quite resistant to elevated temperatures, maximal incorporation of galactose from UDP-galactose occurring at 45° with the proportion of monogalactosyl diglyceride synthesized increasing with temperature. Chang and Kulkarni (1970) also described a similar galactosyl diglyceride synthesizing system in a soluble fraction from spinach chloroplasts, but their preparation apparently differed from that of Mudd *et al.* in requiring no added diglyceride. Ongun and Mudd (1970) also observed mono- and di-galactosyl diglyceride synthesizing capacity in mitochondrial preparations from avocado and cauliflower, but pea root mitochondria were apparently unable to synthesize these lipids from UDP-galactose.

The studies described above gave no indications as to the origins of the diglyceride which acts as initial acceptor of the monosaccharide moiety from UDP-galactose. However, Renkonen and Bloch (1969) studied the incorporation of acyl thiolesters into monogalactosyl diglycerides by crude extracts from green *Euglena gracilis* and found that these extracts catalysed the transfer of acyl groups from thiolesters of both ACP and CoA to monogalactosyl diglycerides in a reaction stimulated by α-glycerophosphate.

The above data are compatible with a pathway for monogalactosyl diglyceride synthesis in chloroplasts involving transfer of fatty acyl groups from ACP (and possibly from CoA thiolesters) to α-glycerophosphate with formation of phosphatidic acids. The products are then dephosphorylated to 1,2-diglycerides before reaction with UDP-galactose.

Ongun and Mudd (1968) and Chang and Kulkarni (1970) also obtained evidence that spinach leaf extracts also catalyse the transfer of monosaccharide from UDP-galactose to di- and tri-glactosyl diglycerides with formation of tri- and tetra-galactosyl diglycerides respectively.

2. Sulphoquinovosyl diglyceride (sulpholipid)

By analogy with the biosynthetic pathway proposed for monogalactosyl diglyceride, Benson (1963) suggested that the plant sulpholipid might be synthesized by transfer to diglyceride of the sulphoquinovose group of a nucleoside diphosphosulphoquinovose, which had been identified in extracts of *Chlorella vulgaris* by Zill and Cheniae (1962). The mechanism as proposed is therefore:

Nucleoside diphosphosulphoquinovose + diglyceride
↓
Sulphoquinovosyl diglyceride + nucleoside diphosphate

As yet, no experimental data have been offered in support of this or any other mechanism for the biosynthesis of the glycolipid, although several groups have speculated on the biological origins of the sulpho-sugar (Benson, 1963; Zill and Cheniae, 1962; Davies *et al.*, 1966).

E. Factors Affecting the Fatty Acid Composition of Individual Lipids

We have already noted (Chapter 4) that the fatty acids in the total lipid extract of a plant tissue are not uniformly distributed among its component lipid classes, many of which may have highly specific fatty acid compositions. Because the pathways established for the formation of many of these lipids have diglyceride and phosphatidic acid as the initial acylated precursors (Fig. 6.2) the question arises as to how such differences in fatty acid composition arise. In practice, several different factors related to cellular structure and metabolism contribute to the overall situation.

1. Separate sites of synthesis—"compartmentalization"

A generalized hypothesis, sometimes offered in explanation of differences in the fatty acid composition of classes of glyceride isolated from a single tissue, is that these lipids are derived from separate pools of diglyceride. However, such arguments alone cannot adequately explain much of the existing analytical and biochemical data.

For example, Tipton and Swords (1966) found that the lecithin in *E. gracilis* is produced by a series of *N*-methylations of endogenous phosphatidyl ethanolamine; despite this these two lipids have dissimilar fatty acid compositions (Nichols and Appleby, 1969). Similarly, there exists a substantial quantity of data (Chapter 4) indicating that mono- and digalactosyl diglyceride fractions from the same cell usually have significantly different fatty acid compositions, even though there is strong biochemical evidence (p. 184) that digalactosyl diglycerides are formed by galactosylation of monogalactosyl diglyceride.

Clearly then, the concept of compartmentalization cannot account for all the known differences in fatty acid composition of glyceride classes from a single cell or tissue, although it can reasonably be applied to certain groups of lipid. Thus those lipids which typify specific organelles of the plant cell, for example mitochondria and chloroplasts, might well be expected to be derived from separate diglyceride pools. It is also conceivable that more than one diglyceride-phosphatidic acid pool may occur within a single subcellular particle, but the technical difficulties involved have so far precluded experimental exploration of this possibility.

2. Specific metabolism of individual glycerides

Lipids with differing fatty acid compositions could also be derived from a single phosphatidic acid pool providing that the enzyme which catalyses the combination of diglyceride with the relevant polar group gives optimum results with diglycerides of specific structure. Such a system would not require the presence of numerous independent diglyceride pools, but could theoretically operate on a single pool comprising a variety of diglyceride structures, specific members of

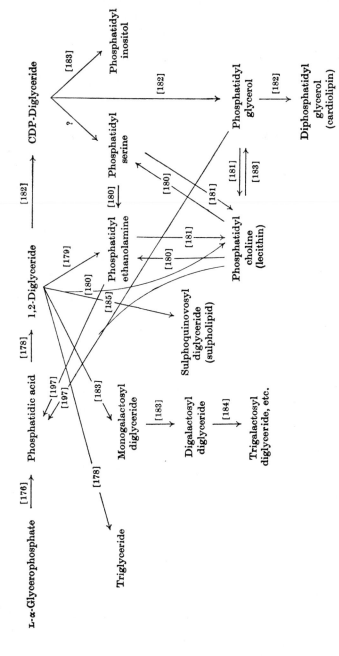

FIG. 6.2. Established and postulated interconversions between major classes of plant glycerides. Numbers in parentheses [ ] indicate the pages on which the appropriate reactions are described.

which would be employed for the synthesis of an individual class of glyceride.

To date the only experimental evidence that lipid synthesis in plants can involve preferential utilization of specific diglycerides is that provided by Mudd *et al.* (1969) who showed that the enzyme system from spinach which catalyses the reaction between a diglyceride acceptor and UDP-galactose to give monogalactosyl diglyceride operates most efficiently when the diglyceride acceptor is highly unsaturated, e.g. dilinolenin. The same principle of selective metabolism could account for the different fatty acid compositions found in other lipid classes which bear a precursor-product relationship towards one another, as occurs with the phosphatidyl ethanolamine and lecithin fractions from *E. gracilis*. Although a requirement for phosphatidyl ethanolamine species of specific fatty acid composition have not yet been shown to be required for lecithin syntheses in plant tissues, Arvidson (1968) has shown that such specificities are operative in the synthesis of phosphatidyl choline in rat liver.

3. Changes in fatty acid structure subsequent to incorporation into lipid

The first direct evidence that changes may occur in the structure of fatty acids subsequent to their incorporation into acyl lipids was obtained by Chung and Law (1964) during studies with the bacterium *Clostridium butyricum*. They found that in this organism phosphatidyl ethanolamine is the required substrate for the conversion of *cis*-vaccenic acid into lactobacillic acid, this conversion taking place after incorporation of the vaccenic acid into the lipid molecule.

More recent studies have suggested that in *Chlorella vulgaris* phosphatidyl glycerol may be the required substrate for the conversion of palmitic acid to *trans*-3-hexadecenoic acid (Nichols *et al.*, 1965b) (p. 158) and that lecithin may play a similar role in the formation of linoleic acid from oleic acid (Gurr *et al.*, 1969) (p. 144) and linolenic acid from linoleic acid (Roughan, 1970). Moreover Nichols (1968) and Safford and Nichols (1970) have suggested that a more general series of transformations of fatty acid structure may occur within the monogalactosyl diglycerides and other lipids of algae.

The mechanisms by which transformations such as those described above occur have not been established, but two alternative general pathways are possible (Fig. 6.3). In *C. butyricum* the addition of a methyl group across the double bond of *cis*-vaccenic acid to give lactobacillic acid occurs while the acyl group is still located within the phosphatidyl ethanolamine molecule, and this kind of mechanism, which involves no intermediate production of partially deacylated lipid and free fatty acid

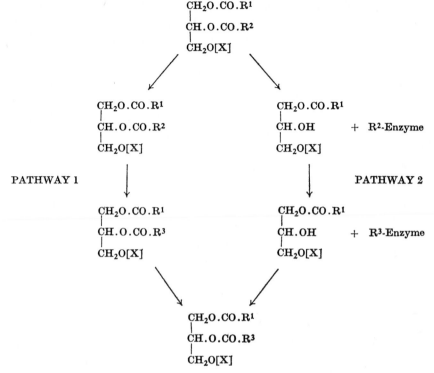

FIG. 6.3. Alternative mechanisms for the involvement of complex lipids in the alteration of the structure of their constituent acyl groups. [X] represents phosphorylated residues or sugars.

(Fig. 6.3) is also believed to operate in the biosynthesis of *trans*-3-hexadecenoic acid from palmitic acid. Bartels *et al.* (1967) noted that 3-hexadecenoic acid is rapidly reduced to palmitic acid when added to cells of *C. vulgaris* suggesting that it is stable only as part of an acyl lipid. Since this acid is usually absolutely specific to the phosphatidyl glycerol fraction of leaves and algae (p. 90) it seems likely that the desaturation of palmitate to *trans*-3-hexadecenoate occurs while the acyl residue is attached to the 2-position of a phosphatidyl glycerol molecule. In this kind of mechanism the acyl lipid must presumably be transiently bound to the desaturating enzyme (Fig. 6.3).

Alternatively the acyl group may first be detached from the lipid and then transferred (possibly via CoA or ACP-bound intermediates) to the relevant site on the desaturase. After desaturation, the acyl group would then return to the parent lipid (Fig. 6.3). Such a mechanism requires lipolytic enzymes capable of releasing acyl groups from the parent lipid and a complementary system which can reacylate the partially de-acylated lipid with the newly formed acid.

We have already noted (Chapter 4) that many plant phospholipids resemble those of animal origin in having the major part of their more saturated acid components situated at the 1-position, with the 2-position containing the more highly unsaturated acids. Any desaturations which occur after the *de novo* synthesis of a phospholipid will therefore occur most frequently on the 2-position, and it is significant that not only has phospholipase A activity been detected in plants (p. 196) but also that Bartels and van Deenen (1966) have shown that spinach leaf homogenates can catalyse the acylation of lyso-lecithin and lysophosphatidyl ethanol-amine with formation of the corresponding diacyl lipids. In the case of the galactosyl diglycerides the studies of Nichols and Moorhouse (1969) and Safford and Nichols (1970) have suggested that in *C. vulgaris* the fatty acids on both the 1- and 2-positions of the monogalactosyl diglyceride fraction are further desaturated subsequent to the *de novo* synthesis of this lipid. Such transformations, should they depend on the removal of acyl groups from the parent lipid, would require an enzyme system capable of removing acyl residues from both the 1- and 2-positions of these glycolipids and the work of Sastry and Kates (1964) has indicated that both classes of enzyme are present in higher plants. Appleby and Nichols (1971) have shown that spinach leaf homogenates can acylate monogalactosyl monoglycerides.

The work of Safford and Nichols (1970) and Mudd *et al.* (1969) have suggested that two types of system could be responsible for the unique fatty acid compositions possessed by the galactosyl diglycerides of plant tissue. The evidence obtained by Mudd *et al.* indicates that the high degree of unsaturation found in the component acids of these lipids in higher plants is due to the preferential utilization of specific classes of diglyceride at the galactosylation stage, whereas evidence obtained with *C. vulgaris* by Safford and Nichols indicated changes in the fatty acid composition subsequent to the introduction of the sugar moiety. It is possible that both pathways are operative in chloroplasts, but the relative importance of each differs according to the class of plant involved.

2. Biosynthesis of Other Acyl Lipids

A. Wax Esters

Kolattukudy (1967b) described the biosynthesis of wax esters by acetone powders prepared from broccoli leaves, and observed that this system catalysed the transfer of palmitic acid from phospholipid into wax esters much more efficiently than would be expected on the basis of free palmitic acid produced from it. This result suggested the importance of a mechanism involving direct acyl transfer from phospholipid to the primary alcohol, and at low concentrations of the free alcohol most of the ester synthesis was apparently mediated by this transesterification mechanism rather than by direct esterification. Nevertheless, fractionation of the water-soluble constituents of the acetone powder clearly demonstrated the presence of an acyl-CoA fatty alcohol transacylase, so that waxy ester synthesis in leaves may proceed by the following mechanism:

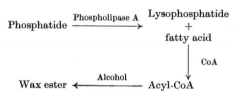

B. Steryl Glucoside Esters

The synthesis of steryl glucosides and their 6-O-acyl esters by plant tissues has been studied in a number of laboratories (Hou *et al.*, 1968; Kauss, 1968; Eichenberger and Newman, 1968; Ongun and Mudd, 1970), who have shown that the steryl glucoside is first formed by enzymatic transfer of the sugar from UDP glucose to free sterol; the glycoside is then acylated to give acylated sterol glucoside. Mitochondria and chloroplasts are both capable of synthesizing these compounds (Ongun and Mudd, 1970).

C. Sphingolipids

There are few published data on aspects of sphingolipid synthesis in plants, although the biosynthesis of phytosphingosine in yeasts has been studied extensively.

D. Cutin

Heinen and Brand (1963) studied some of the enzymatic aspects of cutin synthesis in damaged leaves from *Gasteria verricuosa*, and noted that their preparations possessed enhanced lipoxygenase activity. They

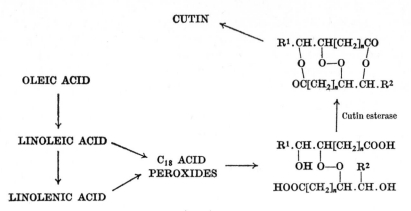

FIG. 6.4. Proposed pathway for cutin biosynthesis (from Heinen and Brand, 1963).

consequently proposed that during cutin synthesis the more highly unsaturated fatty acids produced by the dehydrogenases are oxidized by lipoxygenase to give peroxides, which in turn undergo secondary reactions leading to the formation of polyhydroxy acids. Pairs of these acids are initially linked by hydroxyoxirane bonds and are then more firmly bound by the formation of ester bonds between carboxyl and hydroxyl groups; similar interactions between paired molecules of this sort then result in the formation of larger cutin units of the type described on p. 58 (Fig. 6.4).

CHAPTER 7

Lipolytic Enzymes

It is well established that the bruising or homogenization of plant tissues can lead to partial or complete loss of the normal cellular lipid, often with the concomitant accumulation of hydrolysis products of these lipids. Such phenomena testify to the existence in plants of enzymes capable of hydrolysing a wide variety of lipid structures in a number of different ways, many of which have now been characterized. However, much of the information regarding plant lipolytic enzymes has been obtained from crude or at best only partially purified preparations, so that the differences in substrate specificities and other properties sometimes observed by different investigators, studying apparently analogous systems, are not altogether surprising.

1. Lipase (Glycerol Ester Hydrolase)

Lipase is the trivial name commonly used to describe those enzymes which remove fatty acyl groups from triglycerides. These enzymes are widely distributed in living cells and the mechanisms by which they operate, as well as their chemical and physicochemical requirements for optimum efficiency, have been well characterized for a variety of animal lipases (e.g. Mattson and Volpenhein, 1968).

In general plant lipases have been less rigorously studied than those of animal origin, although the importance of their role in the mobilization of lipid energy reserves during seed germination is well known (p. 274). Thus St. Angelo and Altschul (1964) showed that lipase activity in peanut reached a maximum 13 days after germination whereas in castor bean and cottonseed maximum activity occurred 4 and 2 days after germination respectively.

In some cases, more than one species of lipase have been identified in a single seed. Castor bean, for example, contains an "acid" lipase with a pH optimum at 4·3 (Ory *et al.*, 1962) whereas Yamada (1957) reported that the same tissue also produces a second lipase during germination which has a pH optimum at neutrality. Although St. Angelo and Altschul (1964) were subsequently unable to confirm the presence of this second lipase in castor bean, Ching (1968) found both types of enzyme in

7

Douglas fir seeds, during the germination of which the activity of a neutral lipase (pH optimum 7·1) increased fourfold, while that of an acid lipase (pH optimum 5·2) increased sevenfold.

The cellular location of lipases has been studied by Tavener and Laidman (1970) who showed that these enzymes occur in both the bran and the endosperm of wheat grain, while Ching (1968) found that the major lipase activity in fir seeds is associated with "heavy fat body" and "soluble" fractions; "light fat bodies", mitochondria, protein bodies, microsomes and nuclei contained comparatively little active lipase.

Because the process of germination is normally under the influence of phytochrome, Nyman (1965a, b) examined the effect of light on the development of lipase activity and concluded that in Scots pine (*Pinus silvestris*) lipolysis is not influenced by a light factor. On the other hand, Hacker and Stohr (1966) obtained evidence that the distribution of lipase activity in cotyledonary tissue of *Sinapis alba* during germination is not changed but that the rate of lipolysis is under the control of phytochrome.

A. Castor Bean Lipase

Of all seed lipases, only that of castor bean has been studied in detail with regard to its mechanism of action, substrate specificity and cofactor requirements. It apparently differs from other lipases in requiring a unique cofactor which Ory *et al.* (1964) characterized as a cyclic polymer of ricinoleic acid with linkages through the carboxyl and 12-hydroxyl groups (Fig. 7.1) and which can be separated from the crude enzyme by extraction with butanol.

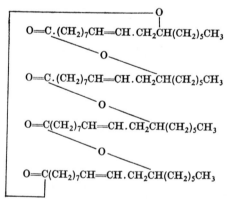

FIG. 7.1. Structure of the cyclic cofactor required for the activation of castor bean lipase (Ory *et al.*, 1964).

The lipase can be prepared by sequential extraction of castor beans with pH 7·5 phosphate-cysteine-EDTA buffer, ether and salt solutions (Altschul et al., 1963) and has a pH optimum at 4·3. The crude enzyme does not require the addition of emulsifiers for maximum efficiency.

A study of the behaviour of the enzyme in in vitro systems showed that hydrolysis of triglycerides is most rapid with triglycerides of saturated fatty acids having chain lengths between C_4 and C_8 (Ory et al., 1962) although its natural endogenous substrate, castor oil, is also hydrolysed rapidly (Ory et al., 1960). Earlier observations from several laboratories that this lipase hydrolyses triglycerides completely, with no accumulation of partial glycerides with fatty acids on the 2-position, led to the conclusion that castor lipase was not specific for position. However, Ory et al. (1969) found that the enzyme specifically catalyses the hydrolysis of fatty acids from the 1- and 3-positions of synthetic triglycerides immediately after achieving proper reaction conditions, and that fatty acids from the 2-position appear in the reaction products only at a much later stage; the same group also demonstrated conclusively that castor lipase does not cleave secondary ester linkages.

Castor bean lipase therefore resembles mammalian lipases in specifically hydrolysing fatty acids from the 1- and 3-positions of triglycerides, and in terms of behaviour towards such substrates can be classed as a glycerol ester hydrolase (EC 3.1.1.3). The earlier observations of all three fatty acid residues being removed by this enzyme are thought to arise from the isomerization of partial glycerides containing 2-acyl groups, which at acid pHs migrate rapidly to the 1- and 3-positions in the presence of this enzyme (Borgström and Ory, 1970). Such a mechanism acting in the natural tissue would ensure the most efficient utilization of available storage fat and would also explain the absence of detectable levels of monoglyceride in germinating seed.

Present evidence (Olney et al., 1968) suggests that the lipase from Vernonia anthelmintica seed may operate in a similar manner to castor bean lipase.

2. Phospholipases

Numerous classes of phospholipid-hydrolysing enzymes—"phospholipases"—have been characterized, and the nomenclature relating to the various members of this general class of enzyme has become extremely confused. In this section we shall adopt the classification system proposed by the 1961 Commission on Enzymes of the International Union of Biochemistry, but the reader should be aware that many papers published both before and after the Commission have employed conflicting notations.

The system approved by the Commission is illustrated in Fig. 7.2 in which phosphatidyl choline is illustrated as a typical substrate, although other classes of phosphatide could have served equally well.

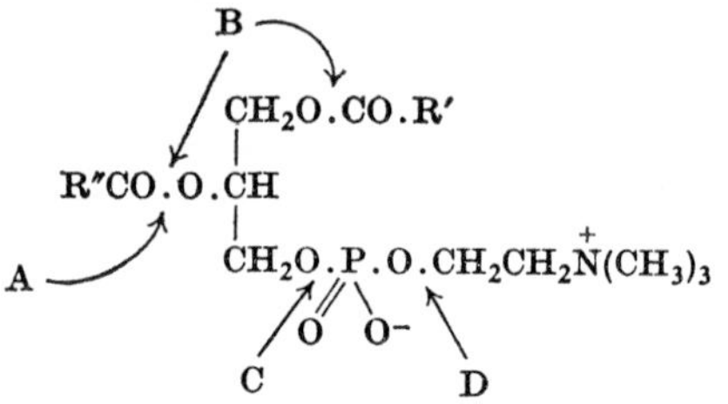

FIG. 7.2. Sites of attack of phospholipases.

A. Phospholipase A (Phosphatide Acyl-hydrolase, EC 3.1.1.4)

Phospholipase A is defined as that enzyme which specifically removes only the acyl residue on the 2-position of a phosphoglyceride (Fig. 7.2), and is found in a wide variety of animal tissues including snake, bee and wasp venoms, pancreas, kidney and bacteria. Direct evidence for the existence of this enzyme in plant tissues is sparse, although Yagi and Benson (1962) have reported that extracts from the green alga *Scenedesmus obliquus* hydrolysed plant glycerophosphatides with formation of the corresponding lysophosphatides. The absence of detectable levels of lysophosphatides in plant extracts does not necessarily argue against the presence of phospholipase A since Bartels (1969) and Appleby and Nichols (1971) have noted that many plant cells degrade mono-acyl complex lipids particularly rapidly.

B. Phospholipase B

Phospholipase B activity has been reported in barley (Acker and Bücking, 1957), while the observation of Miyachi *et al.* (1965) that complete deacylation of plant phospholipids occurs on the cell surface of *Chlorella ellipsoidea* suggests the presence of the enzyme in this class of plant also. Conversely, Appleby and Nichols (1971) were unable to show similar hydrolytic activity in *Chlorella vulgaris*.

Galliard (1970) reported strong phospholipid acyl-hydrolase activity in aqueous extracts of potato tuber, and because no accumulation of lyso-phospholipids was detectable during his experiments, the author proposed that the enzyme involved was of the phospholipase B type, or was possibly a mixture of phospholipases A and B.

C. Phospholipase D
(Phosphatidyl Choline Phosphatidohydrolase, EC 3.1.4.4)

Phospholipase D has so far been found only in plant tissues, and catalyses the removal of choline, ethanolamine, serine and glycerol from their respective phosphoglycerides with the consequent formation of phosphatidic acid. Evidence regarding its behaviour towards other classes of lipid has been somewhat contradictory possibly because of the different states of purity of the enzyme studied in different laboratories. Thus Kates (1956) was unable to obtain hydrolysis of lysolecithin with his enzyme preparation whereas Long et al. (1963) achieved complete hydrolysis of this class of substrate. Similarly, Hack and Ferrans (1959) reported that phospholipase D removed the amino alcohol residue from choline plasmalogens but Lands and Hart (1965) later found that this lipid was completely stable to the enzyme. Despite these occasional reports of phospholipid stability towards phospholipase D, observations such as the rapid conversion of stearoylphosphoryl choline into stearoyl-phosphate and choline (van Deenen and de Haas, 1966) have suggested that it can act on a fairly broad range of substrates, and although it does not appear to attack natural or synthetic cardiolipin in the presence of ether (de Haas et al., 1966) hydrolysis of this lipid occurs in the absence of this solvent (Heller et al., 1965). Moreover, experiments by Bonsen et al. (1965) using stereoisomeric lecithins have confirmed the original observations of Davidson and Long (1958) that the enzyme will completely degrade DL-α and DL-β lecithins as well as the natural L-α- form; phospholipase D therefore does not act completely stereospecifically.

The hydrolytic activity of the enzyme in homogenates or aqueous extracts of different plant tissues is very variable, and in some circumstances is sufficient to cause substantial degradation of endogenous phospholipid within a very short period. Furthermore this activity is catalysed by such solvents as ether and chloroform and Kates (1956) has recommended iso-propyl alcohol as one of the most reliable and convenient solvents for the extraction of leaf phospholipids without significant hydrolysis. The variability of phospholipase D activity in extracts of comparable tissues from different plants may be due to the presence of differing levels of natural inhibitors of this enzyme (Kates, 1954; Tookey and Balls, 1956).

In a major extension of the original studies of Davidson and Long (1958) on the distribution of phospholipase D between individual tissues of various plants, Quarles and Dawson (1969) investigated the activity of the enzyme in subcellular fractions from individual tissues and thereby obtained results which are in apparent conflict with those of Kates (1954) and Davidson and Long. In particular, Kates found that the enzyme was primarily associated with particulate matter, mainly the chloroplast, while Davidson and Long reported that the enzyme was present in both soluble and particle-bound forms in cabbage. Quarles and Dawson, on the other hand, found that all phospholipase D activity appeared in a supernatant fraction from extracts of pea plumules and cabbage leaves. Quarles and Dawson also obtained indirect evidence that the enzyme is not primarily associated with the chloroplast by analyses of such tissues as Jerusalem-artichoke tubers, soya beans and marrow seeds, which were found to be rich in the enzyme yet practically devoid of chloroplasts, and these authors concluded that in general the enzyme is chiefly present in those parts of the plant which contain reserve materials for eventual maturation or the development of new plants.

The physiological importance of phospholipase D is still obscure although Quarles and Dawson believed their results to favour the idea that the enzyme is formed in some types of higher-plant cells during a period of active growth, e.g. in the rapidly growing tip of a root or plumule. Usually this activity is then lost so that the mature stem or leaf is comparatively devoid of activity. However, when this active growth is associated with the formation of a storage tissue and the deposition of reserve substances it is possible that the phospholipase D formed is not always lost but may remain in the tissue until development occurs and the proteins, including the enzyme, are lost by proteolytic activity. Presently available evidence does not suggest any reason why the enzyme should be required by actively growing cells, and since phospholipase D is absent from the actively growing root of potato tuber (Quarles and Dawson, 1969) it does not appear to be a universally essential prerequisite of the growth process.

A second type of reaction catalysed by phospholipase D has also been described, namely the transfer of the phosphatidyl residue from lecithin to water soluble alcohols:

$$\text{Phosphatidyl choline} + \text{ROH} \rightarrow \text{Phosphatidyl-OR} + \text{choline}$$

Yang et al. (1967) presented evidence that both the hydrolase and transferase activities are catalysed by the same enzyme, finding that a constant ratio of these two reactions was preserved during a hundredfold purification of the enzyme. Phosphatidylations of ethanol, ethanolamine and glycerol were readily achieved by incubating phosphatidyl choline and the enzyme in the presence of one of these acceptors, and with glycerol a mixture of phosphatidyl-L-glycerol and phosphatidyl-D-glycerol was formed, indicating that the enzyme reacts specifically with the primary alcohol group. On the other hand, the formation of this mixture shows that the enzyme is not stereospecific and casts some doubt upon the physiological significance of the transfer reaction, as the natural phosphatidyl glycerol is exclusively phosphatidyl-L-glycerol (p. 183).

Similar results were obtained by Dawson (1967) who found that propanol, propanediol, ethylene glycol and diethylene glycol readily accepted the phosphatidyl group from lecithin in the presence of this enzyme, but propane-2-ol, sugars and oxy-acids did not act as acceptors.

The stability of phospholipase D is largely dependent on its degree of purity, being stable after 5 minutes at 55° while in the crude state but becoming more thermolabile on progressive purification (Davidson and Long, 1958). The latter effect can be overcome, however, by the addition of serum albumin to the highly purified enzyme (Dawson and Hemington, 1967).

Methods for the preparation and assay of phospholipase D have been described by Sastry and Kates (1969) and Yang (1969).

3. Glycolipid Hydrolases

A. Galactosyl Glyceride Acyl Hydrolases

Sastry and Kates (1964) have shown that the leaves of the runner bean (*Phaseolus multiflorus*) contain lipases which remove both fatty acids from the galactosyl diglycerides to give water-soluble galactosyl glycerols and free fatty acids. These enzymes are associated with the chloroplasts and are also present in soluble form in the cell-sap cytoplasm, and on the basis of stoichiometric and chromatographic data the following reactions were presumed to be catalysed by these enzymes:

$$
\begin{array}{c}
\text{Monogalactosyl} \rightarrow \text{Monogalactosyl monoglyceride} + \\
\text{diglyceride} \qquad \text{free fatty acid} \\
\downarrow \\
\text{Monogalactosyl glycerol} + \\
\text{free fatty acid}
\end{array}
$$

$$
\begin{array}{c}
\text{Digalactosyl} \rightarrow \text{Digalactosyl monoglyceride} + \\
\text{diglyceride} \qquad \text{free fatty acid} \\
\downarrow \\
\text{Digalactosyl glycerol} + \\
\text{free fatty acid}
\end{array}
$$

In practice, both fatty acid residues were released equally quickly from each lipid so that the intermediate formation of galactosyl mono-glycerides was presumed rather than experimentally confirmed.

Since different pH optima and Michaelis-Menten constants operate for the enzymic hydrolysis of the mono- and di-galactosyl diglycerides respectively, Sastry and Kates concluded that the two lipids are hydrolysed by two distinct lipases. Their enzyme preparations were without effect on triglycerides and phospholipids, and solvents such as ether were inhibitory.

Significant galactolipid-hydrolysing activity was found in the cell-sap cytoplasm of the leaves of three *Phaseolus* species tested, but was absent from similar preparations from soybean, a member of a closely related plant genus. Although these results appear to suggest that these enzymes are confined to the *Phaseolus* genus, it is more probable that galacto-lipases are actually present in tissues from other plants but are prevented from acting on their substrates by the presence of endogenous inhibitors (cf. phospholipase D, p. 197). Thus the observation of Miyachi *et al.* (1965) that incubation of ^{14}C-labelled galactolipids with cells of *Chlorella ellipsoidea* resulted in the accumulation of radioactive galactosyl glycerols in the culture medium suggests that similar enzymes to those in *Phaseolus* leaves are present in this alga.

Galactosyl diglyceride acyl-hydrolase activity observed in aqueous extracts of potato tubers by Galliard (1970) showed some marked differences from the spinach leaf enzyme system. Specifically, the preparation from potato exhibited a much greater activity towards mono-galactosyl diglyceride than towards digalactosyl diglyceride, and monogalactosyl monoglyceride could be detected as an intermediate in the deacylation of the former glycolipid.

Sastry and Kates (1970b) have published detailed methods for the isolation and assay of the galactosyl diglyceride hydrolases.

B. Sulphoquinovosyl Diglyceride Acyl Hydrolases

Yagi and Benson (1962) found that incubation of the plant sulpholipid with extracts of *Scenedesmus obliquus*, *Chlorella* or leaves and roots of alfalfa resulted in the stepwise deacylation of this lipid to give first, sulphoquinovosyl monoglyceride, and then the completely deacylated product, sulphoquinovosyl glycerol. The mechanism of deacylation is therefore clearly similar to that proposed by Sastry and Kates for the galactosyl glycerols, although in this case there was a marked accumulation of the monoacyl intermediate.

Because the *Scenedesmus* extract also hydrolysed phospholipids to the corresponding lyso-compounds but was without effect upon galactosyl diglycerides, the enzyme studied by Sastry and Kates is clearly quite different from that which acts on the sulpholipid.

CHAPTER 8

Biological Degradation of Plant Fatty Acids

1. BETA-OXIDATION OF FATTY ACIDS

A. Introduction

Lipid represents a most concentrated and anhydrous form of biological fuel, and is laid down as such in the seeds of many plants. When they germinate, hydrolysis (p. 193) is followed by breakdown of the fatty acids to provide energy (e.g. as ATP) and material (e.g. as carbohydrate) for the new plant. This catabolic activity is a property of a wide variety of tissues, and the mechanism was first studied in animals. Knoop (1904) originally proposed the theory of β-oxidation after studying the fate of ω-phenylalkanoic acids fed to dogs. It was some fifty years before the process and its individual steps could be substantiated in several laboratories (Lynen and Ochoa, 1953; Drysdale and Lardy, 1953; Lehninger and Greville, 1953; Green, 1954). The established pathway has now been fully described (e.g. Lynen, 1958; Mahler, 1964; Mahler and Cordes, 1966; Green and Allman, 1968a). This involves preliminary activation of free fatty acids to the S-acyl derivatives of CoA (p. 97) and subsequent cleavage with the production of acetyl-CoA molecules from successive pairs of carbon atoms until complete degradation has been achieved (Fig. 8.1). The fragments are further metabolized by a series of accessory reactions: in particular, acetyl-CoA may be fully oxidized to carbon dioxide with conservation of energy as ATP via oxidative phosphorylation associated with the TCA cycle (Fig. 8.2); acetyl-CoA may alternatively be converted to carbohydrate with conservation of carbon via the glyoxylate cycle (Fig. 8.3). The mechanism of β-oxidation has been fully established: it involves dehydrogenation of acyl-CoA to 2-enoyl-CoA; hydration to 3-hydroxyacyl-CoA; dehydrogenation to 3-ketoacyl-CoA; and cleavage of acetyl-CoA. The other product of this cleavage is an acyl-CoA having two carbon atoms per molecule less than the original substrate which may then be subject to the same sequence of reactions (Fig. 8.1). These reactions are tightly coupled, and the sequence is closely associated with the TCA and glyoxylate cycles, so that no metabolite or product of β-oxidation normally accumulates.

The individual enzymes have been isolated and characterized. Mammalian acyl-CoA dehydrogenases have FAD as prosthetic group

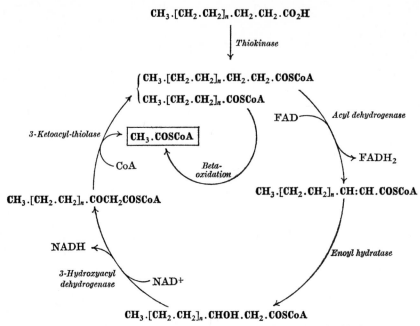

FIG. 8.1. Production of acetate from fatty acids by beta-oxidation.

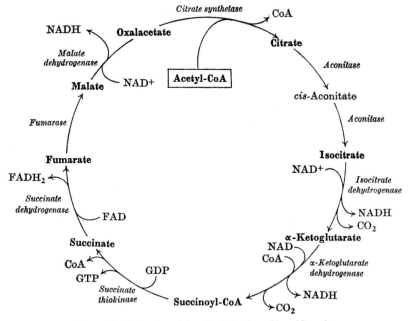

FIG. 8.2. Fate of acetate: the tricarboxylic acid cycle.

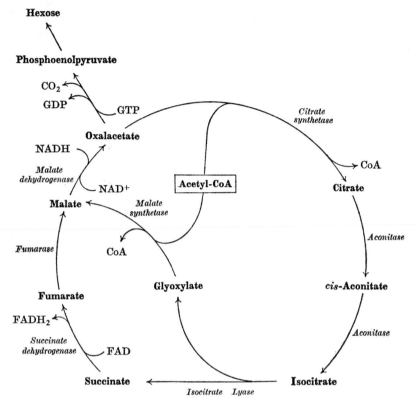

FIG. 8.3. Fate of acetate: the glyoxylate cycle.

and are somewhat selective in chain-length of substrate. Both short-chain (EC 1.3.2.1) and long-chain (EC 1.3.2.2) dehydrogenases convert acyl-CoA specifically to *trans*-2-enoyl-CoA. Enoyl-CoA hydratase (EC 4.2.1.17), also called crotonase, catalyses the conversion of *trans*-2-enoyl-CoA to L(+)3-hydroxyacyl-CoA; this hydration is stereospecific. L(+)3-Hydroxyacyl-CoA dehydrogenase (EC 1.1.1.35) is an NAD-specific enzyme responsible for the formation of 3-ketoacyl-CoA. Finally, acyl-CoA:acetyl-CoA C-acyltransferase, also called 3- or β-ketothiolase, causes the cleavage of acetyl-CoA from the molecule via thiolysis by free CoA; two enzymes of different selectivities have been described, one for long (EC 2.3.1.16) and one for short (EC 2.3.1.9) chains.

This pathway of β-oxidation outlined in Fig. 8.1, describes the mechanism by which the saturated carbon chain is cleaved enzymically, and is based mainly on data from animal systems, where the enzymes are particularly associated with the mitochondria. The same mechanism apparently occurs in plants; it is similar, but not identical, to the reverse of the mechanism of biosynthesis (Fig. 5.7, p. 106), and the two systems are compared and contrasted on p. 119 (Table 5.2).

Much of the fatty acid present in animals and plants is unsaturated; these acids are subject to β-oxidation since the enzymes involved are fairly non-specific with respect to homologous substrates, but the pathway must be modified in some way when a metabolite is reached with a double bond near the carboxyl group. In general terms, β-oxidation of unsaturated fatty acids can proceed as for saturated acids, except that cis and trans 2-enoyl-CoA and 3-enoyl-CoA derivatives would arise after the action of thiolase. The trans-2-alkenoyl-CoA is a normal metabolite (Fig. 8.1) and β-oxidation would continue. Enoyl hydratase converts cis-2-alkenoyl-CoA to D(−)-3-hydroxyacyl-CoA. This is not directly oxidized, since the 3-hydroxyacyl-CoA dehydrogenase present in β-oxidation systems is specific for the L(+) isomer; however Stoffel and Caesar (1965) have shown that animal mitochondria contain a 3-hydroxy-acyl-CoA-3-epimerase which catalyses the interconversion of the D(−) and L(+) isomers; the latter is then available as a substrate for further oxidation (Fig. 8.1). The cis-3-enoyl-CoA is not attacked by β-oxidation enzymes, but can be converted to the trans-2-isomer directly by an enzyme found in rat liver (Stoffel et al., 1964), in guinea-pig liver and in the slime mould Dictyostelium discoideum (Davidoff and Korn, 1964). Struik and Beerthuis (1966) have shown that trans-3-enoyl-CoA, as well as the cis-3-isomer, is converted to the trans-2-isomer by an isomerase (EC 5.3.3.3) isolated from ox liver mitochondria (Rilling and Coon, 1960) and apparently identical with that from rat liver. The trans-2-enoyl CoA so formed fits into the normal β-oxidation pathway, and thus β-oxidation of unsaturated fatty acids in these systems is feasible.

The modified pathway due to the presence of the double bond may affect the rate of β-oxidation: the formation of trans-2-enoyl-CoA from the cis-3-isomer is faster than from the trans-3-isomer (Struik and Beerthuis, 1966). This is reflected in the faster β-oxidation of oleic acid than that of elaidic acid (Anderson, 1967). However cis,cis-linoleic acid is oxidized more slowly than mono-trans- or all-trans-isomers: this may be due to a rate-limiting epimerization of 3-hydroxyoctanoyl-CoA (Anderson, 1968). All geometric isomers of oleic and linoleic acids are readily oxidized in the intact animal (Anderson and Coots, 1967), and double bonds do not therefore present any blockage to the oxidative pathway. An exception is the perfused heart, where accumulation of intermediates can occur: while oleic acid is oxidized normally, elaidic acid gives rise to trans-5-tetradecenoic acid (Willebrands and van der Veen, 1966).

The fact that unsaturated acids are oxidized more rapidly than saturated acids of similar chain length in intact animals and in their isolated mitochondria can be explained by invoking γ-oxidation of the intermediate cis-3-enoates, for which there is some evidence. Instead of isomerization and cleavage of acetate from the resulting 2-enoate, propionate could be released directly; this would be converted to methylmalonate (p. 211). The experiments of Dupont and Mathias (1969) indicate that some methylmalonate is indeed derived in animals directly from linoleate and oleate rather than palmitate, suggesting that unsaturated acids are oxidized partly by a pathway involving propionate.

B. Beta-oxidation in Plants

Early work on intact plants suggested that β-oxidation was not confined to animals. In experiments reminiscent of Knoop's classical work, Grace (1939) showed that the growth-promoting properties of the homologous series of five ω-naphthylalkanoic acids depended on chain length, an alternation of activity occurring as this increased: thus, the ω-(1-naphthyl)-derivatives of acetic, butyric and hexanoic acids promoted the rooting of cuttings, whereas derivatives of propionic acid and valeric acid were less active. Synerholm and Zimmerman (1947) examined seven members of the series of ω-2,4-dichlorophenoxyalkanoic acids, and observed that only those with an even carbon number had the growth-regulating properties of 2,4-dichlorophenoxyacetic acid ("2,4-D"). Fawcett et al. (1954) treated flax seedlings with a series of nine ω-phenoxyalkanoic acids and found that appreciable quantities of phenol were produced only from those acids with odd numbers of carbon atoms. Wain and Wightman (1954) extended this work to demonstrate that a series of some chlorophenoxyalkanoic acids shows typical alternation in growth-regulating properties. All these in vivo results strongly suggest that the side chains are oxidized by successive loss of pairs of carbon atoms.

1. Beta-oxidation in mitochondria

Attempts to demonstrate β-oxidation by plant preparations in vitro were not successful until Stumpf and Barber (1956) found that a mitochondrial fraction from peanut cotyledons 5–8 days after germination would oxidize fatty acid to carbon dioxide, provided that only trace amounts of substrate were present. The reaction was followed by counting $^{14}CO_2$ from acids labelled in various positions; the cofactor requirements were ATP, NAD, CoA, NADP, Mn, a TCA cycle acid and GSH, suggesting that the mechanism of fatty acid oxidation in peanut mitochondria resembles that in animals. Inhibition of butyrate oxidation by malonate and by 2,4-dinitrophenol indicated the involvement of the TCA cycle, and this was confirmed by comparing the rate of appearance of $^{14}CO_2$ from [1-^{14}C]acetic acid and [2-^{14}C]acetic acid (or their precursors, [3-^{14}C]palmitic acid and [2-^{14}C]palmitic acid respectively); moreover, oxidation of [1-^{14}C]butyric acid led to the appearance of label in the TCA cycle acids.

Mitochondrial preparations containing enzyme systems for the oxidation of fatty acids to carbon dioxide have also been prepared from pine seedlings (Stanley and Conn, 1957) and from avocado (Stumpf, 1962). The system evidently causes activation of the fatty acid to the CoA derivative, followed by β-oxidation and total oxidation of the acetyl-CoA formed to carbon dioxide, with the conservation of the energy released to provide for synthetic reactions in the cell; this is also an established function of animal mitochondria.

However, Rebeiz and Castelfranco (1964) observed that their mitochondrial fractions (again from germinating peanut cotyledons) were only irregularly effective, but consistent oxidation occurred if the supernatants were added back to them. They showed that the protein precipitated from this supernatant converted short- and long-chain saturated fatty acids by β-oxidation to acetyl-CoA which accumulated because of the absence of TCA cycle activity. Their results allowed three possibilities: firstly, that the β-oxidation activity was localized in the mitochondria, but was leached out during their isolation; secondly, that it was localized only in the microsomes or hyaloplasm or both (in which case the activity associated with the mitochondria was due to occlusion of extra-mitochondrial material); and thirdly, that it was localized in both mitochondria and in the microsomes or hyaloplasm or both. Later experiments (Rebeiz et al., 1965a) showed that the enzymes responsible for fatty acid activation were particulate and located mainly in the microsomes; and that those responsible for β-oxidation were present in all particulate fractions as well as a high-speed supernatant. However, most of the β-oxidation activity could not be associated with mitochondria, because their fragmentation was unlikely and moreover electron microscopy of the tissue used (peanut cotyledons, 4·5 days after germination) showed that few recognizable mitochondria were present. It was now suggested that the acetyl-CoA produced by extramitochondrial β-oxidation was metabolized via the glyoxylate cycle, which reaches maximum activity at 6 days after germination (Marcus and Velasco, 1960), and which could not be mitochondrial either, and was therefore located in some other membrane system present in the tissue.

This suggestion is confirmed by the observations of Cherry (1963), that mitochondrial function in peanut cotyledons reaches a maximum some three days after the glyoxylate cycle activity, and of Canvin and Beevers (1961) that the TCA cycle is inoperative in castor bean endosperm five days after germination.

Yamada and Stumpf (1965a) estimated β-oxidation activity in soluble, microsomal and mitochondrial fractions from castor beans after up to eight days' germination, and showed that both the mitochondrial and supernatant fractions reached maximum activity on the fifth day, respectively accounting for 2·5 and 97% of the total. Both systems required NADP, NAD, ATP, glyoxylate and manganese for optimal release of $^{14}CO_2$ from [^{14}C]oleoyl-CoA. [^{14}C]Oleic acid was oxidized only when supplemented with a microsomal fraction of maturing castor beans, which contained the appropriate thiokinase. [^{14}C]Palmitic acid similarly yielded $^{14}CO_2$ rapidly if C-1 or C-11 were labelled, but not if the label were at C-2. These results, and the effects of omitting various cofactors, are in accordance with the mechanism proposed by Beevers (1961) and illustrated in Fig. 8.4. It suggests there are two discrete systems involving β-oxidation of fatty acids: the normal energy-producing pathway, localized in the mitochondria, which oxidizes fatty acids to carbon dioxide via the TCA cycle; and the conversion of fatty acids to sucrose via the glyoxylate cycle. In germinating fatty seeds, the glyoxylate cycle initially predominates; this activity is catalysed by "soluble" enzymes which are associated with the fragile "glyoxysomes".

2. Fate of acetate

Improvements in techniques of homogenization of plant tissue and of fractionation of homogenates led to the recognition of new subcellular organelles. "Spherosomes" (Jacks et al., 1967) were shown to be the site of lipid storage in germinating peanut cotyledons, but not of its degradation since no lipase or acyl kinase was present. Breidenbach et al. (1968) isolated three organelle fractions from castor bean endosperm by gradient centrifugation in sucrose: mitochondria (1·19 g/cc), proplastids (1·23 g/cc) and "glyoxysomes" (1·25 g/cc). The latter contained essentially all the particulate malate synthetase, isocitrate lyase, catalase and glycolic oxidase. Malate dehydrogenase, citrate synthetase and aconitase were present in both mitochondria and glyoxysomes. It was proposed that the glyoxysomes were responsible for the metabolism of acetyl-CoA via the glyoxylate pathway, with the net production of succinate. This suggestion was confirmed by Cooper and Beevers (1969a) who studied the distribution of some twenty enzymes of the TCA and glyoxylate cycles between the mitochondria, proplastids, glyoxysomes and supernatant

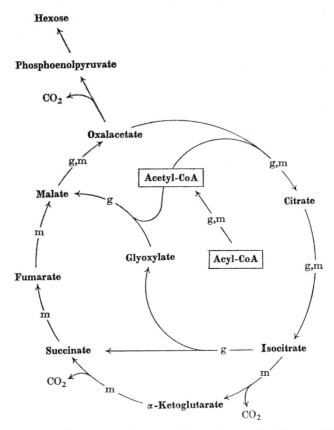

Fig. 8.4. Biological degradation of fatty acids in plants. After Beevers (1961) and Cooper and Beevers (1969a, b). This scheme represents β-oxidation (Fig. 8.1), the TCA cycle (Fig. 8.2) and the glyoxylate cycle (Fig. 8.3). The reactions are localized in the glyoxysomes (g), or the mitochondria (m) or both (g, m). In germinating seeds where fat is converted to sugars, acetyl-CoA yields succinate in the glyoxysomes; succinate is oxidized to oxalacetate in the mitochondria. [1-^{14}C]Acetyl-CoA and glyoxylate yield [1,4-^{14}C$_2$]oxalacetate and thence [1-^{14}C]phosphoenolpyruvate and ^{14}CO$_2$; [2-^{14}C]acetyl-CoA gives rise to [2,3-^{14}C$_2$]oxalacetate followed by [2,3-^{14}C$_2$]phosphoenolpyruvate and unlabelled CO$_2$. In other tissues where fat is required for energy, and which contain no glyoxysomes, acetyl-CoA is fully oxidized to CO$_2$ in the mitochondria.

obtained from castor bean endosperm 5 days after germination. They observed that more than 85% of the total isocitrate lyase and malate synthetase, key enzymes in the glyoxylate pathway (Fig. 8.3), are associated with the glyoxysomes; there was less than 5% mitochondrial activity.

With the exception of aconitase (which was unstable and apparently leached from the organelles during isolation) the activities of the glyoxylate cycle enzymes estimated in the isolated glyoxysomes were more than adequate to account for the observed endogenous rate of acetate metabolism in the intact tissue. Glyoxysomes do not, however, contain either α-ketoglutarate or isocitrate dehydrogenase although both these enzymes are located in mitochondria; hence the only known fate open to isocitrate generated in glyoxysomes is production of succinate and glyoxylate by means of isocitrate lyase. The glyoxylate is metabolized by the malate synthetase which is present, thus utilizing more acetyl-CoA; the apparent absence of NADH-glyoxylate reductase or glyoxylate transaminase activity suggests that this is the only fate of glyoxylate in this organelle. Meanwhile the succinate cannot be further metabolized *in situ*, owing to the absence of succinic dehydrogenase and fumarase, and must apparently be transported to the mitochondria (where these activities are high) before undergoing oxidation to malate.

3. Beta-oxidation in glyoxysomes

Cooper and Beevers (1969a) also noted that in ten tissues tested, four (castor bean endosperm, corn scutellum, water melon cotyledon and peanut cotyledon) converted fat to carbohydrate and contained glyoxysomes possessing glyoxylate cycle activity; six (castor bean hypocotyl, corn root, water melon hypocotyl, avocado, peanut hypocotyl and zucchini squash fruit) did not convert fat to carbohydrate, and contained neither glyoxysome fraction nor glyoxylate cycle activity. The glyoxysome therefore is the organelle responsible for accepting acetyl-CoA and converting it to succinate during the synthesis of carbohydrate from lipid in the germinating seedling. The problem of the supply of acetyl-CoA to the organelle by the enzymes of β-oxidation was investigated by Cooper and Beevers (1969b) and Hutton and Stumpf (1969), who showed that the glyoxysome itself is responsible for much β-oxidation activity. Over 80% of the particulate β-oxidation activity was associated with the glyoxysomes, and virtually none with the purified mitochondria. This could explain why acetate escapes oxidation through the TCA cycle, and is exclusively metabolized via the glyoxylate cycle in this tissue. The glyoxysomal mechanism of β-oxidation appears to be similar to the established mitochondrial pathway (Fig. 8.1), but there are important differences. Glyoxysomes do not re-oxidize NADH (which accumulates *in vitro*) and oxygen is a requirement for glyoxysomal β-oxidation; the production of acetyl-CoA, the reduction of NAD^+ and the uptake of oxygen occur in $1:1:0.5$ stoichiometry. Cooper and Beevers (1969b) suggested that the oxygen uptake must be ascribed to oxidation of a flavoprotein (Fig. 8.5) with the formation of hydrogen peroxide, which would be broken down by the catalase present in the glyoxysome, giving the observed stoichiometry; moreover, they showed that addition of cyanide doubled the oxygen uptake without affecting NAD reduction, which is consistent with the inhibition of the glyoxysomal catalase. Glyoxysomes do not contain the efficient electron-transport system associated with mitochondria, and the mechanism of reoxidation of NADH formed by glyoxysomal β-oxidation is unclear.

Hutton and Stumpf (1969) investigated β-oxidation of fatty acids in maturing as well as germinating castor beans. They estimated the activities of β-oxidation (using various substrates), fatty acid synthetase and several single enzymes in these tissues at different stages of their development. Production of acetyl-CoA from added palmitic acid reached a maximum 28 days after flowering and 3 to 4 days after germination. In both cases, activity was found in the mitochondrial (1·19 g/ml),

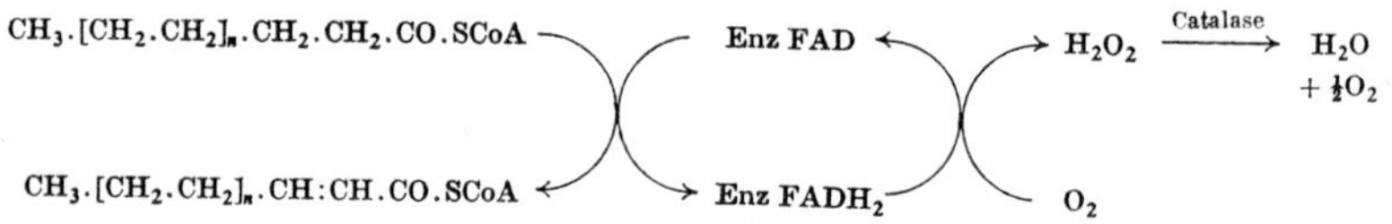

$$CH_3.[CH_2.CH_2]_n.CH_2.CH_2.CO.SCoA \quad Enz\ FAD \quad H_2O_2 \xrightarrow{Catalase} H_2O + \tfrac{1}{2}O_2$$

$$CH_3.[CH_2.CH_2]_n.CH:CH.CO.SCoA \quad Enz\ FADH_2 \quad O_2$$

FIG. 8.5. Proposed mechanism of dehydrogenation of acyl-CoA in glyoxysomes (Cooper and Beevers, 1969b).

cytosomal (1·25 g/ml) and "soluble" enzymes. In the germinating system the cytosomes correspond with glyoxysomes, but their fragility evidently caused extensive leaching of β-oxidation activity into the "soluble" fraction. Thus about 95% of the activity (conversion of palmitic acid to acetyl-CoA) in germinating beans was found to be "soluble", whereas similar tissue assayed by Cooper and Beevers (1969b) on the basis of NAD$^+$ reduction with palmitoyl-CoA as substrate gave about 20% "soluble" activity. It may be that all the β-oxidation activity in systems isolated and found to be "soluble" in fact originated in the glyoxysome or some other fragile extra-mitochondrial organelle.

The function of a system for catabolizing fatty acid in the maturing bean, where the synthesis and storage of large quantities of lipid is taking place, is not clear. Here, however, the cytosomes responsible were clearly not aggregated mitochondria or glyoxysomes, since fatty acid synthetase and acetyl-CoA carboxylase were shown to be present by Hutton and Stumpf (1969). They suggest that the low activity of β-oxidation might provide a scavenging system, preventing the accumulation of any free fatty acid in the maturing seed. The β-oxidation activity did not appear to be controlled by means of limiting thiokinase activity, but rather by a regulation of the separate activities of the component enzymes.

The activation of the long-chain fatty acids (p. 97) occurs in the avocado (Barron and Stumpf, 1962) germinating peanuts (Rebeiz et al., 1965a) and germinating castor beans (Yamada and Stumpf, 1965a); in each case the acyl thiokinase activity is located almost entirely in the microsomes. The acetyl thiokinase in germinating peanuts is soluble (Rebeiz et al., 1965b). Esters of CoA must therefore pass through a mitochondrial or glyoxysomal membrane before β-oxidation can take place. The transport system which operates in some mitochondria involves the biosynthesis of the carnitine derivative which can pass through the membrane, followed by resynthesis of the CoA ester, which cannot; while carnitine is present in plant tissues (Panter and Mudd, 1969) it is not known whether plants contain acyl-CoA-carnitine transferases (Mudd, 1967). However Cooper and Beevers (1969b) have demonstrated that such a system does not operate in the glyoxysome, since added CoA esters are oxidized directly, whereas carnitine derivatives are not.

4. Fate of propionate

The β-oxidation of unbranched fatty acids of even carbon number gives rise to acetyl-CoA only, which can be smoothly metabolized by the TCA and glyoxylate cycles. The β-oxidation of an odd-numbered fatty acid must yield one mole proportion of propionate. Monomethyl-substituted fatty acids also yield propionate during the normal β-oxidation sequence, provided the substituent is at an even-numbered carbon atom. The fate of propionate in plants was investigated by Stumpf and Barber (1956) who observed that their mitochondrial preparations of germinating peanuts released carbon dioxide from added propionic as well as acetic acid. Giovanelli and Stumpf (1957, 1958) demonstrated the unlikelihood of the carboxylation of propionate to methyl malonate or succinate, or of its α-oxidation to pyruvate. They identified β-hydroxy-propionate as an intermediate which accumulated, and proposed that a modified β-oxidation occurred; this was confirmed by examining the fate of radioactivity from specifically labelled substrate. Radioactive carbon dioxide was released more readily from [3-^{14}C]propionate than from [2-^{14}C]propionate; moreover radioactive succinate obtained during [3-^{14}C]propionate oxidation was labelled in the carboxyl groups, whereas that from [2-^{14}C]propionate was not.

This work was extended by Hatch and Stumpf (1962a) who examined tissue slices from wheat (epicotyl and root), safflower (cotyledon, stem and root) and pea (leaf, stem, cotyledon and root). All oxidized added propionate to carbon dioxide; also formed were 3-hydroxypropionate, β-alanine, malonate and methylmalonate. The use of specifically labelled substrate confirmed that the malonic semialdehyde pathway (Fig. 8.6a) operates. While propionate is carboxylated in plant systems to methylmalonate (Hatch and Stumpf, 1962b), the oxidation of this metabolite by means of the vitamin B_{12}-catalysed isomerization to succinate (Fig. 8.6d) does not evidently occur in plants, though it is a major route in animals (Flavin and Ochoa, 1957). Vitamin B_{12} does not occur or does not function in normal plant cells (Stumpf, 1962).

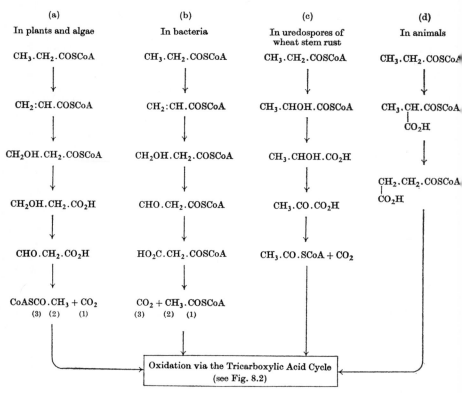

FIG. 8.6. Fate of propionate $CH_3.CH_2.CO_2H$.
(3)   (2)   (1)

The modified β-oxidation pathway via malonic semialdehyde is also the mechanism by which propionate is oxidized by the colourless alga *Prototheca zopfii* (Callely and Lloyd, 1964; Lloyd and Venables, 1967). McConnel and Finlayson (1964) confirmed that propionate was subject to the same fate in wheat plants, but Reisener *et al.* (1963) reported a different pathway in wheat stem rust. In the uredospores, oxidation of specifically labelled propionic acid demonstrated that a modified α-oxidation (p. 213) was operative (Fig. 8.6c). The use of specifically labelled substrate revealed yet another pathway of propionate oxidation in anaerobic bacteria (Vagelos, 1960). Here, the crucial intermediate is the CoA derivative of malonic semialdehyde, and the bacterial pathway is distinguished from the others by the different fates of the individual carbon atoms (Fig. 8.6b).

The oxidation of "unusual" fatty acids has not been well documented, but presumably plants are able to catabolize any acid they synthesize. Phytanic acid (3,7,11,15-tetramethylpalmitic acid) is not directly subject to β-oxidation since the C-3 position is blocked; in animals a preliminary α-oxidation step is obligatory before β-oxidation can occur (p. 219). Presumably the degradation of phytol in plants involves a similar pathway. The unnatural 3,3-dimethyl fatty acids cannot be handled in this way; in animals, ω-oxidation takes place (p. 221).

2. ALPHA-OXIDATION

A. Alpha-oxidation in Germinating Seeds

The degradation of fatty acids in plants by α-oxidation was first investigated by Stumpf and his coworkers. Here oxidation occurs at the 2-position and is followed by cleavage of the 1,2 carbon–carbon bond; it was initially observed when radioactive carbon dioxide was released from specifically labelled palmitate in cell-free preparations of germinating peanuts (Newcomb and Stumpf, 1952). Further investigation showed that microsomal particles released $^{14}CO_2$ from [1-^{14}C]palmitate, [2-^{14}C]-palmitate, [3-^{14}C]palmitate or [11-^{14}C]palmitate (in order of decreasing efficiency); no cofactor requirements were demonstrated with [1-^{14}C]-palmitate, but NAD was necessary in the other cases. Palmitoyl-CoA was oxidized much more slowly than the acid, and the TCA cycle did not appear to be involved (Humphreys et al., 1954). The particulate system was solubilized by treatment with cholate and purified; NAD was now required for $^{14}CO_2$ release from [1-^{14}C]palmitate, and a stimulatory effect was observed on adding boiled extracts of peanut cotyledons (Humphreys and Stumpf, 1955). When Castelfranco et al. (1955) extended this study to the supernatant fractions from homogenates of peanut cotyledons, they found that this stimulation was due to glycolic acid. The crude supernatant fraction released $^{14}CO_2$ from palmitic acid only if the substrate were labelled in the carboxyl group; purified fractions showed an absolute requirement only for glycolic or other short-chain L-α-hydroxyacids, and these could not be replaced by the corresponding D-enantiomers, α-ketoacids, β-hydroxyacids or β-ketoacids. The system caused the oxidative decarboxylation of stearic and myristic as well as palmitic acids; decanoic acid and lower homologues were inert, as was 2-hydroxypalmitate. However, 2-keto[1-^{14}C]palmitate was rapidly oxidized to $^{14}CO_2$, and the suggested pathway of fatty acid oxidation involved conversion to enzyme-bound 2-hydroxyacid, oxidation to 2-ketoacid followed by decarboxylation. Stumpf (1956) now reported that the enzymes responsible for release of $^{14}CO_2$ from [1-^{14}C]palmitate in fact included a peroxidase and demonstrated that the glycolic acid was necessary because its oxidation by traces of glycolic oxidase present yielded the hydrogen peroxide required for the peroxidation of the fatty acid. The peroxidase was also successfully coupled to other systems generating hydrogen peroxide such as glucose–glucose oxidase. Using a bacterial luciferase as a sensitive assay system, the accumulation of long-chain aldehyde was detected during the peroxidation of unlabelled palmitate.

Further work by Martin and Stumpf (1959) established that germinating peanuts contain (a) a peroxidase which decarboxylated a long-chain fatty acid to CO_2 and an aldehyde with one less carbon atom in its chain; and (b) an aldehyde dehydrogenase, which converted this to its corresponding acid by an NAD-specific mechanism. The enzymes were present in mitochondrial, microsomal and cytoplasmic fractions of fresh tissue homogenate, and were responsible for the degradation of fatty acids by successive α-oxidation. The postulated mechanism of the peroxidation of palmitate involved the formation of CO_2 and a pentadecyl hydroperoxide derivative, which would yield pentadecanal. Alpha-oxidation was observed in several plant tissues, and was not due to the commonly occurring plant peroxidases (Stumpf, 1962). It has been invoked as a component of the basal respiration of potato slices (Laties and Hoelle, 1967).

B. Alpha-oxidation in Leaves

Plant leaves contain enzymes which convert traces of free fatty acids to carbon dioxide via an α-oxidation pathway. This has been demonstrated by incubating uniformly labelled oleic acid with leaf systems, and showing that radioactive 8-heptadecenoate, 7-hexadecenoate and lower homologues accumulated (Hitchcock and James, 1964). Oxygen and NAD were required, but in the absence of NAD, 8-heptadecenal accumulated at the expense of 8-heptadecenoate; the metabolic route therefore appeared similar to that proposed by Martin and Stumpf (1959), i.e.

$$R.CH_2.CO_2H \rightarrow R.CHO \rightarrow R.CO_2H \rightarrow etc.$$

TABLE 8.1

Properties of alpha-oxidation systems[a]

	Green leaves	Germinating seeds	Animal systems
Metabolites:			
Oxidation of $R.CH_2.CO_2H$	+	+	+
Formation of $R.CHOH.CO_2H$	+	?	+
Oxidation of $R.CHOH.CO_2H$	+	?	+
Formation of $R.CO.CO_2H$	?	+	+
Oxidation of $R.CO.CO_2H$	?	+	+
Formation of $R.CHO$	+	+	?
Oxidation of $R.CHO$	+	+	?
Formation of $R.CO_2H$	+	+	+
Cofactor requirement:			
Oxygen	+	+	+
NAD	+	+	+
ATP	?	?	+
H_2O_2-generating system	?	+	?
Inhibition:			
By imidazole	+	+	?

[a] After Hitchcock and James (1966).

However, the enzymes responsible were evidently different (Table 8.1), for the leaf systems showed no requirement for hydrogen peroxide, and were particle-bound. Moreover, further investigation showed that the 2-hydroxyacids were intermediate in the α-oxidation of fatty acids in leaves (Hitchcock and James, 1966), thus:

$$R.CH_2.CO_2H \rightarrow R.CHOH.CO_2H \rightarrow R.CO_2H \rightarrow etc.$$

Isotope competition suggested that palmitic acid was oxidized to L-2-hydroxypalmitate (rather than the D-isomer) before decarboxylation (Hitchcock et al., 1968a). This was surprising, since the long-chain hydroxyacids normally present in leaves (and in other sources) are of the D-configuration (p. 23), and indeed the labelled 2-hydroxypalmitate which accumulated during α-oxidation of [1-^{14}C]palmitate was also the D-isomer (Hitchcock et al., 1968b). Hence it appears that during α-oxidation, both enantiomers are formed; the D-hydroxyacid accumulates and is normally incorporated into lipid (particularly cerebroside), while the L-hydroxyacid is immediately oxidized further. Examination of the biosynthesis of the D-2-hydroxypalmitic acid from D- and L-[1-^{14}C,2-^{3}H]-

palmitic acid established that the hydroxylation occurred with overall retention of configuration (Morris and Hitchcock, 1968), since tritium was lost from the D-substrate but retained from the L-substrate. Similarly, it has been shown that tritium from both D- and L-[1-^{14}C,3-^{3}H]-palmitic acid was retained; hence the hydroxylation was not the result of dehydrogenation followed by hydration as during β-oxidation. Investigation of the biosynthesis of pentadecanal from D- and L-[U-^{14}C,2-^{3}H]palmitic acid revealed retention of tritium from the L-substrate, but loss from the D-substrate (Hitchcock and Morris, 1970).

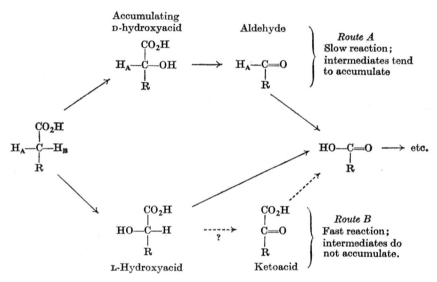

Fig. 8.7. Possible pathways of alpha-oxidation in plants. After Hitchcock and Morris, 1970.

This showed that the aldehyde cannot be derived from the intermediate L-2-hydroxypalmitate, which has lost the hydrogen originally on the L-2-position of palmitate. It appears that the accumulating D-2-hydroxy-palmitate gives rise to any pentadecanal formed, and that any further oxidation to pentadecanoate must be slow compared with the rate of α-oxidation via L-2-hydroxypalmitate. Moreover, the pentadecanal also cannot be derived from a 2-ketopalmitate intermediate since it retains label from L-[2-^{3}H]palmitate. Two pathways of α-oxidation in plant leaves therefore seem to exist (Fig. 8.7). Palmitate may be oxidized via D-2-hydroxypalmitate (which accumulates, ultimately in the sphingo-lipids) and pentadecanal (which can accumulate *in vitro*) to penta-decanoate and CO_2 (Fig. 8.7, route A). Simultaneously L-2-hydroxypalmitate may be an intermediate but this does not accumulate; it may form 2-ketopalmitate during oxidation to pentadecanoate (Fig. 8.7, route B). The evidence for route B rests on isotope competition experiments only; if this interpretation is correct, route B must be responsible for most of the decarboxylation and is therefore the faster route.

The existence of two distinct pathways of α-oxidation is also suggested by experiments with rat kidney homogenates (Levis, 1970). Two soluble oxidases have been separated and shown to be specific for D-2-hydroxy-stearate and L-2-hydroxystearate respectively; the product here is 2-ketostearate, which is decarboxylated in a microsomal fraction (p. 218). A study of the relative contributions of α-, β- and ω-oxidation of fatty acids in rat liver indicated that β-oxidation overwhelmingly predominated (Antony and Landau, 1968). On the other hand, over 80% of the $^{14}CO_2$ evolved by fresh leaves offered [1-^{14}C]palmitate was associated with an imidazole-sensitive system (Hitchcock, 1971); presumably α-oxidation is preferred to β-oxidation in this plant tissue under the conditions used.

C. Other Alpha-oxidation Systems

1. In micro-organisms

In the yeast *Candida utilis*, α-hydroxylation and decarboxylation of long-chain fatty acids take place, but the enzymes responsible are more specific than those in higher plants (Fulco, 1967). Hexacosanoic acid, formed endogenously by chain elongation (p. 119) was converted to 2-hydroxyhexacosanoate. The enzyme seemed quite specific for this chain length, though there might have been some activity with C_{24}, C_{25} and C_{27} substrates; but exogenous hexacosanoic acid added to the cells was not hydroxylated. Apparently some form of activation was necessary, which would entail a novel mechanism of hydroxylation, but since exogenous *trans*-2-hexacosenoate was also not altered, hydration did not appear to be involved. The yeast also contained a relatively non-specific system which caused oxidative decarboxylation of each of the 2-hydroxyacids tested (C_{18}, C_{20}, C_{22}, C_{24} and C_{26}), with the formation of both the unsubstituted acids and aldehydes (C_{17}, C_{19}, C_{21}, C_{23} and C_{25} respectively). This decarboxylation may well be the same as in higher plants.

Evidence that α-oxidation of long-chain fatty acids occurs in bacteria has been reported by Yano *et al.* (1969, 1970a). Palmitic acid was converted to 2-hydroxypalmitate, pentadecanoate and possibly 2-hydroxypentacanoate by growing or resting cells of *Corynebacterium simplex*; this

suggests that the bacterial 2-hydroxyacids (p. 169) arise by hydroxylation and are degraded by an α-oxidation pathway similar to that in
plants.

2. In animals

Long-chain hydroxyacids are found in animal lipids, particularly in
the brain; their biogenesis and degradation involves an α-oxidation
(Bowen and Radin, 1968) similar to the two pathways in plants. These
routes evidently have much in common, but they seem to differ in
details; some of their properties are compared and contrasted in
Table 8.1.

Reports from the laboratories of J. F. Mead and N. S. Radin have
shown that long-chain fatty acids are hydroxylated and decarboxylated
in vivo. Brain preparations *in vitro* were shown to decarboxylate 2-
hydroxystearate and 2-hydroxylignocerate (cerebronate), with the
formation of heptadecanoate and tricosanoate respectively (Levis and
Mead, 1964); there was some evidence that the corresponding ketoacid
acted as an intermediate. Acetone powders of pig brain were indeed shown
to catalyse the oxidative decarboxylation of 2-ketostearic acid (Davies
et al., 1966). Subsequently, ketoacid was shown to accumulate during
oxidation of hydroxyacid in brain preparations under special conditions
(Lippel and Mead, 1968). The accumulation of aldehyde in brain systems
has not, however, been detected (Davies and Radin, 1966). Nevertheless,
MacDonald and Mead (1968) and Lippel and Mead (1968) consider aldehyde may be an intermediate of α-oxidation in these systems, and suggest
that the mechanism involves a ferrous iron-dependent oxidase or
oxygenase, which converts 2-hydroxyacid to 2-ketoacid-enzyme complex, with subsequent decarboxylation by an enzyme which may be a
peroxidase:

$$\begin{array}{ccc} \text{R.CHOH.CO}_2\text{H} & \xrightarrow{\text{Fe}^{2+}} & \left[\begin{array}{c}\text{R.CO.CO}_2\text{H}\\\text{enzyme } \text{H}_2\text{O}_2\end{array}\right] & \xrightarrow{\text{Fe}^{2+}} & \text{R.CO}_2\text{H } \text{CO}_2 \\ \text{enzyme } \text{O}_2 & & & & \text{enzyme } \text{H}_2\text{O} \end{array}$$

Non-nervous tissue also contains these enzymes: for instance, cere-
bronic acid administered orally to rats is metabolized in the intestine by
α-oxidation (Lippel and Mead, 1969). The 2-hydroxypalmitate formed
by catabolism of phytosphingosine in rat liver preparations is subsequently decarboxylated to pentadecanoate (Gatt and Barenholz, 1968).
Long-chain L-2-hydroxyacid oxidase has been localized in the peroxy-
somes of hog kidney (Saga *et al.*, 1969). Levis (1970) has identified two
such oxidases in supernatants from rat kidney homogenates: both are
responsible for the oxidation of DL-2-hydroxystearate to the ketoacid,
which is decarboxylated by a microsomal fraction. They differ in

stereospecificity, each accepting one hydroxyacid enantiomorph only; this suggests two distinct pathways of α-oxidation as in leaves (Fig. 8.7), though there is no indication of ketoacid formation in the plant system.

Lower animals, such as the protozoa *Tetrahymena pyriformis* and *Crithidia fasiculata*, probably also contain the machinery for α-oxidation of fatty acids (Avins, 1968).

Investigations of a rare inherited disorder of the human nervous system has uncovered another aspect of mammalian α-oxidation. Refsum's disease (heredopathia atactica polyneuritiformis) is characterized by an accumulation in all tissues of phytanic acid (3,7,11,15-tetramethyl-palmitic acid), a metabolite of exogenous phytol (Steinberg *et al.*, 1967). Being substituted in the β-position, this acid is not directly catabolized by β-oxidation; a necessary preliminary involves α-oxidation via 2-hydroxyphytanate to pristanic acid (2,6,10,14-tetramethylpenta-decanoic acid) which is subject to successive β-oxidation. (Mize *et al.*, 1969a). Alternations between α- and β-oxidation similarly account for the degradation of 3,6-dimethyloctanoic acid in guinea-pig kidney slices (Stokke, 1969). The enzymic defect in Refsum's disease is the failure of hydroxylation of phytanic acid: patients oxidize [14C]phytanate at much lower rates than normal (Mize *et al.*, 1969b), and they fail to convert 3,6-dimethyl[8-14C]octanoic acid to 14CO$_2$ or to 2,5-dimethyl-[14C]heptanoic acid (Stokke *et al.*, 1967). Enzyme-bound keto-acid or a short-lived aldehyde may be intermediates in α-oxidation of β-branched acids, but direct proof is lacking. Evidence has been presented by Blass *et al.* (1969) that α-hydroxylation of straight-chain and branched-chain fatty acids in mammals may be catalysed by different enzymes.

It is evident that α-oxidation represents a pathway of fatty acid catabolism of wide distribution, though data relating to systems from micro-organisms or animals are not directly relevant to plant metabolism. However, the differences outlined in Table 8.1 may yet prove to be illusory, and a single universal mechanism emerge for α-oxidation as for β-oxidation.

D. Function of Alpha-oxidation

The role of α-oxidation of long-chain fatty acids in plants is far from clear. The process seems wasteful when compared to β-oxidation; it may not be of general importance for energy production unless the cofactors associated with β-oxidation are lacking. The fact that free acid can be metabolized could be advantageous in the absence of kinase activity, but this fact also means that only small quantities of substrate can be dealt with. Hitchcock and James (1964) pointed out that the existence of the oxidizing system in young leaves implied that fatty acids were being translocated, possibly from a more mature leaf. Free fatty acid is a non-specific enzyme inhibitor, and possibly the α-oxidation system functions as a scavenger, thus preventing interference with biologically active proteins by small quantities of adventitious acid.

It may be that metabolites of α-oxidation are necessary precursors of plant materials. Stumpf (1962) suggested that these might include odd-chain fatty acids (which would give rise to propionic acid on β-oxidation) and long-chain aldehydes and alcohols as intermediates in the biogenesis of plant waxes. Aldehydes are unlikely candidates since odd-chain alcohols would presumably result, and these are not widespread. The simple production of CO_2 might be necessary for synthetic reactions if the supply from other sources in the plant were insufficient.

The long-chain D-2-hydroxyacids may be essential for specialized complex lipids such as cerebrosides (p. 54) in plants and animals alike. Their biogenesis evidently occurs by direct α-hydroxylation of the unsubstituted acid, and their degradation involves α-oxidation. Sequential α-oxidation of fatty acids may be a manifestation of the non-specificity of both these systems together, which could control the level of hydroxyacid, though in this case the apparent role of the L-2-hydroxy-acid is confusing. However, it is most likely that the α-hydroxylation reaction is responsible for the accumulation in plants of all unusual fatty acids characterized by a 2-hydroxy group; these include 2-hydroxysterculic and 2-hydroxylinolenic acids as well as the more ubiquitous normal saturated 2-hydroxyacids (p. 168).

Moreover the process of α-oxidation can explain the natural occurrence of some unusual noracids by postulating α-hydroxylation and oxidative decarboxylation of more commonplace acids known to be present. The co-occurrence of linolenate (18:3,9c12c15c), 2-hydroxylinolenate (2h-18:3,9c12c15c; p. 24) and norlinolenate (17:3,8c11c14c; p. 8) in thyme (*Thymus vulgaris*) suggests that α-oxidation intermediates can accumulate. The structures of sterculic, 2-hydroxysterculic and malvalic acids also imply an interrelationship involving α-oxidation (p. 30). The biogenesis of saturated odd-numbered fatty acids such as margarate could involve α-oxidation of even-numbered analogues, but such acids can also be synthesized *de novo*, if for instance propionate takes the place of acetate as a "primer" for fatty acid synthetase (p. 108).

It seems likely that α-oxidation in leaves is an obligatory step in the degradation of phytanic acid, as has been demonstrated in animals (p. 219). The turnover of the phytol moiety of chlorophyll would then require the enzymes of α-oxidation; if these enzymes were non-specific, the observed reactions in leaves of added unbranched fatty acids need have no special significance.

3. OMEGA-OXIDATION OF FATTY ACIDS

Oxidation of the fatty acid chain at the α- or β-positions proceed by hydroxylation or dehydrogenation-hydration respectively, facilitated by the presence of the carbonyl group; however hydroxylation can take place at positions further removed. The presence of γ- and δ-lactones in mammalian systems may be explained by a δ-oxidation pathway of fatty acids, postulated by Dimick *et al.* (1969). Their results suggest that in the mammary gland, lauroyl-CoA is degraded to octanoyl-CoA via 4-dodecenoate, 5-hydroxylaurate and 5-ketolaurate. The unsaturated intermediate may also yield 4-hydroxylaurate, and the hydroxyacids give rise to γ- and δ-lactones in the milk. A number of γ- and δ-lactones are also found in apricots, peaches and pineapples (Tang and Jennings, 1968). These could be formed by a catabolic pathway, such as δ-oxidation, or by β-oxidation of pre-formed long-chain hydroxyacids: for instance, 10-hydroxypalmitate is degraded in plants via 4-hydroxydecanoate (Yamada and Stumpf, 1965b). Another reasonable suggestion would involve the biosynthesis of a 4-enoate *de novo*, involving a mechanism similar to that of the bacterial "anaerobic" pathway (p. 131), by the action of fatty acid synthetase during which one reduction step is omitted; elongation of (say) the intermediate 2-octenoate might not be able to be continued beyond the 4-decenoate stage for steric reasons. This acid could give rise to both γ- and δ-decalactone in fruits.

The biological oxidation of fatty acids at the methyl end of the molecule was first suggested by the accidental discovery of dioic acid acidosis in man after administration of undecanoate as the triglyceride. Verkade (1938) proposed that some acids were oxidized at the ω-position and the dibasic acids produced were then subject to β-oxidation at both ends. Wakabayashi and Shimazono (1963) noted that straight, branched, substituted and unsaturated fatty acids were all subject to ω-oxidation *in vivo*, and demonstrated that cell-free guinea-pig liver preparations contained a hydroxylase that required NADPH and oxygen as cofactors. Preiss and Bloch (1964) observed that rat liver preparations converted stearate to octadecane-1,18-dioic acid, 18-hydroxystearic acid and 17-hydroxystearic acid. These, and analogous products from other substrates, were accidentally encountered during an investigation into desaturation; there was great similarity between the cofactor requirements of ω-hydroxylase and desaturase (p. 134). The ω-hydroxylation of medium-chain fatty acids has been detected in microsomes from livers, kidneys and lungs of mammals, birds and fishes (Ichihara et al., 1969). The participation of cytochrome P-450 was inferred from the inhibitory properties of carbon monoxide (Wada et al., 1968). Lu et al. (1969) have resolved a liver microsomal system into the ω-hydroxylase and three soluble fractions containing cytochrome P-450, a cytochrome P-450 reductase and a heat-stable factor: the latter seemed to be an activating lipid. The system has broad substrate specificity and amounts to a defence mechanism for the safe removal of toxic materials and drugs, since it will act upon fatty acids, hydrocarbons, aromatic compounds and steroids (Das et al., 1968).

Bacteria of the type which grow on hydrocarbons as sole carbon sources might also be expected to oxidize the hydrocarbon end of fatty acid nutrients. Kusunose et al. (1964a,b) indeed demonstrated that when incubated with a cell-free preparation of Pseudomonas oleovorans, octanoate, decanoate, laurate and myristate gave the corresponding ω-hydroxyacid and dicarboxylic acid; palmitate and stearate were similarly oxidized at a slower rate. Hexane, octane and decane were also oxidized, and the bacterial hydroxylation system was composed of three soluble proteins: the ω-hydroxylase; a rubredoxin containing non-heme iron but no labile sulphide; and a NADH-rubredoxin reductase (Peterson et al., 1967). The rubredoxin has been purified (Peterson and Coon, 1968); together with the reductase it acts as an electron carrier during substrate hydroxylation by the ω-hydroxylase. The bacterial enzyme contains no cytochrome P-450, is unaffected by carbon monoxide but is inhibited by cyanide; in these respects it differs from the mammalian system (McKenna and Coon, 1970). It seems reasonable to suppose that a similar system is present in those plants in which ω-hydroxyacids or dicarboxylic acids accumulate (p. 25). The oxidation of hexadecane to palmitate in the leaves of higher plants demonstrates that such oxidations do in fact occur there (Kolattukudy, 1969).

In the absence of experimental data relating to higher plant tissues, the mechanism of fatty acid hydroxylation in yeasts may prove to be relevant. Those of the genus Torulopsis may produce L-(ω −1)-hydroxy-acids and ω-hydroxyacids as extracellular glycosides (p. 25); the ratio of these depends on chain length, the hydroxylation taking place most efficiently when the site of hydroxylation and the terminal carboxyl group are separated by fourteen methylene groups (Jones and Howe, 1968). It was therefore suggested that hydroxylations were catalysed by a single enzyme which inserted hydroxyl (derived from molecular oxygen) by stereospecific displacement of hydrogen attached to either a terminal or a penultimate carbon atom at a preferred distance from a carboxyl group. Evidence in favour of this hypothesis was reported by Jones (1968) who observed the oxidation of [L-17-^{3}H$_1$]stearate, [D-17-^{3}H$_1$]stearate and [DL-17-^{3}H$_1$]stearate to L-17-hydroxystearate and 1,18-octadecandioate. He also concluded that the (ω −1)-hydroxylation of stearate took place without double bond formation and with retention of configuration. Heinz *et al.* (1969) investigated the oxidation of [18-^{2}H$_3$]stearate, [16,18-^{2}H$_5$]stearate, [17-^{2}H$_2$]stearate, [D-17-^{2}H$_1$]stearate, and [L-17-^{2}H$_1$]-stearate to L-17-hydroxystearate; their results also demonstrated that hydroxylation took place with overall retention of configuration, and that unsaturated intermediates were most unlikely. Moreover, incubations of whole cells with 17[^{18}O]-hydroxyoleate and with oleate in the presence of ^{18}O$_2$ or H$_2$^{18}O showed that the oxygen atom of the 17-hydroxyoleate is derived from oxygen and not from water, and that it is not lost on formation of extracellular glycoside. The reaction is apparently due to a mixed-function oxidase, since a cell-free system required oxygen and NADPH, and was strongly inhibited by carbon monoxide, but not by cyanide or azide (Heinz *et al.*, 1970).

4. OXIDATION BY LIPOXYGENASE

A. Autoxidation

Unsaturated lipids can react with atmospheric oxygen without the aid of enzymes: the rate of this autoxidation depends on the number of suitable ethylenic bonds present and is accompanied by degradation and polymerization, with the eventual formation of a complex mixture of oxidized products. Edible oils therefore develop an off-flavour in air, while drying oils autoxidize to a hard polymer. The mechanism of the primary process is the formation of a hydroperoxide by a free radical chain reaction (Fig. 8.8). Such autoxidations are preceded by an induction period associated with the ill-defined initiation reaction; they are catalysed by light and free-radical sources. They are inhibited by "anti-oxidants" such as *tert*-butylhydroxytoluene (BHT) and gallic esters: these react preferentially with any free radicals produced, thus protecting

Initiation: RH → R·

Propagation: R· + O$_2$ → RO$_2^·$
               RO$_2^·$ + RH → RO$_2$H + R·

Termination: R· + R·   → ⎫
               R· + RO$_2^·$ → ⎬ More stable products
               RO$_2^·$ + RO$_2^·$ → ⎭

FIG. 8.8. Mechanism of autoxidation by free-radical chain-reaction.

the unsaturated lipid by breaking the chain reaction. The sequence normally terminates when two free radicals interact to give a more stable product.

Saturated fatty esters are very slowly autoxidized at high temperatures with the production of a mixture of all possible monohydroperoxides; preferential oxidation towards the centre of the molecule was observed (Brodnitz *et al.*, 1968a,b). The autoxidation of monoenoates is a slow process which occurs only after a long induction period; however, mesomeric radicals are evidently involved (Fig. 8.9) since four hydro-

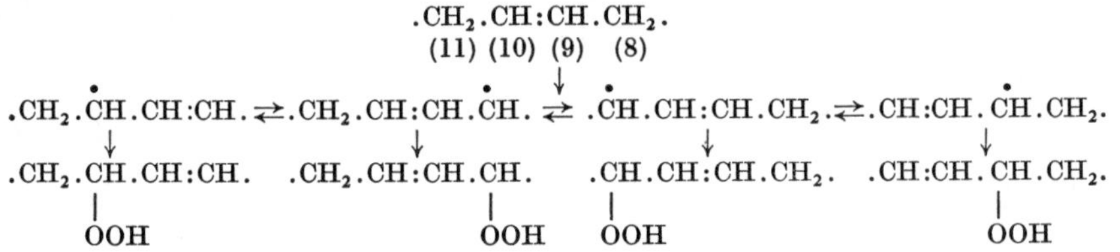

$$.CH_2.CH:CH.CH_2.$$
$$(11)\ (10)\ (9)\ \ (8)$$

$$.CH_2.\overset{\bullet}{C}H.CH:CH.\rightleftharpoons.CH_2.CH:CH.\overset{\bullet}{C}H.\rightleftharpoons.\overset{\bullet}{C}H.CH:CH.CH_2.\rightleftharpoons.CH:CH.\overset{\bullet}{C}H.CH_2.$$

$$.CH_2.\overset{|}{C}H.CH:CH. \qquad .CH_2.CH:CH.\overset{|}{C}H. \qquad .CH.CH:CH.CH_2. \qquad .CH:CH.\overset{|}{C}H.CH_2.$$
$$\overset{|}{O}OH \qquad\qquad \overset{|}{O}OH \quad \overset{|}{O}OH \qquad\qquad\qquad \overset{|}{O}OH$$

FIG. 8.9. Autoxidation of methyl oleate.

peroxides are eventually formed at about equal speeds from oleate (Table 8.2). These are substituted at 8-, 9-, 10- or 11-positions (Ross *et al.*, 1949) and are accompanied by some secondary products.

A methylene-interrupted dienoate is more liable to autoxidation since this methylene group is doubly activated; thus hydroperoxides are similarly formed from methyl linoleate about twenty times more quickly than from oleate after a shorter induction period. The initial product is a mixture of approximately equal amounts of only two conjugated diene hydroperoxides substituted at the 9- or 13-positions; no 11-isomer is detectable (Table 8.2; Fig. 8.10).

Autoxidation at low temperatures leads to the production of mainly *cis,trans*-diene hydroperoxides; at ordinary temperatures thermal rearrangement of these products yields *trans,trans*-analogues in addition (Sephton and Sutton, 1956).

Methyl linolenate is subject to faster autoxidation and requires less induction; secondary reactions also occur quickly. The primary products

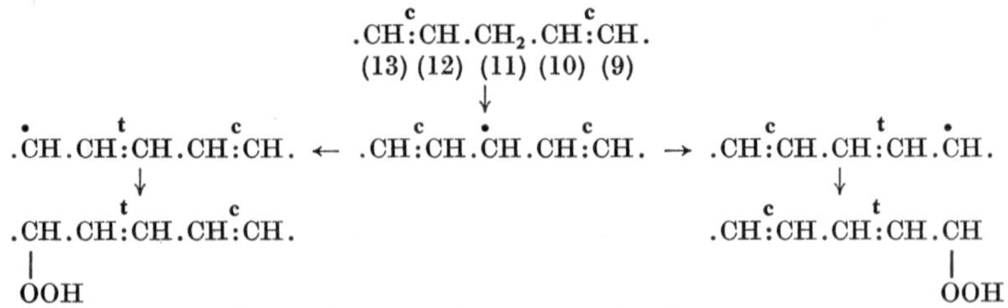

$$.\overset{c}{C}H:CH.CH_2.\overset{c}{C}H:CH.$$
$$(13)\ (12)\ (11)\ (10)\ (9)$$

$$.\overset{\bullet}{C}H.\overset{t}{C}H:CH.\overset{c}{C}H:CH.\leftarrow.\overset{c}{C}H:CH.\overset{\bullet}{C}H.\overset{c}{C}H:CH.\rightarrow.\overset{c}{C}H:CH.\overset{t}{C}H:CH.\overset{\bullet}{C}H.$$

$$.CH.\overset{t}{C}H:CH.\overset{c}{C}H:CH. \qquad\qquad\qquad .\overset{c}{C}H:CH.\overset{t}{C}H:CH.CH$$
$$\overset{|}{O}OH \qquad\qquad\qquad\qquad\qquad\qquad \overset{|}{O}OH$$

FIG. 8.10. Autoxidation of methyl linoleate.

TABLE 8.2

Hydroperoxides formed by oxidation of unsaturated fatty esters

Substrate	Product; autoxidation[a]	Product; photo-sensitized oxidation[b]	Product; lipoxygenase[c]
18:1(9c)	8p-18:1(9e)	9p-18:1(10e)	
	9p-18:1(10e)	10p-18:1(8e)	
	10p-18:1(8e)		
	11p-18:1(9e)		
18:2(9c12c)	9p-18:2(10t12c)	9p-18:2(10e12e)	9p-18:2(10t12c)
		10p-18:2(8e12e)	
		12p-18:2(9e13e)	
	13p-18:2(9c11t)	13p-18:2(9e11e)	13p-18:2(9c11t)
18:3(9c12c15c)	9p-18:3(10t12c15c)	9p-18:3(10e12e15e)	
		10p-18:3(8e12e15e)	
	12p-18:3(9c13t15c)	12p-18:3(9e13e15e)	
	13p-18:3(9c11t15c)	13p-18:3(9e11e15e)	13p-18:3(9c11t15c)
		15p-18:3(9e12e16e)	
	16p-18:3(9c12c14t)	16p-18:3(9e12e14e)	

[a] Ross et al. (1949); Sephton and Sutton (1956); Frankel et al. (1961).
[b] Cobern et al. (1966).
[c] Hamberg and Samuelsson (1967a).

8

are four triene hydroperoxides with *cis,trans*-conjugated diene structures resulting from initial attack at either of the doubly activated methylene groups (Table 8.2). Substitution by oxygen occurs at the 9-, 12-, 13- and 16-positions (Frankel *et al.*, 1961). The more highly unsaturated lipids are subject to faster oxidation in air, presumably by similar mechanisms. The relative stability appears to decrease as a linear function of the number of double bonds present per molecule, but is significantly increased by chromatography on silica or silver nitrate-impregnated silica (Slawson and Stein, 1970). Such autoxidation can lead eventually to a complex mixture of products including monohydroxy-, poly-hydroxy-, epoxy-, hydroperoxy, peroxy- and oxo-derivatives, shorter-chain acids and carbonyl compounds, and polymeric material.

When suspensions of lipids in water are oxidized in air, some products are extracted into the aqueous phase and consequently are not available for further reactions in the oil phase. Products may therefore appear which do not accumulate in the absence of water. From linoleate, Schauenstein (1967) has identified *trans*-2-octanal, *n*-amyl hydroperoxide, 8-hydroperoxyoctanoate, 4-hydroperoxy-2-nonenal, 1-hydroxy-heptan-2-one, 4-hydroxy-*trans*-2-octenal in the water phase.

Light enhances normal autoxidation, but when chlorophyll is also present a sensitized photo-oxidation occurs at a much faster rate. While similar products are observed, the composition of the peroxide mixtures initially formed is different. Cobern *et al.* (1966) characterized the hydroperoxides obtained from oleate, linoleate, linolenate and arachidonate by non-enzymic reaction with oxygen at room temperature in the presence of light and chlorophyll; in each case they identified exactly two products per double bond (Table 8.2). The mechanism of photosensitized autoxidation must therefore be different from normal autoxidation: apparently oxidative attack occurs at all double-bonded carbon atoms with inevitable migration of the double bond.

Lipids already containing conjugated polyene systems are also subject to autoxidation, though here different pathways must operate. Methyl eleostearate (18:3,9c11t13t) does not give hydroperoxides, but rather polymers and oxygenated compounds including 1,4-peroxides together with some 1,2- and 1,6-peroxides (Faulkner, 1958).

B. Lipoxygenase

1. Lipoxygenase-catalysed oxidation

The primary reaction of autoxidation can be catalysed by the enzyme lipoxygenase (linoleate: oxygen oxidoreductase; EC 1.13.1.13; also called lipoxidase). Its activity has been observed in many higher plants, especially legume seeds, cereal grains and oil-seeds; principal sources include soybeans, urd beans, lentils, green peas, peanuts, wheat and sunflower seeds (Tappel, 1963). Leaves are also active (Holden, 1970). The enzyme catalyses the aerobic oxidation of unsaturated fatty acids with a cis-1, cis-4-pentadiene system to conjugated cis,trans-diene hydroperoxides; the enzymic reaction therefore strongly resembles non-enzymic autoxidation discussed earlier. However, the enzymic hydroperoxidation is more specific in substrate requirements than its chemical counterpart (Table 8.3), though both may be followed by

TABLE 8.3

Specificity of lipoxygenase[a]

Substrate	Substrate double bond position[b]	Product (if any) hydroperoxyl group position[b]
1. 18:2(9c12c)	ω6,9	ω6[c]
2. 18:3(9c12c15c)	ω3,6,9	ω6
3. 18:3(6c9c12c)	ω6,9,12	ω6
4. 20:3(8c11c14c)	ω6,9,12	ω6
5. 20:4(5c8c11c14c)	ω6,9,12,15	ω6
6. 20:5(5c8c11c14c17c)	ω3,6,9,12,15	ω6
7. 20:2(8c14c)	ω6,12	none
8. 20:3(5c8c11c)	ω9,12,15	none
9. 20:3(8c11c14c)[d]	ω6,9,12	none
10. 21:3(9c12c15c)	ω6,9,12	ω6
11. 22:3(10c13c16c)	ω6,9,12	ω6
12. 22:3(8c11c14c)	ω8,11,14	none
13. 22:6(4c7c10c13c16c19c)	ω3,6,9,12,15,18	ω6
14. 23:3(9c12c15c)	ω8,11,14	none
15. 18:2(9c12t)	ω6,9	none
16. 18:2(9t12t)	ω6,9	none

[a] 1–14: Hamberg and Samuelsson (1967). 15–16: Privett et al. (1955).
[b] Position counted from the terminal methyl group (ω1-position).
[c] Also a small proportion at ω10.
[d] With a methyl substituent at C-15, i.e. the ω6-position.

complex secondary reactions. Linoleic acid (18:2,9c12c) is a characteristic substrate for lipoxygenase, and the initial products are 13-hydroperoxy-cis-9,trans-11-octadecadienoate and 9-hydroperoxy-trans-10,cis-12-octadecadienoate (Table 8.2; Fig. 8.11).

The trans-analogues of linoleate (18:2,9c12t; 18:2,9t12t) are not acceptable (Privett et al., 1955).

Other lipoxygenases exist which catalyse the same reaction with trilinolein (Koch, 1968) or methyl linoleate (Guss *et al.*, 1968); however the enzyme active against free acids (salts), and known simply as "lipoxygenase", has been studied most extensively.

The oxidation of linoleic acid catalysed by lipoxygenase can cause the non-specific co-oxidation of other compounds present, including carotenoids (Tookey *et al.*, 1958) and digalactosyl diglycerides (Guss *et al.*, 1968). In the latter case, the polar lipid was not oxidized in the enzyme preparation unless linoleic acid in catalytic quantities was also added; oxygen uptake then exceeded that required for the linoleate alone. Since this cooxidation occurred in soybean but not in active wheat extracts,

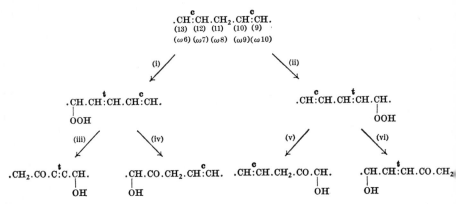

Fig. 8.11. Oxidation of linoleic acid by lipoxygenase. After Gardner (1970). Reaction (i) is associated particularly with soybean lipoxygenase; reaction (ii) with corn lipoxygenase. Linoleate hydroperoxide isomerase from corn catalyses reactions (iii) to (vi); reactions (iv) and (v) lead to the major products. Flax isomerase catalyses reaction (iv).

it cannot be due to autoxidation of galactolipid catalysed by linoleate hydroperoxide. Linoleate could not be replaced by oleate, whose physical effects on the system could be comparable. The co-oxidation might be due to the linoleate acting as a co-enzyme in the mixed system, rather than as a substrate. The co-oxidation of carotene appears to be due to a second heat-sensitive enzyme (carotene oxidase) rather than to lipoxygenase itself, which is relatively heat-stable (Kies *et al.*, 1969).

Fractionation by electrophoresis demonstrated that lipoxygenase exists in several multimolecular forms (Hale *et al.*, 1969) which are difficult to separate. The existence of these isoenzymes may explain differences in oxidation of various substrates by a crude preparation, or indeed of the same substrate by different purified preparations (Guss *et al.*, 1968). One isoenzyme, lipoxygenase-2, distinct from the well-established lipoxygenase-1, has been isolated from soybeans by Christopher *et al.* (1970).

Lipoxygenase is inhibited by the anti-oxidants which are active against lipid autoxidation, including α-tocopherol, nordihydroguaiaretic acid, propyl gallate and hydroquinone. They are not inhibited by reagents which complex with metals or combine with prosthetic groups of other oxidative enzymes, such as cyanide, azide and fluoride. Lipoxygenases from urd bean and mung bean were sensitive to thiol reagents, which had no effect on the enzyme from other sources (Siddiqi and Tappel, 1957); evidently some lipoxygenases contain an essential thiol group. Indeed, of the 637 amino-acid residues of pea lipoxygenase, seven corresponded to cysteine, which is absent from soybean lipoxygenase (Eriksson and Svensson, 1970).

Hamberg and Samuelsson (1967a) tested some fourteen unsaturated fatty acids as substrates for crystalline soybean lipoxygenase, and concluded that there was a requirement for a *cis*-ω6, *cis*-ω9-diene grouping, i.e. for a doubly-activated methylene group at the ω8 position from which hydrogen is lost. In these cases, oxygen was introduced at the ω6 position (at C-13); linoleate was oxidized primarily to 13-hydroperoxy-*cis*-9,*trans*-11-octadecadienoate, though a small proportion of product was oxygenated at the ω10-position, i.e. at C-9 (Table 8.3). Dolev *et al.* (1967a) have however reported that their crystalline lipoxygenase is completely specific for oxidation at the ω6-position of linoleic acid, though crystalline preparations from the same supplier catalyse attack at both the ω6- and ω10-carbon atoms. Flaxseed lipoxygenase also yields mainly ω6-hydroperoxides from linoleate or linolenate; substitution at the ω10-position (20% and 12% respectively) also occurs, and this could be due to autoxidation (Zimmerman and Vick, 1970). However, enzymic ω10-hydroperoxylation is more likely (Veldink *et al.*, 1970a) in which case even the crystalline enzyme is a mixture of two or more lipoxygenases, in addition to peroxidases which are often present. On the other hand, lipoxygenase from corn germ oxidizes linoleic acid primarily at the ω10-position, i.e. to D-9-hydroperoxy-*trans*-10,*cis*-12-octadecadienoic acid (Gardner and Weisleder, 1970).

Holman *et al.* (1969) have tested all the methylene-interrupted *cis*,*cis*-isomers of linoleic acid as substrates for purified soybean lipoxygenase. They report that the natural 9,12-isomer (ω6) was best; 50% of this activity was exhibited by the 13,16-isomer (ω2), while other isomers showed less activity (<25%) particularly if the double bond was near the carboxyl group. In the presence of calcium ions, all these activities were enhanced; the same general pattern emerged, though the specificity was somewhat broader. When nineteen unsaturated acids were offered to crude lipoxygenase activated by calcium, a pattern emerged (Table 8.4) which suggested that the rate of oxidation was not correlated with double bond positions with respect to the carboxyl group; the presence of double bonds at the $\omega 6$ (and $\omega 9$) positions usually resulted in efficient oxidation. Crepenynic acid (18:2, 9c 12a) was a poor substrate, implying that the enzyme does not accommodate acetylenic analogues, while 2-methylarachidonate was not oxidized at all.

TABLE 8.4

Rates of oxidation of polyunsaturated fatty acids by lipoxygenase[a]

Substrate	Substrate double bond position[b]	Relative oxidation rate[c]
1. 18:2(9c 12c)	$\omega 6,9$	100
2. 22:6(4c 7c 10c 13c 16c 19c)	$\omega 3,6,9,12,15,18$	93·6
3. 20:5(5c 8c 11c 14c 17c)	$\omega 3,6,9,12,15$	91·4
4. 20:2(11c 14c)	$\omega 6$	86·9
5. 19:3(10c 13c 16c)	$\omega 3,6,9$	83·3
6. 18:3(9c 12c 15c)	$\omega 3,6,9$	79·0
7. 20:4(5c 8c 11c 14c)	$\omega 6,9,12,15$	76·8
8. 20:3(11c 14c 17c)	$\omega 3,6,9$	55·7
9. 21:4(6c 9c 12c 15c)	$\omega 6,9,12,15$	44·6
10. 18:3(6c 9c 12c)	$\omega 6,9,12$	39·3
11. 18:3(5c 8c 11c)	$\omega 7,10,13$	39·1
12. 18:2(9c 15c)	$\omega 3,9$	32·6
13. 18:4(6c 9c 12c 15c)	$\omega 3,6,9,12$	27·6
14. 17:3(5c 8c 11c)	$\omega 6,9,12$	25·6
15. 17:2(9c 12c)	$\omega 5,8$	19·3
16. 16:3(6c 9c 12c)	$\omega 4,7,10$	9·7
17. 19:2(10c 13c)	$\omega 6,9$	3·5
18. 18:2(9c 12a)[d]	$\omega 6a,9$	3·8
19. 20:4(5c 8c 11c 14c)[e]	$\omega 6,9,12,15$	0·0

[a] Holman et al. (1969).
[b] Position counted from the terminal methyl group ($\omega 1$-position).
[c] The oxidation of each substrate by crude soybean lipoxygenase in the presence of Ca was measured by increase in absorption at 234 mμ; the initial rates of oxidation were expressed as a percentage of those of linoleate standards.
[d] Crepenynic acid.
[e] With a methyl substituent in the 2-position (i.e. 2-methylarachidonic acid).

Lipoxygenase-catalysed hydroperoxylation of linoleate is stereospecific. The product is optically active (Privett et al., 1955); optical rotatory dispersion spectroscopy showed that the 13-hydroperoxyacid belonged to the L-series, since L-13-hydroxystearate and L-2-hydroxyheptanoate were derived from it (Hamberg and Samuelsson, 1967a).

Similarly, the other product of hydroperoxylation has been identified as D-9-hydroperoxyoctadecadienoic acid (Veldink et al., 1970a; Gardner and Weisleder, 1970). Both products are therefore indeed formed enzymically and not by non-enzymic autoxidation; moreover, the different configurations of the products would be expected if the substrate is oriented as a flat structure on the enzyme and is attacked by oxygen from one direction at the $\omega 6$ or $\omega 10$-position. The oxygen so incorporated comes from the gas phase and not the aqueous phase (Dolev et al., 1967b).

Using cis-8, cis-11, cis-14-eicosatrienoic acids stereospecifically labelled with tritium at the 13-position, Hamberg and Samuelsson (1967a) demonstrated that only hydrogen of the L-configuration is removed from the 13-position during enzymic conversion to L-15-hydroperoxy-cis-8,cis-11,trans-13-eicosatrienoic acid. Similarly, they showed that the single hydrogen atom originally at the 15-position is retained.

2. Mechanism of oxidation by lipoxygenase

The detailed mechanism of lipoxygenase catalysis is obscure. Free radicals can be detected during the reaction; the co-oxidation of other substrates and inhibition by antioxidants confirm the existence of free-radical intermediates (Tappel, 1963). If the mechanism is analogous to that proposed for non-enzymic autoxidation (p. 223), then differences must be postulated to account for the specific nature of the enzymic pathway. The enzyme contains no metal, and no cofactor or prosthetic group appears to be involved. Its low activity with acids whose carboxyl groups are hindered by methyl substitution or unsaturation nearby suggests enzyme binding through this group; this may be via a bound calcium atom, since calcium enhances enzymic activity (Holman et al., 1969). Evidence in favour of this concept is provided by Koch (1968) who observed that lipoxygenases which act on trilinolein are not stimulated by calcium; here binding through calcium is unlikely, but presumably some other interaction occurs. Enzyme–substrate complexes must also be stabilized by lipophilic interaction, which would have the effect of extending the carbon chain of the substrate on the enzyme surface. Indirect confirmation of lipophilic substrate–enzyme interaction is provided by an investigation into the inhibition of lipoxygenase by saturated monohydric alcohols (Mitsuda et al., 1967). This is reversible, and increases with increasing chain length; it has partially competitive and purely non-competitive mixed character, suggesting that non-specific binding of the alcohol to lipophilic regions of the enzyme near the catalytic site discourages formation of the specific enzyme–substrate complex and also prevents breakdown of the complex once formed. The enzyme surface could indeed have islands or clefts of lipophilic character, since of about 881 amino-acid residues in the protein, 307 are non-polar (Holman *et al.*, 1969). The surface must also have some geometric characteristics which ensure specific orientation of the substrate, and so explain the specific nature of the enzymic reaction. Similar concepts are invoked in the consideration of enzymic desaturation of fatty acids (p. 146).

   The mechanism of hydroperoxylation may involve the creation of the enzyme-bound free radical (Fig. 8.12) by the ejection of hydrogen from

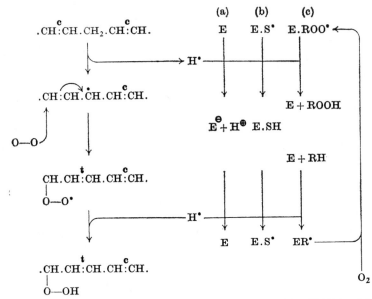

FIG. 8.12. Possible mechanism of oxidation by lipoxygenase. After Siddiqi and Tappel (1957). The possible fates of the hydrogen atom involve: (a) Enzyme (E) acting as electron sink; (b) Enzyme (ES*) acting as free radical; (c) Enzyme-bound substrate (E.RH) participating in a chain reaction.

the activated methylene group of the enzyme–substrate complex; this could be loss of a hydrogen ion while the enzyme momentarily holds the electron (Fig. 8.12a) or loss of a hydrogen atom by the participation of the enzyme as a free radical (Fig. 8.12b), for instance at a thiol group (Siddiqi and Tappel, 1957). The isotope effect observed by Hamberg and Samuelsson (1967a) during the loss of hydrogen from the 8-position of labelled eicosatrienoic acid implies that this occurs before the establishment of a carbon–oxygen bond. An allylic shift of one electron is now possible with the formation of a conjugated system, followed by (or simultaneously with) attack of enzyme-bound molecular oxygen at the radical centre. Finally the peroxy radical is stabilized. If the enzyme is acting as an electron sink, the electron temporarily held is now replaced at the radical centre; provision of a hydrogen ion completes the product (Fig. 8.12a). If a hydrogen atom is being stored during the reaction, this is now required (Fig. 8.12b). Alternatively the peroxy radical may remove hydrogen from the activated methylene group of another substrate molecule nearby: this would represent a lipoxygenase-regulated chain reaction (Fig. 8.12c).

This type of mechanism does not explain why, when two mesomeric radical centres are possible (at $\omega6$ and $\omega10$), only one product is some-

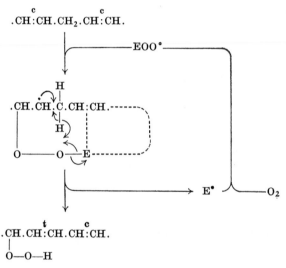

FIG. 8.13. Possible mechanism of oxidation by lipoxygenase. After Dolev *et al.* (1967a). In this model, the free radical adds to the substrate rather than abstracting hydrogen from it.

times formed. It may be assumed that the enzyme-bound oxygen is positioned to react at only one radical centre; otherwise a different mechanism proposed by Dolev *et al.* (1967a) must be considered. Here the enzyme E is activated by oxygen to form a free radical $EOO^{\bullet}$ which reacts specifically with the substrate at the $\omega6$-position rather than abstracting hydrogen from it (Fig. 8.13). The intermediate free radical rearranges to give the product directly with the liberation of the enzyme, perhaps as a free radical $E^{\bullet}$ ready for more oxygen to regenerate the activated form $EOO^{\bullet}$. Most lipoxygenases contain no prosthetic group, and an understanding of the detailed mechanism of oxygen binding or hydrogen transfer awaits further research.

3. Function of lipoxygenase

Lipoxygenase itself is not present in animal tissues or micro-organisms, and its function in plants is not clear. Enzymic oxygenation of unsaturated fatty acids occurs during prostaglandin biosynthesis in animals (Hamberg and Samuelsson, 1967b) but the peroxides formed by lipoxygenase in plants cannot be required for a similar metabolic role. They can, however, be isomerized by an enzyme found in flax seed (Zimmerman, 1966; Veldink et al., 1968, 1970c). Of the two hydroperoxides obtained from linoleate, only the 13-hydroperoxy-cis-9,trans-11-octadecadienoate was an acceptable substrate for the flax isomerase; the product was 13-hydroxy-12-keto-cis-9-octadecenoate (Fig. 8.11). In this α-ketol, only the oxygen at C-12 originated at C-13 of the hydroperoxide; the conversion seems to involve a cyclic intermediate with cleavage of the carbon–oxygen bond at C-13 (Veldink et al., 1970b). The product has the ketol structure of ascorbic acid, and could function in an electron transport system. The isomerase from corn germ evidently accepts both the linoleate hydroperoxides (Gardner, 1970). Its natural substrate is 9-hydroperoxy-trans-10,cis-12-octadecadienoate, which is converted to 9-hydroxy-10-keto-cis-12-octadecenoate; some 13-hydroxy-10-keto-trans-11-octadecenoate is also formed. When soybean lipoxygenase is used to generate 13-hydroperoxy-cis-9,trans-11-octadecadienoate, this is converted by the corn isomerase to 13-hydroxy-12-keto-cis-9-octadecenoate and 9-hydroxy-12-keto-trans-10-octadecenoate (Fig. 8.11).

The hydroperoxides also decompose enzymically (Gini and Koch, 1961) or spontaneously; secondary products include 9-keto-11,13- and 13-keto-9,11-octadecadienoates (Vioque and Holman, 1962). If the hydroperoxide is formed by lipoxygenase in vivo, one would expect it to be rapidly dissipated in a controlled manner, otherwise chain reactions could develop during which considerable damage to amino-acid residues of nearby proteins might occur, resulting in enzyme inactivation and membrane disruption (Roubal and Tappel, 1966). Hydroperoxide isomerases may function to provide protection from hydroperoxide accumulation (Veldink et al., 1970c).

The linoleate hydroperoxides are the exact peroxy-analogues of the hydroxyacids coriolic acid (13h-18:2,9c11t) and α-dimorphecolic acid (9h-18:2,10t12c). This suggests that lipoxygenase or a similar enzyme in certain seeds may be involved in the biosynthesis of conjugated hydroxydienoic and of conjugated trienoic acids; such a possibility is considered on p. 161. Lipoxygenase may have a function in catalysing the co-oxidation of polyphenols, carotenoids and chlorophyll (Tappel,

1963), of digalactosyl diglycerides (Guss *et al.*, 1968) or of glutathione (Mapson and Moustafa, 1955), or in providing oxidized polymerized lipid for plant cutin (p. 191). Since lipoxygenase activity is only evident when a plant is cut or damaged, it may be responsible for the breakdown of lipids to physiologically active compounds such as traumatic acid (2-dodecen-1,12-dioic acid), which has been claimed to induce cell division after the wounding of plant tissues (p. 276).

CHAPTER 9

Lipid and Fatty Acid Metabolism During Organogenesis and Senescence

Changes in the morphological character and metabolic behaviour of plant cells are invariably accompanied, (and in some cases possibly stimulated) by changes in their lipid content and metabolism. Many such examples have been referred to in preceding chapters but it is appropriate at this stage to consider specifically some of the more important kinds of cellular development in plants in relation to the corresponding changes in lipid metabolism which accompany them.

1. Seed and Fruit Maturation

Maturation or ripening of fruits normally occurs throughout a period of several weeks or months after flowering and is associated with considerable increases in the size and weight of the tissues. In many cases much of this weight increase is due to the accumulation of fat in the endosperm or embryo (germ) of the seed, and in some cases, also in the fleshy fruit pericarp (p. 60). The rate at which fat accumulates is seldom constant throughout the period of maturation and it is not uncommon for a large proportion of the total fat content of a mature seed to accumulate during a comparatively small part of the ripening period. Moreover the component fatty acids which constitute the oil of the mature seed are not usually synthesized at constant rates throughout this period so that at any time before full maturation the oil may have a composition quite different from that of the ripe tissue.

Weber (1969) described a detailed study of the lipids and fatty acids of the maturing maize grain in which individual grains were analysed at various stages between 10 and 90 days after pollination. The least mature grains studied contained 3% total fat, of which polar lipid constituted 70% and only some 10% was triglyceride; at full maturity the fat content of the seed was nearly 14%, of which 92% comprised triglyceride (Table 9.1). In all varieties of grain studied, the rate of lipid synthesis was greatest between 15 and 45 days after pollination, and throughout the maturation period the major class of lipid synthesized was triglyceride with polar lipids comprising a steadily decreasing proportion of the total

TABLE 9.1

Variations in the lipid composition of maize grain (*Zea mays*, var. Illinois high oil) during maturation
(from Weber, 1969)

Days after pollination	100 kernels Wet wt g	100 kernels Dry wt g	Oil % dry wt	% of total lipids by weight Hydrocarbons + sterol esters	Tri-glycerides	Free fatty acids	Sterols	Partial glycerides	Polar lipids
10	8·3	1·1	3·0	4·7	10·1	0·5	6·8	7·1	70·8
15	15·6	2·5	5·6	4·2	41·1	0·5	5·0	4·1	45·1
30	26·7	11·5	10·9	1·6	78·4	1·4	4·3	4·7	9·5
45	31·6	18·0	13·7	2·0	84·0	0·5	3·2	4·3	6·0
60	30·6	19·0	13·8	1·5	88·1	0·4	2·5	3·0	4·5
75	33·4	23·4	13·4	1·3	92·0	0·4	1·5	0·9	3·9
90	31·8	23·8	13·8	1·0	94·0	0·3	1·3	1·1	3·9

grain lipid. At no stage were free fatty acids present at significant levels. Weber also observed that there were large changes in the fatty acid composition of the triglycerides between 10 and 45 days after pollination, the relative proportions of palmitic, linoleic and linolenic acid falling and those of oleic acid increasing with maturation.

Similar studies on seeds of *Crambe abyssinica* and the *Brassica napus* varieties "Golden" and "zero-erucic" have been described by McKillican (1966), who found that maximum rates of oil accumulation occurred between 10 and 20 days after flowering (DAF) in *Crambe* and between 20 and 30 days after flowering in the seeds from the two varieties of *Brassica* (Table 9.2). As in corn grain, the proportion of phospholipids and glycolipids in the total seed fat decreased with increasing maturity of the tissue, but unlike maize grain the oils from the immature seeds of *Crambe* and *Brassica* both contained appreciable levels of free fatty acid. The erucic acid (13-docosenoic acid) which typifies the oils from *Crambe* and "Golden" *Brassica* was present in the oils of the less mature seeds in reduced proportions but its level increased at a rate inversely proportional to the decrease in the level of oleic acid, its natural precursor

TABLE 9.2

Fatty acid composition of triglyceride fractions from seeds of *Crambe abyssinica* and *Brassica napus* (rape) during maturation (McKillican, 1966)

Seed	16:0	16:1	18:0	18:1	18:2	18:3	20:0	20:1	22:1
			Weight % of total fatty acids						
Crambe									
10 DAF[a]	5·3	1·2	2·4	33·9	13·3	7·3	1·1	11·2	24·3
20 DAF	2·7	0·3	1·0	16·2	7·3	5·1	0·3	3·4	61·6
30 DAF	3·4	0·6	1·3	20·6	9·3	6·1	0·8	3·0	55·0
mature	2·8	0·5	0·8	16·3	11·0	8·7	0·6	3·5	54·9
Zero-erucic rape									
10 DAF	13·2	0·8	6·4	9·3	51·6	16·8	0·4	—	—
20 DAF	7·5	0·9	4·3	61·1	19·5	3·6	1·1	1·1	—
30 DAF	5·7	0·4	3·3	56·2	23·5	8·8	0·7	1·3	—
mature	4·3	0·3	2·2	58·3	22·6	10·9	0·6	0·9	—
Golden rape									
10 DAF	8·5	1·3	5·1	31·5	29·3	9·2	1·0	6·8	6·7
20 DAF	5·1	0·5	2·5	25·7	17·8	7·8	1·1	13·1	25·5
30 DAF	4·2	0·5	1·2	19·3	16·2	10·4	—	11·2	36·7
mature	3·4	—	1·4	14·6	14·7	10·3	0·6	10·2	44·8

[a] DAF = days after flowering.

(p. 129). Maturation of seeds from zero-erucic rape is characterized by increases in the proportion of oleic acid which parallel those of erucic and eicosenoic acids in "Golden" rape and *Crambe* (Table 9.2). On a weight per seed basis, phospholipids plus glycolipids reached a maximum at 20 DAF in *Crambe* and 30 DAF in rape, and the fatty acid compositions of these fractions were characterized by greater changes between 10 DAF and maturity than occurred in other lipid classes. The erucic acid content of the complex lipids in mature *Crambe* was similar to that of the corresponding triglyceride fraction, but in "Golden" rape the phospholipids and glycolipids contained only about 10% erucic acid at maturity.

Maturation of castor beans is accompanied by similar changes of lipid composition to these described above, oil formation commencing 21 days after flowering with two-thirds of the total oil accumulating during the next 20 days and the remaining third being formed in the final period of 20 days (Canvin, 1963). In this case, the fatty acid which typifies the seed-oil, ricinoleic acid, was not present in the very young seed but appeared 12 days after flowering and represented 90% of the total oil fatty acids after 36 days. The fatty acid composition of the oil then remained constant for the final 25 days of maturation.

Roehm and Privett (1970) examined in detail the changes in triglyceride composition which occur in soybean during maturation and noted that the major changes in fatty acid composition of the triglycerides occurred during the first 52 days after flowering. During this period linolenic acid decreased from 34% to 12%, the percentages of linoleic and oleic acids increased, stearic remained fairly constant and palmitic acid decreased slightly. Although changes occurred in percentage and amount of each triglyceride species as isolated by argentation-TLC (p. 290) the positional distribution of fatty acids remained virtually unchanged throughout maturation.

Similar but less detailed studies on the lipid composition of maturing oil seeds include those by Huber and Zalik (1963) on flaxseed and by Sims *et al.* (1961) on flaxseed and safflower seed.

The only comparable data for the maturation of fruits possessing oil-rich exocarps are those published by Crombie (1956) and Crombie and Hardman (1958) for the fruit of the West African oil palm (*Elaeis guineensis*), and these papers reveal some interesting interrelationships between the synthesis of kernel and exocarp oils. In her first paper, Crombie described the oil content of the fruit kernel during a period of 8 to 20 weeks after pollination, showing that the palm kernel oil accumulates at a more even rate than is usual with oil seeds, the major part appearing between 12 and 18 weeks after pollination with full fruit maturation taking approximately 20 weeks (Table 9.3). Crombie and

TABLE 9.3

Changes in the lipid content of the fruit of the oil palm during maturation
(from Crombie, 1956 and Crombie and Hardman, 1958)

Weeks after pollination	Wt. of lipid (mg/fruit)			
	Tree 6/140		Tree 6/173	
	Seed	Exocarp	Seed	Exocarp
8	—	4·81	—	4·22
10	32·7	3·74	12·6	—
11	—	4·48	27·4	—
12	—	2·72	53·4	3·64
13	—	—	175·5	—
14	—	4·84	198·4	—
15	175·2	4·14	294·1	—
16	—	2·78	441·0	4·35
19	—	25·74	512·5	76·16
20	341·5	848·0	569·9	1194·97[a]

[a] Lipid was incompletely extracted from this sample.

Hardman (1958) later showed with similar tissues that very little oil accumulates in the fruit exocarp during the first 18 or 19 weeks after pollination, i.e. while most of the kernel oil is accumulating, but is then synthesized very rapidly, a major proportion of the palm oil being formed within a single week (Table 9.3).

The rate of incorporation of labelled metabolites such as [^{14}C] acetate into fatty acids by seed tissues of different states of maturity usually confirms the supposition that those periods during which fat accumulates most quickly are those in which fatty acid and triglyceride synthetase activities are usually greatest. Such studies also show that at seed or fruit maturity the capacity for further synthesis is either very low or non-existent. For example, although ricinoleic acid constitutes 90% of the total fatty acids in mature castor beans the enzymes for the biosynthesis of this acid are active only during a brief period in the development of the seed (Yamada and Stumpf, 1964). Similarly McMahon and Stumpf (1966) confirmed that most of the capacity for saturated and unsaturated fatty acid biosynthesis in developing seeds from Safflower (*Carthamus tinctorius* L.) occurs between 10 and 25 days after flowering.

There have been few descriptions of the synthesis of the different classes of complex lipid during seed maturation although Singh and Privett (1970) have published a short report on the incorporation of ^{33}P into soybean phosphatides. Disodium hydrogen [^{33}P] phosphate or sodium [1-^{14}C] acetate were injected into soybean seeds approximately

30 days after flowering, the beans harvested and extracted at appropriate intervals up to 24 hours after injection and the lipids extracts examined radiochemically. The results (Table 9.4) showed that after 15 minutes more than half the label was present in phosphatidic acid, the proportion of label in this fraction then decreasing sharply during the next three hours, as might be predicted from the established intermediary role of this compound in fat synthesis (p. 176). Phosphatidyl inositol was also rapidly labelled but its proportion of the total lipid-bound ^{33}P remained fairly constant after the first hour. This study also indicated that N-acyl phosphatidyl ethanolamine may not only play an important role in seed germination (p. 244) but is also actively involved in the lipid metabolism associated with seed maturation, since an initially sharp and then a more slow but steady increase in the proportion of label in this lipid during soybean maturation was observed.

TABLE 9.4

Distribution of ^{33}P activity among lipid components of immature soybean after injecting $Na_2H^{33}PO_4$ (Singh and Privett, 1970)

Lipid component	^{33}P activity (% of total counts/min) Time of harvesting the seeds after the administration of the isotope:						
	15 min	30 min	1 h	2 h	4 h	12 h	24 h
X₁	8·9	8·2	2·4	2·4	1·0	1·2	1·2
Phosphatidyl inositol	12·3	18·4	21·0	26·5	26·2	23·3	22·7
Phosphatidyl serine	2·3	2·7	1·4	0·9	0·6	0·3	0·5
Phosphatidic acid	58·2	45·5	33·4	10·9	8·1	5·9	5·3
Phosphatidyl choline	3·0	3·4	5·8	12·6	9·8	14·6	13·6
Phosphatidyl ethanolamine	4·7	5·7	7·4	12·9	18·6	12·1	10·3
Phosphatidyl glycerol	3·1	5·2	6·1	6·8	5·0	6·1	3·9
N-Acyl phosphatidyl ethanolamine	7·0	10·5	22·0	26·2	30·3	35·8	42·4

Singh and Privett also found that the major glycolipid present, digalactosyl diglyceride, did not accumulate a measurable quantity of radioactivity during the 24 hours following incubation with [^{14}C]acetate, despite the fact that all phospholipids became strongly labelled during the same period.

2. SEED GERMINATION AND SEEDLING DEVELOPMENT

One of the most prominent features of germination is the rapid mobilization of reserve materials accompanied by their concurrent utilization. This metabolic activity consists mainly of catabolic processes occurring in reserve organs such as the cotyledons of leguminous plants and of anabolic processes in growing tissues of seedlings, such as hypocotyls and radicles. In oil-bearing seeds metabolism of storage lipid provides the main source of energy for the early cellular development and in general the onset of germination is concomitant with a large increase in lipase activity in the storage tissue (e.g. Ching, 1968; Laidman and Tavener, 1969; Drapron et al., 1969) resulting in the release of free fatty acids which then become available for the catabolic processes by which energy is released.

Many studies of the changes in lipid composition and metabolism which accompany the germination process have involved analysis of the whole, unfractionated tissue. In consequence these results reflect the effects of both the catabolic processes which occur in the seed and also those of the lipid metabolism associated with the development of new organs such as roots, shoots and cotyledon leaves.

Zimmerman and Klosterman (1965) described the changes in the lipid composition of flax seed between germination and 90 hours after the onset of this process. They found (Table 9.5) that during the first 18 hours there was only a slight decrease in the level of triglycerides but that thereafter this fraction was metabolized more rapidly, so that after 90 hours germination the oil content had decreased by 53%. The authors emphasized that the changes described were the net changes in lipid composition in the whole seedling and pointed out the possibility that triglycerides could have been catabolized in one tissue of the seedling at the same time that the same class of lipid was being synthesized in

TABLE 9.5

Weight of total lipid, triglycerides, free fatty acids and phospholipids during 90 h of germination of flax seed (Zimmerman and Klosterman, 1965)

Germi-nation h	Seedling length mm	Dry wt g	Total lipid g	Tri-glycerides g	Phospho-lipids g	Free fatty acids g
0	0	0·5477	0·1900	0·1735	0·0055	0·0003
18	0–0·5	0·5385	0·1934	0·1699	0·0059	0·0003
36	4–5	0·5197	0·1837	0·1589	0·0076	0·0010
54	12–15	0·5214	0·1650	0·1409	0·0078	0·0058
72	25–40	0·5263	0·1287	0·1021	0·0080	0·0095
90	35–48	0·5489	0·1041	0·0815	0·0069	0·0100

another tissue. The constant increase in free fatty acid levels in the germinating tissues from 18 hours onwards suggested that lipase activity was greater than that of the enzymes involved in the oxidative breakdown of the fatty acids, and indicated that the rate hydrolysis of the triglycerides by lipase was not limiting the rate of fatty acid oxidation.

Zimmerman and Klosterman noted that these free fatty acid fractions accumulated increasing numbers of fatty acids of odd carbon number (e.g. 15:0, 17:0, 17:1, 17:3) and saturated fatty acids of comparatively large carbon number (20:0, 22:0, 24:0, 26:0), none of which was present in detectable concentrations at the earliest stages of germination. They ascribed the presence of the odd chain length acids to the α-oxidation (p. 213) of the acid of next higher carbon number (i.e. 16:0, 18:0, 18:1 and 18:3 respectively) while the appearance of the long chain saturated acids was attributed to the development of active fatty acid synthetase systems in the seedling tissues.

The comparatively minor changes which occurred in the fatty acid composition of the triglyceride fraction throughout the germination period studied suggested that no particular class of fatty acid or glyceride was preferentially utilized for the generation of the energy required for the induction of the germination process.

In contrast with the comparatively minor changes which occurred in the composition of the triglycerides, the fatty acid composition of the phospholipid and glycolipid fractions underwent considerable changes during the germination of flax seed. These changes can almost certainly be attributed to the *de novo* synthesis of these complex lipids which invariably accompanies the formation of new plant cells, and because the cells formed during the germination process have different morphology and functions from those in the original seed it is not surprising that their component lipids should have a different fatty acid content. Triglycerides are not synthesized to a significant extent at this stage of plant growth, so that the fatty acid composition of this fraction in the whole seedling tissue would not be expected to alter to the same extent.

Similar studies on germinating seeds of the water melon (*Citrullus vulgaris*) and West African oil palm (*Elaeis guineensis*) were described some years earlier by Crombie and coworkers (Crombie and Comber, 1956; Hardman and Crombie, 1958; Boatman and Crombie, 1958) and in general respects their results were comparable with those already described for rapeseed. Thus, in both cases germination was accompanied by a rapid loss of storage fat in which no specific class of fatty acid or triglyceride was lost preferentially, nor did free fatty acid levels become particularly high. Later stages of the germination process were accompanied by the synthesis of lipids having a highly unsaturated fatty acid composition similar to those commonly associated with the leaf and stem tissues of mature plants.

McMahon and Stumpf (1966) studied germinating safflower seedlings for their capacity to incorporate [^{14}C] acetate into fatty acids, and in confirmation of the conclusions derived from purely analytical studies described above, found that most of the fatty acid synthesizing capacity was confined to the developing cotyledons and progressively increased with time, whereas the activity in the hypocotyl tissue was much lower and decreased with time.

Quarles and Dawson (1969) observed that when pea seeds were germinated the cotyledons of the developing plant lost a considerable part of their phospholipid complement over an 11-day period. In the dark this loss was 60%, while in the light it was 48%, and consideration of the individual phospholipids showed that these losses were not uniform. Phosphatidyl choline and phosphatidyl ethanolamine, the quantitatively major lipids present, decreased by approximately the same degree as the total phospholipid fraction, whereas there was a greater percentage depletion of phosphatidyl inositol than that of total phospholipid; the small quantities of cardiolipin and phosphatidyl serine present in the original seed were not lost from the cotyledon during growth. The concentrations of phosphatidic acid present in the cotyledons showed no evidence of an increase as would be expected if the phosphoglycerides disappearing were broken down by phospholipase D. A further phospholipid, identified in subsequent studies as N-acyl phosphatidyl ethanolamine (Dawson et al., 1969) was present in the dry pea cotyledons to the extent of about 5% of the total phospholipids, but rapidly disappeared when the seeds were hydrated in the initial stages of germination so that within 17 hours only 20% of the original concentration of this lipid remained. The physiological significance of this rapid metabolism is not clear, but Dawson et al. (1969) pointed out that if phosphatidyl ethanolamine is produced by the catabolism of its N-acyl derivative there could be a pronounced change in the physical characteristics of the seed membrane at this point, since the phospholipid would pass from the acidic to the zwitterionic form and at the same time lose a hydrophobic residue.

The data obtained by Quarles and Dawson was based purely upon the estimation of total lipid levels, and absolute rates of phospholipid synthesis were not evaluated. This aspect of seed germination was however studied in germinated seeds and etiolated seedlings of the mung bean (*Phaseolus radiatus*) by Katayama and Funahashi (1969) who traced the incorporation of label from [^{32}P] orthophosphate into the lipids of the seed cotyledon and into the hypocotyls and radicles of the seedlings. The greater part of the incorporated activity was observed in the phosphatidyl choline and phosphatidyl ethanolamine fractions in the cotyledons, and in the phosphatidyl ethanolamine fractions of the hypocotyls and radicles.

3. Chloroplast Development

The changes in lipid composition and metabolism which accompany the development of photosynthetic capacity in plants have been studied in a wide variety of tissues including *Euglena gracilis*, a protist which when cultured in the absence of light contains neither chlorophyll nor structures which one can call chloroplasts (Wolken, 1959; Rosenberg and Pecker, 1964) and has a metabolism similar to that of non-photosynthetic protozoa. When placed in the light in media containing only inorganic salts, *Euglena* rapidly develops active chloroplasts and metabolizes in a manner comparable with that of a green alga.

Dark-grown (etiolated) *Euglena* contains very low levels of sulpholipid (Rosenberg and Pecker, 1964) and galactosyl diglycerides (Rosenberg *et al.*, 1966), the cellular location of which has not been determined although proplastids and "plant-type" mitochondria (p. 65) certainly contain these lipids. Illumination of dark-grown *Euglena* in a purely mineral medium initially results in a lag phase during which there is no detectable formation of chlorophyll or chloroplasts and a sharp drop in the cellular lipid and fatty acid content occurs (Fig. 9.1). This initial loss of fatty material is largely associated with the cellular utilization of the

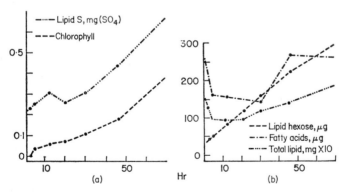

Fig. 9.1. The relative quantities of chlorophyll and sulpholipid (a) and lipid hexose, fatty acid and total lipid (b) in greening cells of *Euglena gracilis* (Rosenberg and Pecker, 1964).

waxy ester fraction which occurs only in dark-grown cultures of this organism (Rosenberg, 1963) and which apparently serves as a store of chemical energy in the etiolated cell. Following this lag phase, detectable chlorophyll synthesis ("greening") occurs once the shorter-chain saturated fatty acids (which are particularly associated with the waxy ester fraction) have dropped to a very low, static level. Sulpholipid accumulates before measurable amounts of chlorophyll, and subsequently the rate of sulpholipid synthesis closely parallels that of chlorophyll; on the basis of these findings Rosenberg and Pecker (1964) suggested that the glycolipid might be necessary for the orientation and possibly the functions of chlorophyll. In contrast, the total galactosyl diglyceride level increased at a linear rate from the onset of illumination (Fig. 9.1a) and these lipids are apparently synthesized at rates independent of the corresponding chlorophyll level.

In a subsequent study, Rosenberg and Gouaux (1967) showed that during greening the monogalactosyl diglyceride content of *Euglena* rose from approximately 2 μmoles in 100 mg of dark-grown cells to 27 μmoles in the same weight of fully green cells, while the corresponding digalactosyl diglyceride level increased from 1 μmole to 11 μmoles. The digalactosyl diglycerides accumulated more rapidly than the monogalacto compounds at first, but this accumulation rate diminished long before greening of the cell was complete. In both classes of lipid there occurred a marked increase in the degree of unsaturation of their component C_{16} and C_{18} fatty acids and the authors suggested that accumulation of galactosyl diglycerides with specific fatty acid compositions may be required in the building of chloroplasts in *Euglena*.

The leaves of dark-grown seedlings of higher plants differ from etiolated *Euglena* in containing much higher levels of sulpholipid and galactosyl diglycerides, which are probably largely associated with the prolamellar bodies (proplastids) characteristic of dark-grown leaves and which are the immediate structural precursors of chloroplast lamellae. Nevertheless, analytical and radiochemical studies by Appelqvist *et al.* (1968a) have shown that the light-stimulated conversion of proplastid to chloroplast during the greening of barley leaves involves the *de novo* synthesis of monogalactosyl diglyceride species containing linolenic acid, while Thomas and Stobart (1970) found that greening of tissue cultures of *Kalenchoe crenata* is accompanied by general glycolipid synthesis, that of the sulpholipid maintaining a close relationship to the synthesis of chlorophyll. Present evidence therefore suggests that changes in the lipid composition of greening leaves are compatible with those obtained for greening *Chlorella* and *Euglena* (Nichols, 1965a; Shibuya and Hase, 1965), namely that during this period sulpholipid synthesis closely parallels that of chlorophyll and there is a more regular accumulation of unsaturated galactosyl diglycerides.

The effect of greening on phospholipid synthesis has been less intensively studied than that of glycolipid synthesis but in their studies on tissue cultures of *Kalenchoe* Thomas and Stobart (1970) showed that phosphatidyl glycerol was the only phospholipid to increase in illuminated dark-grown callus, a result consistent with the fact that only this phospholipid is present in major proportions in chloroplasts (p. 70).

In their work with *Euglena*, Rosenberg and Pecker (1964) showed that greening was accompanied by the initial loss of C_{13} and C_{14} acids which was rapidly balanced by the accumulation of polyunsaturated fatty acids of the C_{16} and C_{18} series. Similar increases in the level of polyunsaturated acids with increasing photosynthetic activity has been noted in numerous algae and higher plants, and the relationships between chloroplast development and fatty acid synthesis have been described for numerous higher plant tissues. For example, Wallace and Newman (1965) studied the effect of light on the lipid composition of bean leaf plastids and showed that, as in algae, chloroplast development was accompanied by the increased synthesis of galactosyl diglycerides and linolenic acid, while Nichols *et al.* (1967) confirmed that the increase in linolenic acid levels during illumination of etiolated castor leaves occurs within the galactosyl diglyceride fractions (Table 9.6). In addition to the increase in levels of polyunsaturated fatty acids during greening,

TABLE 9.6

Light induced changes in the fatty acid composition of castor leaf lipids
(Nichols *et al.*, 1967a)

	% of total fatty acids					
	16:0	16:1(9c)	16:1(3t)	18:1	18:2	18:3
Etiolated						
Monogalactosyl diglyceride	7	3	—	7	14	67
Digalactosyl diglyceride	14	2	—	8	21	51
Phosphatidyl glycerol	38	5	—	11	20	14
Sulphoquinovosyl diglyceride	29	5	—	15	12	27
Phosphatidyl choline	15	2	—	11	37	24
After 20 h illumination						
Monogalactosyl diglyceride	4	—	—	3	4	88
Digalactosyl diglyceride	15	2	—	7	7	65
Phosphatidyl glycerol	41	—	13	12	13	15
Sulphoquinovosyl diglyceride	39	3	—	6	10	33
Phosphatidyl choline	16	2	—	8	41	25

there also appears to be a close parallel between chloroplast development and the accumulation of *trans*-3-hexadecenoic acid. Nichols (1965a) and Nichols *et al.* (1967a) observed that dark-grown algal and leaf tissue contain no detectable quantities of this acid but that during chloroplast formation it becomes a major component of the phosphatidyl glycerol fraction (Table 9.6), although subsequent work showed that the lamellae of blue-green algae do not accumulate this acid (Nichols *et al.*, 1965a).

4. Maturation of Green Tissue

There is some evidence that the lipid composition of healthy leaves can change significantly over periods of several weeks or months, although most studies have been confined to changes in fatty acid composition only. As far as one can generalize on the basis of relatively little data, it appears that there is usually an overall decrease in the degree of unsaturation of leaf fatty acids with increasing maturity of the tissue. Thus Hawke (1963) analysed the fatty acids of "new growth" and mature rye-grass and found that the major acids in the new growth were linolenic (75%), linoleic (8%), and palmitic (12%) whereas the mature grass contained 65% linolenic, 12% linoleic and 16% palmitic acids. Similar results were obtained by Newman (1962) for the leaves from different bush bean nodes and by Klopfenstein and Shigley (1967) for *alfalfa* leaves. The latter authors showed that the decrease in linoleic acid and increase in palmitic acid was particularly reflected in changes in the fatty acid composition of the sulpholipid fraction.

5. Lipid and Fatty Acid Metabolism during "Steady-state" Photosynthesis

Descriptions of experiments with seeds or photosynthetic tissues together account for a very large proportion of the total literature on fatty acid and acyl lipid metabolism in plants. Much of the work relating to the biosynthesis of individual classes of fatty acid and lipids in leaves and algae have already been discussed in Chapters 6 and 7 and this section is mainly concerned with studies aimed primarily at elucidating the inter-relationships which exist between the synthesis and metabolism of individual classes of lipid in photosynthetic tissue.

A. General Metabolism in Leaves and Algae

Studies of general lipid and fatty acid metabolism in leaves and algae have been based most frequently upon the incorporation of label from comparatively unspecific metabolites such as $^{14}CO_2$ and [^{14}C] acetate. These metabolites have the advantage that their carbon atoms may be incorporated into all classes of acyl groups and lipids but have the concomitant disadvantage that they may also be incorporated into other residues in the lipid molecule such as glycerol, sugars and amino-alcohols; consequently the distribution of label between individual lipids can only be interpreted in meaningful terms when the distribution of ^{14}C activity in the different structural moieties of each lipid are also established. For this reason the use of acetate is often preferable to that of $^{14}CO_2$ because under certain conditions the former metabolite has a high if not absolute specificity for incorporation into the more lipophillic type of molecule such as fatty acids, and Yung and Mudd (1966) have defined the optimum conditions for the incorporation of label from [^{14}C] acetate into the lipids of *Chlorella pyrenoidosa*. Nevertheless, results using $^{14}CO_2$ as a source of label can be extremely valuable when the distribution of isotope between the structural elements of individual lipid classes is measured, and by application of this technique to studies on the metabolism of *Chlorella pyrenoidosa*, Ferrari and Benson (1961) established the high metabolic activity of phosphatidyl glycerol and monogalactosyl diglyceride during steady-state photosynthesis, noting that the component fatty acids of these fractions were labelled more rapidly than those of any other lipid. The water-soluble moieties of the galactosyl diglyceride and phosphatidyl glycerol fractions also accumulated label appreciably faster than those in any other class of lipid, the sulpholipid and phosphatidyl inositol fractions becoming labelled only slowly. On the basis of the rapid turnover of label in the sugar component of the galactosyl diglycerides, Ferrari and Benson proposed an active role for these lipids in carbohydrate transport in chloroplasts, but subsequent studies of UDP-galactose metabolism in two laboratories (Ongun and Mudd, 1968; Chang and Kulkarni, 1970) indicated that the galactosyl residues of spinach leaf glycolipids do not turnover appreciably.

In similar studies using [^{14}C] acetate, Nichols *et al.* (1967b) showed that in *Chlorella vulgaris* the rate of labelling of the component fatty acids of individual lipids decreases in the order:

Phosphatidyl glycerol > Monogalactosyl diglyceride > Phosphatidyl choline > Sulpholipid > Phosphatidyl inositol > Digalactosyl diglyceride > Phosphatidyl ethanolamine

indicating that, as in *C. pyrenoidosa*, the rates of synthesis and turnover of fatty acids in the phosphatidyl glycerol and monogalactosyl diglyceride fractions during photosynthesis are greater than those in other lipid classes. Comparable results were obtained by Sastry and Kates (1965) who incubated *Chlorella vulgaris* with [^{32}P] glycerophosphate and found that the component phospholipids were labelled in the order

Phosphatidyl glycerol > Phosphatidyl choline > Phosphatidyl ethanolamine > Phosphatidyl inositol

From this information it is evident that during steadystate photosynthesis in *Chlorella* species monogalactosyl diglyceride and phosphatidyl glycerol are synthesized faster than the other cellular lipids, and that their component fatty acids exhibit the greatest turnover rates.

Nichols (1968) compared the patterns of incorporation of label from [^{14}C] acetate into the component lipids of *C. vulgaris* with those of two blue-green algae, *A. cylindrica* and *A. nidulans* and showed that the relative metabolic importance of individual lipid classes can vary according to the class of alga involved. Thus the fatty acids of the digalactosyl diglyceride and sulpholipid fractions of the blue-green algae were labelled much more rapidly than those of the corresponding *Chlorella* fractions, whereas the activities incorporated into the acids of the phosphatidyl glycerol fraction from *A. cylindrica* and *A. nidulans* were relatively lower than that in the green alga (Fig. 9.2). These variations could not be adequately explained by differences in the fatty acid composition of the algae and their component lipids.

The results of Nichols *et al.* (1967b) obtained from studies on heterotrophic cultures of *Chlorella* were interpreted as indicating that some degree of acyl transfer had occurred between different classes of lipid during the phase of cellular development investigated, but in the later studies of Nichols (1968) which involved photoautotrophic algal cultures the author concluded that such mechanisms could have occurred to only a very limited extent. Roughan (1970) has nevertheless suggested that the lecithin fraction of squash leaves may provide linolenic acid residues for the synthesis of the cellular galactosyl diglycerides. This suggestion was based on the observation that ^{14}C activity incorporated into the lecithin fatty acids diminished with time, while that in the glycolipids rose by a similar extent, although other possibilities, such as loss of activity from the phospholipid by oxidative breakdown of the acyl groups, could not be ruled out by the data reported.

On the basis of several detailed studies of fatty acid metabolism in *Chlorella* (Nichols *et al.*, 1967b; Nichols, 1968; Nichols and Moorhouse, 1969; Gurr *et al.*, 1969; Safford and Nichols, 1970) it has been concluded that major changes in the fatty acid composition of individual lipid classes can occur subsequent to their *de novo* synthesis; in other words the lipids are not merely acceptors of the end products of fatty acid synthetase and desaturase systems but may be intermediates and/or cofactors

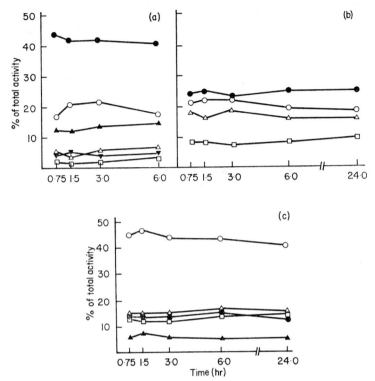

Fig. 9.2. Distribution of radioactivity among the acyl lipids of *C. vulgaris* (a), *A. cylindrica* (b) and *A. nidulans* (c) during incubation with [^{14}C] acetate under similar conditions (Nichols, 1968). ○, Monogalactosyl diglyceride; △, digalactosyl diglyceride; □, sulphoquinovosyl diglyceride; ●, phosphatidyl glycerol; ▲, phosphatidyl choline; ▼, phosphatidyl ethanolamine. Triglyceride and pigment fractions accounted for the remaining activity in all cases.

in some or part of these systems. These ideas, and the evidence from which they are derived, are discussed more fully on pages 144–188.

In studies with the phytoflagellate *Euglena gracilis*, Constantopoulos and Bloch (1967) showed that the intensity of the light source can have a profound effect on the lipid metabolism of this organism, and found that by increasing the light intensity on their cultures from 120 to 610 foot-candles the chlorophyll and total lipid content of the cell declined, whereas the proportion of 4,7,10,13-hexadecatetraenoic acid and α-linolenic acid in the total fatty acid rose sharply. The increased content of the two polyunsaturated acids was most pronounced in the chloroplast lipids, particularly the monogalactosyl diglyceride fractions, and the finding that Hill reaction activity and lipid unsaturation show the same responses to changes in light intensity led the authors to discuss the possible role of polyunsaturated fatty acids in photosynthetic oxygen evolution (see p. 271).

The majority of studies of lipid metabolism during photosynthesis have been carried out on whole cell systems and comparable data for isolated chloroplasts is scanty. Moreover, since comparatively little is known of the lipid metabolism associated with other plant organelles such as microsomes and mitochondria (see below), which certainly contain some of the lipid classes found in chloroplasts, the results from experiments such as those described above are not easily interpreted in terms of lipid metabolism in the photosynthetic apparatus alone. That lipid metabolism in the leaves of higher plants may also differ in several important respects from that observed in algae is suggested by the differences in fatty acid composition established for these systems and in apparent differences in the desaturases operating on saturated fatty acids in the two classes of tissue. Unfortunately, few data are available regarding general lipid metabolism in leaves apart from that of Roughan (1970) for squash leaves, in which he found the monogalactosyl diglyceride fraction to become labelled from [^{14}C] acetate relatively far more slowly than is commonly observed for green algae.

B. Cuticular Waxes

Concomitant with the formation of normal cellular constituents, the development of higher plant leaves is accompanied by the deposition of a complex mixture of lipophilic compounds, known as cuticular wax, on the epidermal cells of leaves, fruits and stems. Smaller quantities of waxy materials have also been reported in almost all other parts of higher plants, and in algae (Kolattukudy, 1970). While many of the components of these waxes are not acyl lipids in the true sense of the word, their formation is so closely related to acyl lipid and fatty acid metabolism that it will be pertinent to summarize here some studies into cuticular wax biosynthesis.

Detailed analyses of the surface wax of cabbage leaves (*Brassica oleracea*) by Purdy and Truter (1963a,b,c) revealed that some 65% of the wax consisted of seven straight-chain aliphatic compounds all containing 29 carbon atoms. The remaining 35% comprised C_{12} to C_{24} fatty acids of mainly even chain length, C_{12} to C_{28} primary alcohols, and esters of acids and alcohols of similar structure to those present in the free acid and alcohol fractions. Among the C_{29} compounds were the C_{29} paraffin, nonacosane, and 15-nonacosanone, which were originally isolated from cabbage by Channon and Chibnall (1929) who proposed that the paraffin and ketone must be interrelated metabolically because of their structural similarities. Chibnall also suggested that condensation of 2 molecules of pentadecanoic acid followed by decarboxylation of the product could give rise to 15-nonacosanone which in turn could be reduced to the corresponding C_{29} paraffin, but it is only comparatively recently that Kolattukudy has tested this and alternative hypotheses by biochemical studies.

Kolattukudy (1965) modified the original hypothesis of Channon and Chibnall to account for the apparent inability of plants to synthesize n-C_{15} acids directly, which he suggested could arise by α-oxidation of

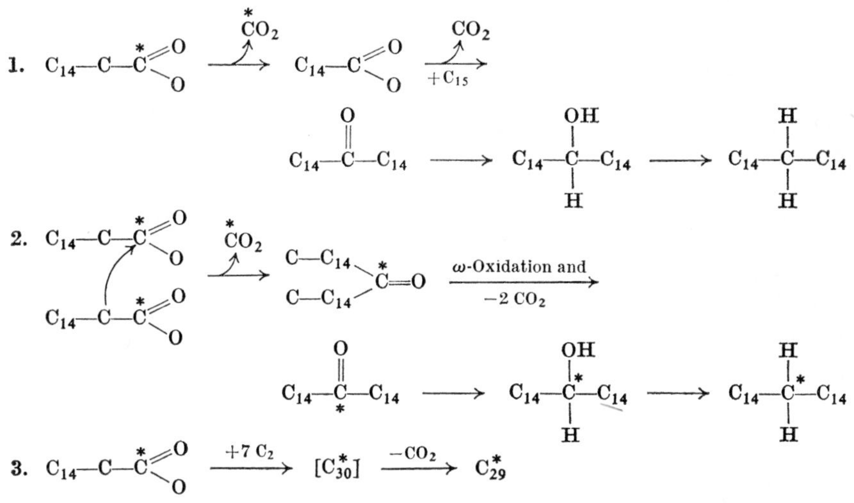

FIG. 9.3. Alternative mechanisms proposed for the biosynthesis of the C_{29} hydrocarbon of cabbage leaf wax (Kolattukudy, 1970). $\overset{*}{C}$ indicates radioactive carbon atom.

palmitic acid (Fig. 9.3.1). An alternative scheme which could operate without involving fatty acids of odd carbon number was originally proposed by Kreger (1948) and involves the condensation of two palmitate molecules followed by a double ω-oxidation (Fig. 9.3.2). These two alternative schemes both require the intermediary formation of a ketone and (probably) a secondary alcohol, which are classes of compound frequently found in wax fractions.

According to pathway 1 (Fig. 9.3) the n-C_{15} acid would be expected to be incorporated into the C_{29} paraffin at least as fast as palmitic acid, which should be the precursor of the C_{15} acid. However, Kolattukudy (1966) found that the C_{16} acid was more readily incorporated into the C_{29} paraffin of *Brassica oleracea* than was the C_{15} acid. Furthermore if pathway 1 were operative the carboxyl carbon of palmitic acid would be lost during the α-oxidation process, so that $[1\text{-}^{14}C]$ palmitate would not be expected to yield radioactive paraffin; in fact the carboxyl-labelled substrate labelled the paraffin as efficiently as the universally-labelled acid. Also, $[16\text{-}^{14}C]$ palmitate and $[1\text{-}^{14}C]$ palmitate were equal in their incorporation of radioactivity into the hydrocarbon (Kolattukudy, 1968) indicating that the entire carbon chain of the acid must be incorporated into it. These and other results (Kolattukudy, 1965; 1966; 1968) show that neither pathways 1 nor 2 are possible, and that the 15-nonacosanone and 15-nonacosanol are not likely to be precursors of nonacosane. Present evidence now favours an elongation-decarboxylation pathway (Fig. 9.3.3) in which saturated fatty acids of intermediate chain-length are lengthened by successive C_2 units until the chain length reaches the vicinity of C_{30} when the fatty acid is decarboxylated. The C_{30} chain length is apparently preferred for decarboxylation in *Brassica oleracea* and many other plants where the major hydrocarbon is C_{29} paraffin, and in other plants where the major paraffin is C_{31} or C_{33} a similar complex with a slightly different specificity may operate. The main experimental support for this mechanism lies in the observation that higher fatty acids, particularly palmitic and stearic acids, are incorporated directly into the paraffin without prior breakdown (Kolattukudy, 1966), stearic acid being incorporated at least twice as fast as palmitate. However, attempts to obtain direct evidence for a decarboxylation have so far failed. For instance, exogenous labelled $n\text{-}C_{30}$ acid did not label $n\text{-}C_{29}$ paraffin in broccoli leaf slices or homogenates (Kolattukudy, 1970) although such exogenous substrates might not have had access to the relevant metabolic site.

Moreover, a time course study of the incorporation of label into these two classes of compound failed to indicate a typical precursor product relationship (Kolattukudy, 1967a) since the amount of label present in the acids did not diminish for several hours after the incorporation of label into paraffins ceased. These results suggested that the labelled very long chain acids were those that were diverted from the site of paraffin synthesis and trapped into a pool from which they were no longer available for paraffin synthesis. In support of this hypothesis, these very long acids were in fact found as minor components of the total phospholipid fraction of the broccoli leaf lipids. In addition trichloroacetate strongly inhibited the incorporation of acetate into paraffin but not into the more common fatty acids of the major leaf lipids, indicating that this inhibitor acts on the elongation stage of paraffin synthesis.

Recent analyses of the chemical composition of wax from the leaf surfaces of normal and "glossy" mutant *Brassica oleracea* (Macey and Barber, 1970) showed that genes of at least two different loci responsible for the glossy character result in a severe reduction in the concentrations of the C_{29} paraffin and the C_{15} acid. The authors pointed out that their results indicated an apparent relationship between the C_{15} acid and the C_{29} paraffin but might be reconciled with the elongation-decarboxylation pathway by assuming that α-hydroxypalmitate, shown to be an intermediate of α-oxidation (Hitchcock and James, 1966), is the substrate which is elongated to C_{30} and subsequently decarboxylated. On this view the pentadecanoate found in the free fatty acid fraction could be considered as a by-product of α-hydroxy acid production, the latter acid being a specific substrate for the elongation system concerned with the production of C_{29} compounds.

On the basis of evidence available to date Kolattukudy (1970) has proposed a scheme for the interrelationships between fatty acid, paraffin and wax ester synthesis in *Brassica oleracea* leaves (Fig. 9.4). *De novo* fatty acid synthesis takes place in the chloroplasts which provide the substrate, palmitic acid, for paraffin synthesis. Fatty acids for paraffin synthesis may also be synthesized in the epidermal layer of cells themselves. The C_{16} acid becomes the substrate for the elongation-decarboxylation complex which elongates the acid to the appropriate chain length and then decarboxylates the product releasing the paraffin. After each successive addition of C_2 units some elongated acids dissociate from the complex and are rapidly esterified into phospholipids from which they may be subsequently transferred to wax esters (p. 191).

6. LIPID METABOLISM IN PLANT MITOCHONDRIA. AGEING

The phenomenon of wound respiration which occurs when a dormant storage organ is sliced and incubated aerobically("ageing"), is associated with the development of new enzyme activities (Click and Hackett, 1963) and an increase in the number of mitochondria and other membranous components such as endoplasmic reticulum (Lee and Chasson, 1966; Jackman and Van Steveninck, 1967). Since lipids comprise an integral part of all known membrane systems it is predictable that "ageing" should be accompanied by an increased rate of synthesis and metabolism of lipoidal compounds which are required for the development and function of mitochondrial membranes.

Thus Willemot and Stumpf (1967a) studied the development of fatty acid synthetase activity in ageing discs of potato tuber, and found that during the first hours of the ageing process the fatty acid synthetase activity increased 6- to 12-fold. In the same period the tissues synthesized increasing quantities of polyunsaturated acids (especially linolenic acid)

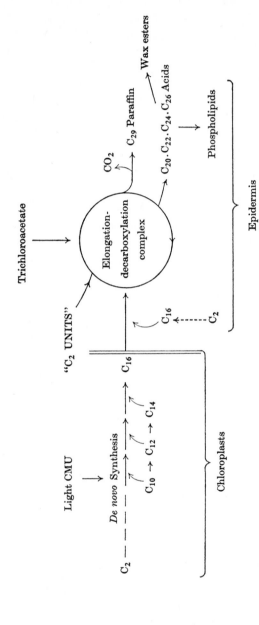

Fig. 9.4. Proposed scheme for the biosynthesis of surface lipids in cabbage leaves (Kolattukudy, 1970).

and long chain (C_{20}–C_{24}) fatty acids which were not synthesized by the discs immediately after slicing; the same authors subsequently demonstrated that these changes in fatty acid synthetase activity paralleled a temporary rise of protein and RNA synthesis (Willemot and Stumpf, 1967b). Using a similar potato system to that studied by Willemot and Stumpf (1967b), Tang and Castelfranco (1968) subsequently showed that the major classes of lipid formed during ageing included lecithin, lysolecithin, phosphatidyl ethanolamine and sterol, all of which are major lipids of plant mitochondria.

Galliard *et al.* (1968b) studied the system of lipid synthesis which developed during the ageing of peel discs from pre-climacteric apples and showed that it resembled that of potato tuber in many respects. Lipid synthesis from [^{14}C]acetate increased rapidly during the first four hours of ageing and thereafter dropped steadily, but in the presence of cycloheximide (an inhibitor of protein synthesis) there was a considerable repression of lipid synthesis during this period. Also comparable with results from studies with potato slices was the observation that the ageing apple tissue accumulated label preferentially into phospholipid fractions whereas the galactosyl diglycerides, which are major components of chloroplasts but are also present in minor proportions in mitochondria, were not labelled appreciably. Galliard *et al.* also noted a large increase in incorporation of label into a lipid fraction having the chromatographic characteristics of free fatty acid, which they considered to indicate an uncoupling of integrated lipid synthesis during ageing. In addition the increased lipid synthesizing capacity in ageing apple peel was accompanied by an increased capacity to oxidize fatty acid and acetate to CO_2.

Evidence that much of the increase in lipid synthesizing activity in ageing tissue is specifically associated with the formation and metabolism of new mitochondria was obtained by Abdelkader *et al.* (1969) who isolated mitochondrial fractions from ageing potato discs following their incubation with labelled substrates (*in vivo* studies), and also incubated mitochondria with similar substrates *following* isolation of the organelle from the tissue (*in vitro* studies). The *in vivo* experiments showed that a large proportion of the lipids synthesized during ageing accumulated in the tissue mitochondria, particularly in the lecithin, phosphatidyl ethanolamine and phosphatidyl glycerol fractions, and that the synthesis of these lipids was accompanied by the rapid synthesis of saturated, mono- and di-unsaturated fatty acids. A considerable increase in the synthesis of lipids and fatty acids by mitochondria *in vitro* during ageing was also observed, but there were very considerable differences in the distribution of label between individual lipid and fatty acid classes

9

obtained in these studies and those from the corresponding *in vivo* experiments. In particular there was a considerably lower incorporation of label into the lecithin and polyunsaturated fatty acid fractions in the *in vitro* studies than occurred *in vivo*. These results, coupled with those from a later study (Abdelkader and Mazliak, 1970) in which microsomal fractions from similar tissues were shown to be the major sites for lecithin and polyunsaturated acid synthesis, led the authors to the proposition that plant mitochondria are not capable of the *de novo* synthesis of phospholipids or polyunsaturated acids, but that these components are synthesized by the cellular microsomes and are then rapidly translocated to the site of mitochondrial membrane synthesis.

There is little evidence to indicate exactly how slicing tuber tissue results in the increased rates of respiration, lipid and protein synthesis typical of "ageing" tissue. Willemot and Stumpf (1967b) have suggested that immediately after slicing one or more repressors may be removed and the resulting de-repression may lead to the synthesis of short-lived mRNA's. These mRNA's would then direct synthesis of new proteins towards

(1) lipid synthetase systems, and as a result, an increased lipid synthesis
(2) structural proteins able to form, in combination with the lipids, membranous structures, and
(3) enzymatic proteins associated with these membranes in structures able to support an activated metabolism.

7. SENESCENCE

Although there have been comparatively few detailed studies of the changes in lipid composition and metabolism which occur during leaf senescence, there is little doubt that this process is accompanied by an extremely rapid loss of these substances, particularly those of the chloroplast.

In studies of senescence in cucumber cotyledons, Draper (1969) observed that during the phase of rapid yellowing there was a rapid accumulation of linolenic acid in the free fatty acid fraction and a rapid decline in some complex lipid levels, particularly those of the galactosyl diglycerides and sulpholipid. Over the same period the lecithin and phosphatidyl ethanolamine levels remained constant while that of sterol glucoside rose. Such changes are consistent with the morphological data obtained by Butler (1967) for senescence in similar tissue, who showed that the phase of rapid yellowing is accompanied by a steady degeneration of the chloroplasts whereas mitochondria become smaller but remain morphologically intact until a very late stage in senescence. It is therefore not surprising that the major lipids of the mitochondrion (phosphatidyl choline and phosphatidyl ethanolamine) are degraded much more slowly than those of the photosynthetic apparatus.

The mechanisms involved in the disappearance of cellular lipid during senescence have not been established experimentally, but it is likely that the first stages involve the degradation of endogenous lipid by lipases (such as those described in Chapter 7) to yield free fatty acids and water-soluble components such as glycerol, glycerophosphate, glycosyl glycerols and free amines. The free fatty acids so produced are then broken down catabolically to CO_2 in a process which is so rapid that the overall accumulation of free acid during senescence is comparatively slight in comparison to the amount of acyl lipid degraded. The enzymes involved in this catabolism could be the α- and β-oxidases described in Chapter 8, and also lipoxygenase. Although the latter enzyme initially attacks only unsaturated acids containing cis-1,4-pentadiene systems, e.g. linoleic and linolenic acids, the products of these reactions might then promote chemical oxidation of other acids in the tissue. Apparent support for the pre-eminence of lipoxygenase activity during senescence is provided by the evidence of several workers (e.g. Newman, 1966) that linolenic acid levels fall much faster than those of other acids over this period, although this tendency might equally well be explained by the fact that the galactosyl diglycerides are particularly rapidly degraded, thus providing a high level of linolenic acid in the free fatty acid pool available for catabolism. Holden (1970) has confirmed that linoleic acid peroxidizing activity is widely distributed in leaves and that this activity is due to a lipoxygenase-type enzyme.

Catabolic breakdown may not be the only *immediate* fate of free fatty acids released during senescence, and Grob and Csupor (1967) noted that intermediate stages of senescence in the leaves of *Acer plantanoides* are characterized by the accumulation of fatty acid esters of xanthophylls, terpene alcohols and other alcohols, presumably by reactions between the alcohols and free fatty acids derived from lipid degradation. Eventually even these fatty acids are catabolized.

8. EFFECT OF ENVIRONMENT ON FATTY ACID SYNTHESIS AND METABOLISM

We have already described (p. 152) the significant effects of ambient temperature on fatty acid metabolism in many tissues, and the nutritional status of the environment in which the plant grows can also have a great influence on fatty acid and lipid composition.

A. Leaves

The fatty acid and lipid composition of leaves can be profoundly affected by a variety of environmental conditions but in most cases it is possible that the effects may be largely of an indirect nature and originate from changes induced in the relative proportions of different cellular organelles.

The majority of effects so far reported have been similar in kind, namely that in cases of mineral deficiency the level of polyunsaturated fatty acids, particularly linolenic acid, are depressed below those in similar tissues receiving a balanced mineral intake. The conditions which produce these changes, such as deficiencies of iron (Newman, 1962, 1964), manganese (Bloch and Chang, 1964; Constantopoulos, 1970) and nitrogen (Wallace and Newman, 1965) are those which also inhibit or depress the rate of chloroplast development. Consequently the depressed levels of linolenic acid noted in vegetative tissues suffering from such deficiencies may be due to the presence of smaller concentrations of chloroplasts in these tissues, although an alternative possibility is that the development of chloroplasts depends on the synthesis of linolenic acid in optimal quantities and that repression of polyenoic acid synthesis results in slower accumulation of the photosynthetic apparatus. If the second alternative is operative then such nutritional deficiencies may be affecting lipid synthesis *directly*.

Evidence that nitrogen levels may exert some direct influence on fatty acid synthesis in chloroplasts has been described by Newman (1966) who found that nitrogen-deficient chloroplasts contained a higher ratio of saturates to unsaturates than did similar organelles isolated from tissues grown on a complete medium. Moreover, nitrogen-deficient chloroplasts incorporated labelled acetate into linolenic acid less rapidly than did chloroplasts from plants grown under normal conditions.

B. Oil-seeds

The effects of variations in the mineral nitrition of oil seed crops on the fatty acid composition of their seeds have been widely studied. Many of the more pronounced effects reported by earlier workers such as Schmalfuss (1937) with flax have not been confirmed by subsequent studies in which more modern analytical techniques have been employed (Appelqvist, 1968) and in general the effects of mineral nutrition on the lipid and fatty acid composition of seeds are rather small when compared with those observed in the corresponding vegetative tissues.

Thus Howell and Collins (1957) investigated the polyenoic acid content of soybeans from plants grown under a variety of conditions and found little or no consistent effect due to levels of nitrogen, phosphorus, potassium, sulphur, trace elements or additions of manure or plant residues, although the addition of iron chelates to the soil caused a slight increase in the linolenic acid content of the soybeans. Yermanos *et al.* (1964) reported that iron chelates had no effect upon the iodine value of safflower oil, but noted a slight decrease in unsaturation with increased levels of nitrogen dressing.

Van den Driessche (1961, 1964) varied both the relative concentrations of cations (K^+, Ca^{2+} and Mg^{2+}) and anions (NO_3^-, SO_4^{2-} and PO_4^{3-}) as well as the ionic strength of the nutrient solution applied to soil-free cultures of cottonseed plants, and concluded that no statistically significant effect of nutrients on the iodine value or linoleic acid content occurred in cotton seeds.

In a detailed study Appelqvist (1968) recorded the effect of nitrogen, phosphorus and potassium nutrition on the lipid composition of rape-seed (*Brassica napus*) grown in soil-free culture. High levels of nitrogen caused an increase in seed weight, but a decrease in oil content, so that the average amount of oil per seed remained constant. Small decreases in palmitic and eicosenoic acid content, and increases in oleic and erucic acid were observed under similar conditions of high nitrogen, suggesting that a small decrease in the extent of elongation of oleic acid to erucic acid occurs in seeds from plants with low levels of nutritional nitrogen. The effects produced by phosphorus, potassium and sulphate at varying concentrations were even less significant than those produced by nitrogen.

9. GENETIC CONTROL

In comparatively recent years there have been found certain species of oil-seed crops within each of which occur varieties that produce oils of atypical fatty acid content. Examples include the strains of rape *Brassica napus* and *Brassica campestris* which produce no erucic acid (Stefansson and Hougen, 1964; Downey, 1964) and the oils of low iodine number in Safflower and Sunflower (Horowitz and Winter, 1957; Knowles and Mutwakil, 1963; Putt *et al.*, 1969). In rape, safflower, flax and sunflower breeding trials have indicated that the composition of the oil is under genetic control (Downey and Harvey, 1963; Knowles and Mutwakil, 1963; Yermanos and Knowles, 1962; Putt *et al.*, 1969). More specifically, it is apparent that rapeseed varieties contain a single gene which controls the chain elongation of oleic acid to erucic acid (Appelqvist, 1969) whereas in safflower there are three genes at one chromosome locus which govern the proportions of linoleic and oleic acids while at a second locus there have been identified two genes governing levels of stearic acid (Krzymanski and Downey, 1969; Knowles, 1969).

The compositional differences which occur within species due to differences in genotype are far greater than those created by environmental differences such as those of climate and nutrition, and offer great opportunities for the breeding of crops for oils of specific nutrition.

CHAPTER 10

The Role of Lipids in Plant Metabolism

The ubiquity of acyl lipids in plant tissues is some indication of the essentiality of these compounds to plant cell physiology, and consideration of the wide variety of lipid structures involved suggests that their roles may be many and varied. A large number of specific physical, physico-chemical and chemical functions have now been proposed for the lipids of biological systems and although the experimental evidence offered in support of such theories often falls short of being definitive it is the purpose of this chapter to summarize the principal roles established or proposed for plant lipids and to discuss where relevant the experimental evidence on which these proposals are based.

1. Plant Lipids as Components of Membranes

A. Structural Role

Biological membranes are the fundamental unit of transport and govern all permeability processes whether they be simple diffusion, exchange diffusion, facilitated diffusion or active transport against a prevailing electrochemical gradient. They are also the sites of many of the important enzymic reactions of cells.

All known membranes contain protein and lipid in proportions which vary considerably according to the membrane source. In myelin, for example, the ratio of protein to lipid is 0·25, whereas in bacterial membranes it is approximately 3·0 (Korn, 1968). Such generally high proportions of lipids show that these compounds must inevitably play an essential role in determining membrane structure, although the precise manner by which they do this is, in most cases, far from clear. Nevertheless, considerable progress has been made in the past few years in establishing the nature of the interactions between membrane proteins and lipids, and there is now a great deal of evidence for non-ionic, hydrophobic bonding between lipids and proteins in membranes. This evidence that membranes owe their structural integrity to such weak hydrophobic interactions as well as protein–protein interactions has led to a complete revision of the bimolecular leaflet as an acceptable model for all membranes. To accommodate this new evidence the latest membrane models place the hydrophilic groups of the amphipathic lipids and proteins at the aqueous interfaces, the whole structure being stabilized by non-polar interactions between proteins and lipids as well as hydrophobic interactions within the two classes of membrane components, i.e. protein–protein and lipid–lipid. Lipoproteins have been released from some membrane preparations and this, together with studies of the effectiveness of various solvents in removing lipids, suggest that the firmness of binding between individual proteins and lipids may vary. As yet there is no definitive evidence for covalent bonding between lipid and protein in membranes.

Analyses of a variety of membranes have demonstrated striking differences in the chemical nature and quantitative distribution of lipid classes which may be related to differences in the specific properties and functions of the interfaces, such as permeability. This relationship is underlined by the fact that for a given membrane the composition is normally very constant and specific.

1. Chloroplasts

The chloroplasts of most higher plants are discoidal or ellipsoidal in side view, usually measuring 1–$10\,\mu$ across and 2–$3\,\mu$ in height. Most of the pigments and acyl lipids of the chloroplasts are located in lamellae

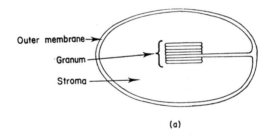

Outer membrane →
Granum
Stroma

(a)

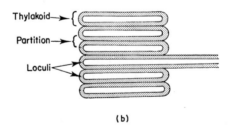

Thylakoid →
Partition →
Loculi

(b)

Fig. 10.1. General structure of the chloroplast and chloroplast granum.

which are usually piled one upon the other to form "grana"; these are embedded in a granular stroma which is enclosed within a double membrane envelope (Fig. 10.1). The lamellar system is composed of tightly aggregated lipoprotein membranes termed thylakoids which contain 45% protein and 55% lipophilic substances, and it is this internal lamellar system which is the site of energy transduction in photosynthesis.

The detailed structure of the grana and the arrangement of acyl lipids, pigments and proteins within these structures has been the subject of extensive studies in several laboratories, and an exhaustive review of all the experimental data and the hypotheses derived therefrom is beyond the scope of this book. We shall mainly consider those theories in which the involvement of different classes of acyl lipid have been specifically invoked.

The models of Weier and Benson (1967) and Branton and Park (1967) both conform to the now generally accepted view that the thylakoids consist of continuous layers of asymmetric globular lipoprotein subunits. Weier and Benson suggest that these subunits comprise a protein core surrounded by components which are determined by the environment of the membrane, and because the stroma and loculi (Fig. 10.2) contain aqueous materials it is proposed that the membranes bordering these spaces bind the surface active glycolipids with their polar moieties associating with the aqueous phase. Freeze-etching studies by Branton and Park also indicated that the membrane is composed of a central component of protein coated by lipids. Since the Weier and Benson model has the polar moieties of the glycolipids projecting into the stroma, it follows that the lipid–protein interaction must be largely through the acyl residues of the lipids, and these authors propose that the highly unsaturated acyl residues normally present in the galactosyl glycerides associate with the internal hydrophobic regions of the lamellar lipoprotein units, this association being controlled by interactions between the planar olefinic systems of the acyl groups and the π-orbitals of the aromatic amino-acids of the membrane protein. Experiments comparing the ability of lipids of different structure and fatty acid composition to reassociate with membrane protein also indicated that the major forces binding lipid to protein are hydrophobic in character (Ji and Benson, 1968).

The Weier and Benson model also requires most of the chlorophyll to be associated with the central hydrophobic area of the partition (Fig. 10.2) with the phytol chain interacting with hydrophobic regions within the protein subunit. It is consequently possible that the phytol groups of chlorophyll and the acyl groups of the glycolipids could interact with one another since both are considered to penetrate this hydrophobic region,

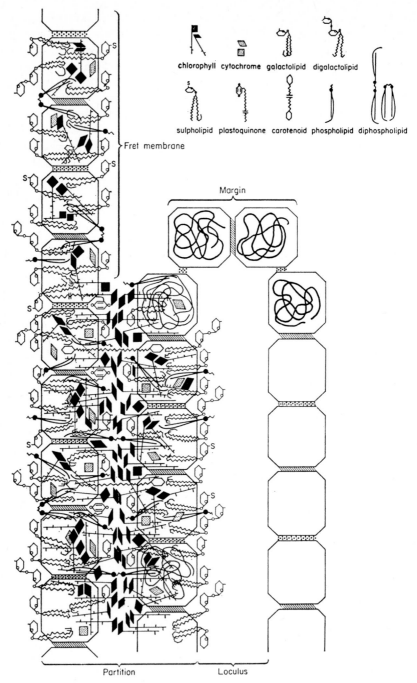

FIG. 10.2. Arrangement of lipids in chloroplast membranes as proposed by Weier and Benson (1967).

and Rosenberg (1967b) has suggested that the *cis* double bonds of poly-unsaturated fatty acids create pockets in their spatial configuration into which the methyl branches in the phytol group of chlorophyll might fit. This would permit closer and more stable hydrophobic association between the fatty acid residues of the galactosyl diglycerides and the phytol groups, thereby facilitating the formation and stabilization of the film of chlorophyll required to provide an efficient photoreceptive surface. The hypotheses of Rosenberg and of Weier and Benson both suggest essential physico-chemical functions for the polyenoic acid residues in chloroplast lipids and may explain why the galactosyl diglycerides of leaves and higher algae invariably contain major proportions of these acids (p. 92). Some blue-green algae contain small and sometimes even negligible proportions of polyenoic acids (p. 78) and it may be significant that the photosynthetic organelles of this class of alga are not normally arranged in the closely packed grana structures typical of higher plants.

On the other hand, the considerably reduced levels in linolenic acid observed in a *Scenedesmus* mutant blocked in the Hill reaction do not result in a detectable change in its chloroplast lamellar structure (Bishop, 1962). Similarly, removal of large amounts (about 50%) of linolenic acid from lamellar systems from barley chloroplasts did not change the membrane structure as observed by electron microscopy (Appelqvist et al., 1968b). These results indicate that the integrity of chloroplast thylakoids is not necessarily dependent on a *high* content of linolenic acid.

In studies of lipid synthesis in greening cultures of *Euglena gracilis*, Rosenberg and Pecker (1964) noted that the plant sulpholipid accumulates before measurable amounts of chlorophyll and that subsequently the rate of synthesis of this glycolipid closely parallels that of chlorophyll. On the basis of these findings these workers suggested that the sulpholipid might be necessary for the orientation and possibly the functions of chlorophyll. Studies of chlorophyll–sulpholipid interactions *in vitro* (Trosper and Sauer, 1968) also support this possibility.

2. Mitochondria

In both plants and animals those processes of cellular respiration which are responsible for the controlled oxidation of respiratory substrates take place within the mitochondrion, which also conserves the energy thus released in forms usable for the conduct of the energy requiring functions of the cell (p. 203).

Isolated mitochondria that meet rigid biochemical requirements for purity and integrity have shown a striking uniformity of general morphology in all tissues investigated, and electron microscopy has shown that mitochondria are bounded by two membranes (Fig. 10.3) usually referred to as the outer (or limiting) membrane and the inner membrane (Bonner, 1965). On the basis of microscopic studies which showed that mitochondrial membranes consist of globular subunits with no continuous lipid bilayer structures, Sjostrand and Barajas (1970) proposed a model for mitochondrial membranes in which their structure is maintained primarily through hydrophobic interaction between proteins, lipids and proteins, and lipids. A predominantly hydrophobic milieu would be found in the interior of the membrane and both non-polar amino-acid side chains exposed at certain regions of the surfaces of the protein molecules and the hydrocarbon tails of the phospholipids would contribute to the non-polar character of the membrane interior.

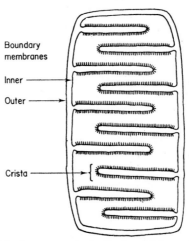

Boundary membranes

Inner

Outer

Crista

FIG. 10.3. Arrangement of mitochondrial membranes.

Thus in general principle the types of interaction proposed for the stabilization of chloroplast and mitochondrial membranes are not dissimilar, although the detailed chemical structure of the proteins and lipids involved are very different. For example, the major, general class of lipid in mammalian mitochondria is phospholipid (e.g. Stoffel and Schiefer, 1968) and although plant mitochondria contain minor quantities of glycolipids (p. 65), phospholipids are the major lipids of these organelles also. The detailed lipid compositions of the inner and outer membranes of *plant* mitochondria have yet to be established, but it is well known that the inner and outer membranes of mammalian mito-chondria differ chemically and enzymically. For example, the outer membrane contains on a protein basis, 2 to 3 times more phospholipid than the inner membrane, and although phosphatidyl choline and phosphatidyl ethanolamine are the major lipids of both fractions the major qualitative differences in composition lie in the distribution of cardiolipin, which is almost exclusive to the inner membrane, and phosphatidyl inositol, which is most abundant in the outer membrane (Parkes and Thompson, 1970).

Indications that some classes of phospholipid may be more important than others for the maintenance of mitochondrial membrane structure were obtained by Awasthi et al. (1969) who exposed beef-heart mitochondrial christae (electron-transport particles) to phospholipase A which digested the endogenous phospholipids in the order phosphatidyl ethanolamine, phosphatidyl choline and cardiolipin. Electron microscopy indicated no changes in membrane structure following digestions sufficient to cause almost complete hydrolysis of phosphatidyl ethanolamine and phosphatidyl choline. The lyso-lipids and cardiolipin which remained after the digestion were apparently adequate to preserve membrane structure but their removal from the partially digested membrane caused noticeable damage to the membrane structure.

B. Involvement in Membrane Function

1. Membrane permeability and transport

(a) *Ion permeability.* The idea that phospholipids may be involved in the transport of ions across membranes is frequently propounded in a variety of ways, and the fact that membrane lipoproteins bind cations in a specific manner is well documented. Whether the primary binding sites are protein carboxylate groups or phospholipid phosphate groups is open to examination, although evidence favouring involvement of lipid phosphate groups is found in Rothstein's (1968) observations of transport inhibition by uranyl ions. Benson (1968) envisages active transport of cations to result from conformational changes whereby an electrostatically bound ion or group of ions is transferred through the lipoprotein membrane by movement of the protein chain and the binding groups, thereby placing the bound ion in the new environment where it is released.

Consistent with these views are results obtained by studies with synthetic bilayers, which have many physico-chemical properties in common with natural membranes, and which led Hopfer et al. (1970) to conclude that the charge on the polar head of membrane lipids plays an important role in controlling the ion-selective permeability of the bilayer and hence possibly that of the natural membranes in which such lipids occur. In particular negatively charged lipids showed pronounced selectivity with regard to cations whereas membranes formed from an uncharged glycosyl glyceride exhibited only slight cation selectivity.

Therefore if lipids are in fact directly implicated in cation transport in natural membranes then it will be those with an overall negative charge which will most probably be involved, and in the chloroplast only phosphatidyl glycerol and the sulpholipid could then be expected to have this function.

Much work nevertheless remains to be done on the chemical nature of the carrier involved in the membrane before it can be concluded with any certainty that phospholipids or other negatively charged lipids are directly implicated in the selective processes of ion transport.

(b) *Sugars and other metabolites.* An interesting proposal for the mechanism for transport of sugars through membranes has been advanced by Benson (1963) on the basis of studies in which the sugar moieties of algal galactosyl diglycerides showed rapid turnover of label following incubation of the organism with $^{14}CO_2$ (Ferrari and Benson, 1961). In this theory glucose epimerizes to galactose which reacts with diglyceride giving galactosyl diglycerides, in which form the sugar is moved across the membrane and is then released by β-galactosidase activity. Some doubt whether this mechanism can be an important one in photosynthetic tissues has been cast by subsequent studies by Chang and Kulkarni (1970) and Roughan (1970) who were unable to detect significant turnover of the galactose residues in the galactosyl glycerides of spinach and pumpkin leaves.

Indirect evidence that the galactosyl diglycerides might facilitate membrane permeability towards sugars and other uncharged metabolites by a mechanism possibly different from that proposed by Benson has been obtained by studies on the permeability of spontaneously formed liquid crystals in salt solutions (liposomes). These structures appear to be suitable model systems for the study of permeability properties and de Gier and coworkers (1968) observed that the permeability of lecithin liposomes towards glycerol and erythritol increased markedly with increasing degree of unsaturation of the lipid acyl groups. These results, and those obtained with natural systems (van Deenen and de Gier, 1964) indicate that permeability of membranes towards water-soluble non-electrolytes may depend primarily on the degree of unsaturation and chain length of the hydrocarbon chains of the constituent lipids. The importance of the high degree of unsaturation of the galactosyl diglycerides may therefore be the facilitation of the very active transport activities of chloroplast membranes, and it is interesting to note that Brody *et al.* (1970) found that the addition of linolenic acid to chloroplasts and subchloroplast particles had pronounced effects on light-induced adsorbtion changes which could be interpreted as reflecting an increase in the permeability of the thylakoid membrane.

2. Role in enzyme activity

A substantial quantity of evidence has now accumulated which points to the essentiality of certain classes of lipid for the functioning of many of the enzymes located in biological membranes. In most cases the exact function of the lipids concerned in these enzymic reactions is not clearly understood, and at least two distinct types of involvement can be envisaged. In the first the presence of the lipid is necessary for maximum enzymic activity although the lipid itself undergoes no chemical change, and in the second the lipid participates as a carrier of the substrate and fulfills the classical role of cofactor. In some situations both types of involvement may operate.

Much of the evidence available at present has been obtained from studies of the role of phospholipids in mitochondria of mammalian origin, but the structures and functions of mitochondria from mammals and plants are sufficiently similar that one may reasonably infer that lipid function may be similar in particles of either origin. In addition, several proposals have been made for the function of lipids in chloroplast metabolism specifically.

(a) Electron transport. Most cells which photosynthesize with concomitant evolution of oxygen contain a high proportion of cis-poly unsaturated fatty acids which are present in particularly high proportions in the major acyl lipids of the chloroplast, i.e. the galactosyl diglycerides. These data suggested to Erwin and coworkers (1964) that acyl lipids containing linolenic acid might be essential cofactors in the electron transport systems involved in the photosynthetic evolution of oxygen (the Hill reaction) and this theory was supported by several groups of workers (e.g. Chang and Lundin, 1965; Appelman et al., 1966) and later extended by others to account for Hill-reacting systems which synthesize C_{20} polyunsaturated acids or γ-linolenic acid but no α-linolenic acid (Kates and Volcani, 1966; Patton et al., 1966; Nichols and Wood, 1968). More direct evidence was obtained by Chang and Lundin (1965) who found that galactosyl diglycerides stimulated the rate of photoreduction of cytochrome c by intact chloroplasts.

The major objection to this hypothesis hinges on the observation by Holton et al. (1964) and Kenyon and Stanier (1970) that some blue-green algae such as Anacystis nidulans do not accumulate polyenoic acids, although there is some doubt as to the true status of this last organism and its ability to carry out the Hill reaction.

An alternative explanation for these facts as offered by Benson (1964) is that since oxygen production and linolenic acid synthesis occur together, it may be the oxygen released which controls linolenate synthesis rather than *vice versa*. It has been established that the synthesis of unsaturated acids in photosynthetic tissue requires oxygen (p. 143) and Benson has suggested that the galactosyl diglycerides may be desaturated while adsorbed in the quantasomes of the lamellar membrane. This concept is consistent with the mechanisms of glycolipid metabolism and desaturation proposed by Nichols *et al.* (p. 190).

Far more convincing is the evidence for the essentiality of lipids in the electron transport systems of mitochondria in which each of the four complexes of the electron transport chain is made up of a combination of proteins and lipids. The individual proteins in each of these complexes maintain their fine structure and some activity in the absence of lipid, but the overall integrated activity of the complex requires the reintroduction of lipid into the complex (Fleischer *et al.*, 1967) and it has been established that there is a requirement for phospholipid in three segments of the electron transport chain, viz. from succinate to Coenzyme Q, reduced Coenzyme Q to cytochrome *c* and reduced cytochrome *c* to oxygen (Green and Fleischer, 1963). This requirement of lipid for the electron flow has been interpreted by Green and Fleischer in terms of providing a medium of low dielectric constant. The exact nature of the phospholipid reinserted into the electron transport complexes to restore activity seems less important than a general requirement for phospholipid although a definite requirement for some acidic phospholipid has been reported by Green and Fleischer (1964). Even this specificity may be due more to the physical problems of getting lipid material into and bound by the mitochondrial particles than to any specific role for acidic phospholipid, since it has now been shown that lipid composition may have a great effect on the physical properties of lipid preparations.

(*b*) *Energy transduction and mitochondrial swelling.* The mitochondrion is known to be capable of cyclical contraction or swelling and it is believed that these conformational changes are fundamental to energy transfer in the mitochondria (Green *et al.*, 1968). Work from several laboratories has suggested that phospholipids and fatty acids may be intimately involved in these processes.

(i) Phosphatidyl inositol and mitochondrial contraction. Ohnishi and Ohnishi (1962) isolated from liver mitochondria a protein fraction having ATP-ase activity and which was capable of an ATP-induced contraction, while Vignais *et al.* (1963a) found that similar contractile proteins from mitochondria could restore ATP-induced contraction in mitochondria which had lost this property. They also discovered that the ability to promote mitochondrial contraction disappeared when the lipid was extracted from these compounds, showing subsequently that phosphatidyl inositol was the only class of lipid capable of restoring contractibility in aged mitochondria in the presence of ATP and Mg^{2+} (Vignais et al., 1963b). Other phospholipids, including the anionic phosphatidic acid and cardiolipin were ineffective, thus indicating a specific role for phosphatidyl inositol in this mitochondrial membrane process.

(ii) Fatty acids, phospholipids and mitochondrial swelling. The addition of free fatty acids to isolated mitochondria causes swelling in this organelle and at the same time uncouples oxidative phosphorylation by inhibiting the translocation of adenine nucleotides through mitochondrial membranes (Wojtczak and Zaluska, 1969). Studies by Wojtczak and Lehninger (1961) indicated that the swelling of rat-liver mitochondria induced by calcium ions can be ascribed to the enzymic liberation of fatty acids from endogenous phospholipid and that subsequent ATP-induced contraction paralleled the reincorporation of the free fatty acids into phospholipid. Similarly Wojtczak et al. (1963) found that formation of [^{32}P]phosphatidic acid from labelled glycerophosphate paralleled the contraction process. These studies point to an intimate relationship between phospholipid metabolism and the swelling–contraction cycle of mitochondria which possibly possess a very active system which couples the hydrolysis of phospholipid into the corresponding lyso-compound and free fatty acid with the reverse reaction (p. 190) in which lysophospholipids are reacylated to give the corresponding diacyl phospholipids.

(c) *Fatty acid desaturase systems.* Jones and coworkers (1969) found that a solubilized fatty acid desaturase system from rat liver microsomes lost activity when the preparation was partially delipidized by extraction with aqueous acetone, and this activity was largely restored when the extracted lipid was restored to the system. Although Gurr et al. (1970) were unable to repeat these findings, the observation of Jones et al. illustrates the likelihood that some of the enzyme systems involved in various aspects of lipid metabolism will themselves depend on lipids as necessary structural elements.

This kind of lipid participation in acyl desaturases has not yet been observed in plant tissues although there is now good evidence that lipids may act as essential cofactors, and even substrates, in reactions leading to changes in the structure of their component acids. Some of this evidence that plant acyl lipids play an intermediary role in the synthesis of unsaturated fatty acids has already been described (Chapter 6). Not all classes of lipid are likely to be involved in this type of role, and on the basis of rates of incorporation and turnover of label from [^{14}C]acetate added to *Chlorella vulgaris* Nichols *et al.* (1967) suggested that phosphatidyl glycerol, phosphatidyl choline and monogalactosyl diglyceride are particularly involved in fatty acid synthesis and metabolism in this way with the other cellular lipids playing a more specifically structural role. Subsequent studies by Nichols (1968) indicated that the relative importance of different classes of lipid in this type of involvement may differ according to the evolutionary status of the plant involved.

It is not clear whether the acyl desaturations in question occur within the lipid itself or whether the acyl groups are first transiently detached from the lipid prior to the dehydrogenation (p. 189). In the bacterium *Clostridium butyricum* synthesis of the cyclopropanoid acid lactobacillic acid from vaccenic acid occurs while the precursor and product acids remain part of the phosphatidyl ethanolamine molecule, and similar mechanisms appear to be involved in the synthesis of cyclopropane and cyclopropene acids in some plant tissues (Chung and Law, 1964; Hooper and Law, 1965).

(*d*) *Other enzymes.* One of the best established examples of a lipid cofactor requirement in plant metabolism is the characterization of a cyclic tetramer of ricinoleic acid as an essential cofactor for the activation of castor bean lipase (p. 194) although this function is an unique one and does not appear to be generally operative in seeds. In addition the cellulose synthetase system from oat seedlings has a non-specific requirement for phospholipid for optimum activity (Pinsky and Ordin, 1969) while several membrane-bound enzyme systems common to mammalian and plant cells have been found to require phospholipids for optimum activity. These include succinic dehydrogenase (Cerletti *et al.*, 1967), GTP-activated fatty acid thiokinase (Sartorelli *et al.*, 1967), NADH-cytochrome *c* reductase (Jones and Wakil, 1967), glucose-6-phosphatase (Dultera *et al.*, 1968), ATP-ase (Martinosi *et al.*, 1968) and UDP-glucuronyl transferase (Graham and Wood, 1969). All these systems seem to require phospholipids primarily as structural elements and do not appear to have absolute requirements for phospholipids of specific structure, although mammalian mitochondrial β-hydroxybutyrate dehydrogenase activity, which is lost when depleted of lipid, is restored only by the addition of sulphydryl compounds and lecithin. Other mitochondrial lipids are ineffective or inhibiting (Jurtshuk *et al.*, 1961).

2. Other Functions

A. Energy Storage

Germinating seeds require a source of carbon skeleton precursors and also a source of energy (ATP) to assemble these precursors into the carbohydrates, proteins, fats etc. needed for the elaboration of the cell. Both requirements are effectively supplied by acyl lipids, usually in the form of triglycerides. When seeds germinate there is a marked increase in the level of endogenous lipase activity (p. 194) resulting in a rapid fall in the concentration of triglyceride, and the released fatty acids are then catabolysed to ATP and acetyl-CoA by a system largely involving β-oxidation (p. 205). This acetyl-CoA is then converted to sucrose which is freely water-soluble and is consequently easily transported to other parts of the tissue where it is converted into many of the compounds required by the developing plant.

The conversion of acetyl CoA to sucrose is effected by a series of reactions which constitute the glyoxylate cycle (Fig. 8.3, p. 203), and Beevers (1961) intensively studied the many aspects of this cycle in plant tissues. He showed that in some 25 different plants examined, both isocitratase and malate synthetase were confined to tissues rich in lipid, and were particularly active in these tissues from germinating seedlings. In general, no activity can be found in the ungerminated seed, in the maturing seed or in parts of the plant which do not convert lipid to carbohydrate. During the first few days of germination the enzyme activities rise dramatically but after most of the lipid has been utilized the activities of the two enzymes rapidly decline and glyoxylate cycle activity ceases. It is about this time when the plant is beginning to photosynthesize.

Although the lipid involved in energy generation is most commonly triglyceride, there occur a few classes of seed such as those of *Simmondsia californica* (Green *et al.*, 1936) and *Murraya koenigi* (Kartha and Singh, 1969) which contain little or no triglyceride and whose energy during the process of germination is largely derived from wax esters and hydrocarbons. Wax esters in dark-grown cells of *Euglena gracilis* play a similar role as reserve material (Rosenberg, 1967a) and Miyachi and Miyachi (1966) have suggested that the plant sulpholipid might serve as an emergency reserve of carbon and sulphur in plant cells. The occurrence of large quantities of galactosyl diglycerides in the seeds of some grasses (Smith and Wolf, 1966) suggests that in some circumstances these lipids may also serve as reserves of metabolic energy, a possibility which is strengthened by the observations of Rosenberg (1967b) that the levels of these compounds diminish precipitously when healthy green cells of *Euglena gracilis* were starved of light and nutrients, until a ratio of one molecule of chlorophyll to two molecules of galactosyl diglyceride was reached. Beyond this point chlorophyll and galactosyl diglycerides disappeared together and the organism began to become dormant and to disintegrate.

B. Cellular Repair Mechanisms

In 1922 Haberlandt reported that damaged plant cells produce a "wound hormone" which accelerates the enlargement of cells around the affected area, and 17 years later English and coworkers (1939) demonstrated that the active principle involved is 1-decene-1,10-dicarboxylic acid, which they named traumatic acid:

$$\overset{t}{HO_2C.CH\!=\!CH(CH_2)_8CO_2H}$$

More recently the work of Hall and Morris (1970) showed that the traumatic acid generated by bruised bean tissue is derived from free linoleic and linolenic acids, and these authors proposed a mechanism for this conversion (Fig. 10.4) which involves initial attack on the polyenoic acid by lipoxygenase and isomerization by hydroperoxide isomerase (Zimmerman, 1966) of the hydroperoxide formed. This mechanism is consistent with the observations of Heinen and Brand (1963) who noted

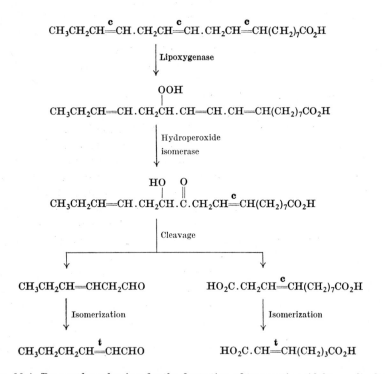

Fig. 10.4. Proposed mechanism for the formation of traumatic acid from α-linolenic acid (Morris and Hall, 1970).

that lipoxygenase and fatty acid dehydrogenase activity increase significantly when plant leaves are damaged.

An additional function for acyl lipids containing linoleic or linolenic acids may therefore be to provide suitable substrates for the formation of traumatic acid in damaged tissue, since plant cells do not accumulate significant quantities of free fatty acids and lipoxygenase does not normally react with lipid-bound fatty acids. Bruising the tissue presumably liberates lipolytic enzymes which free the polyunsaturated acids from endogenous phospholipids or glycolipids after which the liberated acids would then be converted into traumatic acid by the mechanism described (Fig. 10.5).

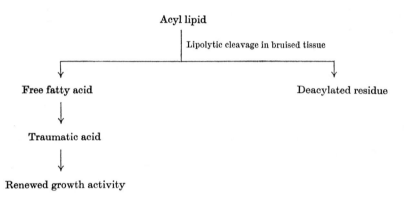

FIG. 10.5. Possible sequence of events following the bruising of plant tissue.

C. Cutin Formation

Cutin and waxes form the major constituents of the cuticle layer which covers the epidermal cells of the leaves, fruits and other tissues of higher plants and although the physiological significance of the cuticle is only partly understood it is certain that it plays a very important role in transpiration (Hall and Jones, 1961). An important secondary role for the lipids of leaves and fruits may therefore be their availability for incorporation into the cutin and wax fractions of the cuticle layer which according to Lee and Priestley (1924) is built up by the excretion of the required precursors through the peripheral plant cell walls. At the surface, when in contact with atmospheric oxygen and possibly with the aid of enzymes diffusing into the cuticular complex from epidermal cells, these unsaturated molecules oxidize and polymerize forming cutin according to the mechanisms already described (p. 191) (Mazliak, 1968).

D. Ethylene Production

In higher plants ethylene has a variety of important physiological functions and is now frequently assigned the status of a true plant hormone (Pratt and Goeschl, 1969).

Two substrates have been implicated as precursors of this compound in plant tissue, namely methionine and linolenate, and although tracer and enzymic studies make it clear that methionine can serve as a precursor of ethylene in a number of plant tissues (Mapson and Wardale, 1967) studies by Galliard et al. (1968a) have emphasized the possibility of linolenate also being a precursor. These workers found that cell-free extracts of apple skin can produce ethylene (and ethane) enzymically when supplemented with linolenic acid and ascorbate, in a reaction which appears to involve the initial formation of epoxidized linolenic acid by the action of lipoxidase. Although the degree of conversion of [14C]linolenic acid to ethylene was 10–20 times less than that obtained with [14C]methionine, Galliard et al. believed that the rate of conversion of linolenic acid to ethylene by extracts was related to the true ethylene production in vivo of the tissue from which the extracts were prepared.

Subsequent work by Mapson et al. (1970) failed to detect, in tissue slices of a variety of plants, the presence of a natural system for the conversion of labelled linolenate to labelled ethylene, although the fatty acid did appear to stimulate the conversion of methionine into ethylene in apple tissue. Mapson et al. consequently concluded that linolenate must play a secondary but essential role in ethylene biosynthesis in apple. It was proposed that this role might be the replacement of a glucose oxidase system by a lipoxygenase system. The fatty acid hydroperoxide produced by this latter enzyme is known to be capable of replacing hydrogen peroxide in the generation of ethylene from methionine or its derivatives (Mapson and Wardale, 1968) and some preliminary studies by these workers have indicated that in tomato a lipoxygenase rather than a glucose oxidase is involved in the synthesis of ethylene.

CHAPTER 11

The Analysis of Plant Lipids

The past decade has seen a rapid growth in our understanding of the chemical nature of the lipids and fatty acids synthesized by the plant kingdom and of the biochemical processes concerning their formation and utilization by the cell. To a large degree credit for this advance in knowledge must be due to the development of a variety of analytical techniques, many of them chromatographic, which have made the fractionation and analysis of plant lipids and their component acids a comparatively simple, routine operation.

In many cases the development of these newer techniques has taken place originally in laboratories in which mammalian chemistry and biochemistry is the major interest, and methods of such origin are often not applicable to studies of plant lipids without some degree of modification. In this chapter we review the more important methods of lipid and fatty acid analyses putting particular emphasis on those techniques which have been successfully applied to plant tissues.

1. EXTRACTION AND STORAGE

The initial problem in the determination of the lipid content of a plant tissue occurs in selecting the best method of extraction; insufficient attention to this crucial aspect of the analytical operation may result in a variety of errors stemming either from the incomplete extraction of the cellular lipids or from non-physiological alterations in lipid structure occurring at this stage.

As a general rule tissues from higher plants should be extracted as soon after isolation from the parent plant as is practicable, since storage at above-freezing temperatures can result in alterations in the structure of endogenous lipids (van der Veen and Olcott, 1967). These alterations cannot always be satisfactorily avoided by deep-freezing the tissues since this operation may result in partial breakdown of cellular structure by ice crystals, and on thawing endogenous lipases released by this cellular breakdown can effect the hydrolysis of acyl lipids by the mechanisms described in Chapter 7. Usually, the presence of significant quantities of free fatty acid and/or phosphatidic acid in extracts from plant tissues is indicative of some degree of non-physiological autolysis of cellular lipid.

Although it is not usually difficult to obtain quantitative extraction of lipids by maceration of fresh tissue with any one of a variety of solvents, most of which contain a fairly high percentage of alcohol to effect dehydration of the tissue, care must be taken in selecting the correct combination in order to minimize lipase activity during the operation. In leaf tissues the enzymes normally most active during extraction are phospholipase D and those which catalyse the hydrolysis of the galactosyl diglycerides, particularly the former. Kates (1956) found that phospholipase D activity is activated by many of the organic solvents commonly used in extraction mixtures, including chloroform and diethyl ether, but noted that maceration of the tissue with *iso*- or *n*-propyl alcohols effectively eliminated lipolytic activity. Of these two alcohols, the *iso*-compound is the more convenient in use because it has the lower boiling point, and a procedure normally satisfactory for leaves involves maceration of the tissue in 20 volumes of cold *iso*-propanol followed by re-extraction of the residue with chloroform–methanol (2:1 v/v).

The tendency for lipids to become lysed during extraction varies considerably from tissue to tissue, phospholipase D activity being apparently greater in some leaves than in others; this is probably due to the presence of varying levels of natural inhibitors of this lipase (Davidson and Long, 1958). These natural inhibitors are usually removed in the course of the preparation of subcellular organelles, such as chloroplasts, and as a consequence extreme caution must be taken in the isolation of chloroplasts and the extraction of their component lipids.

In general the pitfalls described above are most commonly encountered during the analysis of leaf tissue. Other tissues from higher plants appear to possess lower levels of lipolytic activity, while most unicellular algae present few extraction problems and shaking the cells with about 20 volumes of chloroform–methanol (2:1 v/v) usually gives quantitative recovery of endogenous lipid. Seed lipids can also be satisfactorily removed by maceration of the tissue with chloroform–methanol or, when a high concentration of triglyceride is present, with petroleum. However, freezing and thawing of seed tissue frequently results in release of substantial quantities of free fatty acids.

There also occur a small minority of plant tissues from which it is apparently impossible to remove all endogenous lipid by treatment with the more common solvent combinations. Some classes of algae in particular are extracted only with difficulty and Nichols and Appleby (1969) found that extraction of cells of the yellow alga *Monodus subterraneus* with chloroform–methanol for 24 hours at room temperature removed little chlorophyll and only 50–75% of the total endogenous lipid. In this case there was no indication that a specific class of lipid was preferentially retained by the cells since the fatty acid compositions of the total extractable lipid fraction and that not extracted were almost identical.

It is also difficult to extract all lipids from cereals and their products (e.g. wheat or wheat flour) and special solvent systems have been devised for this specific purpose. Of these, the water-saturated butanol system proposed by Mecham and Mohammad (1955) has been most widely employed although Inkpen and Quackenbush (1969b) have found the chloroform–ethanol–water (200:95:5 v/v) is equally effective and more convenient to use. However, Inkpen and Quackenbush also demonstrated that neither solvent system completely removes all lipid and fatty acid from cereals and cereal products and that up to one-sixth of the total fatty acid content of wheat flour could be isolated only following acid hydrolysis of the solvent extracted residue. Significantly the composition of these more strongly bound fatty acids was very different from that of the corresponding extractable lipid fraction, which suggested the possibility that the strongly bound lipid may be structurally different from that which is more easily extracted. This possibility was directly substantiated by Wren and Merryfield (1970) who found that wheat starch granules gave measurable yield of hydrolysate lipid even after exhaustive extraction with methanol or dioxan. This "firmly bound" lipid was nevertheless readily released when the granular structure was disrupted by gelatinization in boiling water or 80% dioxan at 30° or by erosion in pure dimethyl sulphoxide, and mainly consisted of lysolecithin which did not comprise a significant proportion of the more easily extractable lipid fraction.

Plant lipid extracts frequently possess acyl residues of fairly high degrees of unsaturation which may render them vulnerable to oxidative deterioration. While extracts from many tissues will be stabilized by the presence of natural antioxidants such as the tocopherols, partially purified fractions from these extracts may not be so stabilized and it is advisable that they be stored at low temperatures (−5°C) in solvent containing a suitable level of added antioxidant. Wren and Szczepanowska (1964) have found that BHT antioxidant (butylated hydroxy toluene or 2,6-di-tert-butyl-cresol) at levels of 0·005% with respect to the solvent is extremely effective in this respect.

2. FRACTIONATION AND ANALYSIS OF PLANT LIPIDS

A. Preliminary Separations

Many plant tissues contain lipid mixtures of such complexity that any single analytical method is insufficient to separate all the lipid classes present, and a combination of complementary techniques is frequently required. In particular a preliminary bulk separation into two or more groups of lipid classes may considerably facilitate the ultimate resolution of the mixture.

Traditionally this initial operation comprised a precipitation of phospholipids by the addition of 4 volumes of cold acetone to a solution of mixed lipids in diethyl ether, triglycerides and other neutral lipids remaining in solution. It is now generally recognized that in many circumstances this technique is very inefficient because of the ability of large proportions of triglycerides to co-solubilize more polar lipids into acetone, and because glycolipids and some phospholipids, such as highly unsaturated lecithins, are naturally soluble in this solvent. Similarly, partial segregation of phospholipid mixtures by precipitation with ethanol to give lecithin (soluble) and cephalin (insoluble) fractions are now known to yield grossly impure and cross-contaminated preparations.

More satisfactory bulk separations of "neutral lipids" (hydrocarbons, sterols, sterol esters, mono-, di- and tri-glycerides and free fatty acids) from phospholipids and glycolipids may be obtained by passing a solution of plant lipids in diethyl ether over a short column of silicic acid. Only the last two classes of lipid are retained by the adsorbent under these conditions and these can be eluted subsequently by chloroform–methanol mixtures (Nichols, 1964). A simple countercurrent distribution procedure based on a solvent system devised by Carter et al. (1961a) has also been employed by Nichols (1964) to effect a quantitative separation of "neutral" and "polar" plant lipids. Nichols employed separating funnels in a 3-funnel 6-withdrawal system in which the two phases were petroleum (br. 80–100°C) and 95% aqueous methanol, neutral lipids accumulating in the petrol phase and phospholipids and glycolipids in the methanol phase. This method is much faster than the alternative chromatographic procedure but has the minor disadvantage that free fatty acid, if present, distributes fairly evenly between the fundamental and withdrawn phases.

If comparatively small quantities of mixed lipids are to be separated, as when fatty acid analyses of general groups are required, then preparative thin layer chromatography on silicic acid using chloroform as the mobile phase, is the quickest and most convenient method available. With this solvent triglycerides have an R_f of about 0·5 whereas phospholipids and glycolipids do not move from the origin.

Bulk separations such as those described here are particularly valuable in studies concerned with the nature of phospholipids and glycolipids in tissues containing high proportions of triglycerides, as in oil-seeds.

Preliminary separations of certain lipid classes may be achieved in some circumstances by chemical methods. For example sphingolipids which are stable to alkali may be prepared from lipid extracts by removing ester-lipids with alkali, an approach which was utilized by Carter et al. (1962) to isolate phytoglycolipids from extracts of oil-seeds.

B. Separation of Neutral Lipid Classes

1. Column chromatography

Early chromatographic separations of the different classes of neutral lipid were originally carried out on silica gel columns, but the separations obtained with this adsorbent are somewhat inferior to those obtained by Carroll (1961) who devised a method which employed activated magnesium silicate (Florisil) as stationary phase and hexane–ether mixtures, with stepwise increases in polarity, for elution. Hydrocarbons, sterol esters, triglycerides, free sterols, diglycerides, monoglycerides and free fatty acids are separated in that order, that is to say in the same order as that obtained with silicic acid except that free fatty acids are eluted after monoglycerides. In this respect Florisil has a distinct advantage over silicic acid which often gives a poor resolution of triglycerides from free fatty acids. Other advantages are that it requires no pre-washing or other preliminary treatment apart from deactivation with water, that the columns are quickly and easily packed, and that the relatively coarse particles of Florisil permit excellent flow rates. One disadvantage is that phospholipids are less easily recovered quantitatively from this adsorbent than from silicic acid. Carroll (1963) subsequently reported that washing Florisil with acid gives a silicic acid, which, when packed as a column, retains the excellent flow rate characteristics of the silicate columns but on which separations of the type obtained with ordinary silica gel are possible.

2. Thin-layer chromatography

Although silicic acid impregnated filter papers were widely employed in the past for the fractionation of neutral lipid mixtures, their use has been largely superseded by thin-layer chromatography (TLC) on silica gel which gives superior resolutions to those obtained with impregnated papers and also permits the use of highly sensitive non-specific spray reagents such as sulphuric acid, which cannot be used with papers.

Most of the major classes of neutral lipid can be separated by unidimensional TLC on silicic acid using one (Vogel et al., 1962; Brown and Johnson, 1962) or more than one (Vogel et al., 1962; Skipski et al., 1965) development. By double development TLC on gypsum-free silicic acid using isopropyl ether–acetic acid (24:1 v/v) and light petroleum–ethyl ether–acetic acid (90:10:1 v/v) as consecutive solvents, Skipski *et al.* (1965) separated hydrocarbons, cholesterol esters, fatty acid methyl esters, triglycerides, free fatty acids, 1,3-diglycerides, 1,2-diglycerides, sterols and polar lipids in that order, the latter class of lipid remaining at the origin (Fig. 11.1).

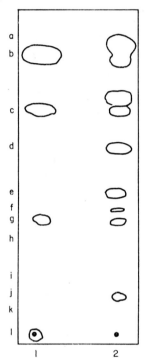

FIG. 11.1. Separation of neutral lipid classes by double development TLC (from Skipski *et al.*, 1965). Chromatogram of total lipids extracted from human serum. Lipid extract (1); mixture of reference compounds (2). Identification of components: a, hydrocarbons; b, cholesterol esters; c, triglycerides; d, fatty acids; e, 1,3-diglycerides; f, 1,2-diglycerides; g, cholesterol; h, i, unknown; j, monoglycerides; k, unknown; l, phospholipids (at origin). Solvent systems: isopropyl ether–acetic acid (24:1 v/v) followed by petrol–ethyl ether–acetic acid (90:10:1 v/v).

C. Separation of Phospholipids and Glycolipids

1. Column chromatography

Although column chromatography on silicic acid often yields good separations of the polar lipid fractions from many animal tissues, only partial fractionation of the phospholipids and glycolipids commonly found in plant extracts can be obtained with this adsorbent (Wheeldon,

1960). Far superior separations of the polar lipids of plants can be obtained by a combination of column chromatography on DEAE cellulose (acetate form) and either TLC (Nichols and James, 1964) or column chromatography on silica gel (Allen et al., 1964) (Fig. 11.2). The method devised by Nichols and James is essentially as follows. The acetate form of DEAE cellulose powder prepared according to the method of Rouser et al. (1963) is poured into a glass column as a slurry in glacial acetic acid, and is then thoroughly washed with methanol and

Solvent	Volume [ml]	Tubes	Lipids present
CHCl₃	400	1–3	Neutral lipids, U1 + U2
		4–20	U3
		21–40	Monogalactosyl diglyceride
CHCl₃–CH₃OH (95:5 v/v)	200	41–59	Phosphatidyl choline / Cerebroside / Sterol glycoside
CHCl₃–CH₃OH (90:10 v/v)	200	60–77	Digalactosyl diglyceride
CHCl₃–CH₃OH (60:40 v/v)	200	78–99	Phosphatidyl ethanolamine
CHCl₃–CH₃OH (2:1 v/v, saturated with conc. NH₄OH)	200	—	Phosphatidyl inositol / Phosphatidyl glycerol / Cardiolipin / Phosphatidic acid / Sulpholipid

Fig. 11.2. Fractionation of narcissus bulb lipids on DEAE-cellulose (Nichols and James, 1964). 120 mg lipid (1·2% of total dry bulb weight) applied to DEAE cellulose column (12 mm × 340 mm). 10 ml fractions collected. U1, U2, U3 indicate unidentified lipids.

then chloroform. The mixture of lipids to be fractionated is then applied to the column as a solution or suspension in pure chloroform, the load/stationary phase ratio not exceeding 50 mg/g. Lipid classes are then eluted by stepwise increases of methanol in chloroform, the course of the fractionation being followed by TLC. Acidic lipids are eluted with ammoniated chloroform–methanol and can then be fractionated by preparative TLC on silicic acid using chloroform–methanol–acetic acid–water (85:25:8:3 v/v) as mobile phase, individual components being located with dichlorofluorescein and the pure compounds then displaced from the adsorbent with chloroform–methanol (1:1 v/v).

2. Thin-layer and impregnated paper chromatography

(a) General. Of the various solvent systems employed for the separation of plant phospholipids and glycolipids by TLC, the one composed of

Di*iso*butylketone–acetic acid–water (40:25:3·7 v/v) gives good resolution of most major classes, but this mixture has the disadvantage of running comparatively slowly and the ketone is not easily removed from the chromatograph (Nichols, 1963), and Pohl *et al.* (1970) have found that acetone–benzene–water (91:30:8 v/v) gives almost equally good separations. A chloroform–methanol–acetic acid–water (170:30:20:7 v/v)

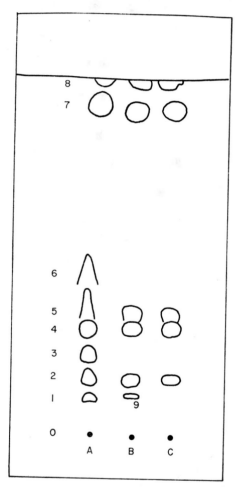

Fig. 11.3. Separation of phospholipids and glycolipids by one-dimensional TLC (from Nichols *et al.*, 1965a). Chromatogram of total lipids extracted from *Chlorella vulgaris* (A), *Anabaena variabilis* (B) and *Anacystis nidulans* (C). Identification: 1, phosphatidyl inositol; 2, sulpholipid; 3, lecithin; 4, digalactosyl diglyceride; 5, phosphatidyl glycerol; 6, phosphatidyl ethanolamine; 7, monogalactosyl diglyceride; 8, neutral lipid; 9, unidentified lipid. Solvent system: chloroform–methanol–acetic acid–water (170:30:20:7 v/v).

system devised by Nichols *et al.* (1965a) also gives good results (Fig. 11.3) although the resolution between phosphatidyl glycerol and digalactosyl diglyceride is sometimes poor. Clayton *et al.* (1970) obtained resolution of some minor lipids of wheat flour on thin layers of silica gel using chloroform–acetone–methanol–acetic acid (73:25:1·5:0·5 v/v), which separated steryl esters, diglycerides, free fatty acid, 6-*O*-acyl monogalactosyl diglyceride, monogalactosyl diglyceride and *N*-acyl phosphatidyl ethanolamine in that order of decreasing R_f. Silicic acid impregnated filter papers also give good separations of major plant lipids when a diisobutyl ketone–acetic acid–water (40:25:5) mixture is employed as mobile phase (Sastry and Kates, 1965), although the technique has the disadvantage that sensitive non-specific corrosive reagents (p. 293) cannot be used.

When detailed qualitative analyses of lipid extracts are required, it is often desirable to employ two-dimensional thin layer chromatography (2-D TLC) which not only yields clearer-cut separations than are obtained by TLC in one dimension but also permits larger quantities of lipid mixtures to be applied to a single chromatogram, thereby facilitating the detection and identification of minor components. Nichols (1964) reported the separation of phospholipids and glycolipids from several plant tissues using 2-D TLC on silicic acid in which chloroform–methanol–7N ammonium hydroxide (65:30:4) and chloroform–methanol–acetic acid–water (170:25:25:4) were used as the successive mobile phases (Fig. 11.4). Lepage (1964) achieved similar separations with an alternative system which has the slight disadvantage of utilizing a rather involatile solvent system based on diisobutylketone, and more recently Singh and Privett (1970) achieved good resolution of the complex lipids of immature soybean using slight modifications of the systems employed by Nichols.

For detailed qualitative analyses of lipid mixtures an alternative approach is to effect a bulk separation of "neutral" polar lipids from acidic lipids by treatment with DEAE-cellulose followed by unidimensional TLC of the two lipid groups.

(*b*) *Inositol lipids.* The complex inositol-containing phosphoglycolipids found in seeds (p. 55) are not easily isolated or fractionated by the more standard techniques described above. Carter and Kisic (1969) separated mixtures of phosphatidyl inositol, phytoglycolipids and ceramide phosphate polysaccharides from each other and from the more common glycerophosphatides by a series of simple countercurrent distributions, first as the naturally-occurring calcium and magnesium salts and subsequently in the sodium salt form. The solvent employed was *n*-hexane–water-saturated *n*-butanol–95% methanol. The separations obtained were checked by analysis of each fraction by paper

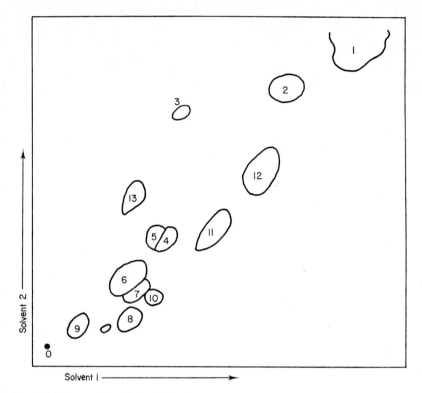

Fig. 11.4. Separation of phospholipids and glycolipids by two-dimensional TLC (from Nichols, 1970). Chromatogram of total lipids extracted from *Ochromonas danica*. Identification of components: 1, neutral lipids; 2, monogalactosyl diglyceride; 3, cardiolipin; 4, phosphatidyl glycerol; 5, phosphatidyl ethanolamine; 6, digalactosyl diglyceride; 7, sulpholipid; 8, lecithin; 9, phosphatidyl inositol; 10, 11, 12, 13, unidentified lipids; O, origin. Solvent systems: 1, chloroform–methanol–7N ammonium hydroxide (65:30:4 v/v); 2, chloroform–methanol–acetic acid–water (170:25:25:4 v/v).

chromatography on formaldehyde impregnated papers according to the method of Horhammer *et al.* (1959) using *n*-hexane–acetic acid–water (4:1:5 v/v, upper layer) as mobile phase.

3. Analysis of water-soluble derivatives

Before TLC became established as the principal technique for determining the lipid composition of plant tissues, Benson *et al.* (1958, 1959a, b) had developed paper chromatography of the water soluble products of lipid deacylation as an elegant and highly informative analytical technique. In this method the lipid preparation is reacted with 0·1N methanolic KOH at 37° for 15 minutes after which it is neutralized with an ion-exchange resin and the water-soluble hydrolysis products

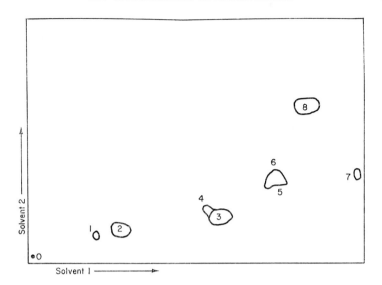

FIG. 11.5. Separation of products of deacylation of phospholipids and glycolipids by two-dimensional paper chromatography (from Kates and Volcani, 1966). Chromatogram of deacylated lipids of *Navicula pelliculosa*. Identification of components: deacylation products from 1, phosphatidyl inositol; 2, sulpholipid; 3, digalactosyl diglyceride; 4, phosphatidyl glycerol; 5, phosphatidyl ethanolamine; 6, monogalactosyl diglyceride; 7, lecithin; 8, neutral glycerides. Solvent systems: 1, phenol–water (100:38 v/v); 2, butanol–propionic acid–water (142:72:100 v/v).

concentrated in a small volume of water. Portions of this solution are then submitted to two-dimensional paper chromatography on heavy filter paper using phenol–water (3:1 v/v) for the first development and butanol–propionic acid–water (14:7:10 v/v) for the second (Fig. 11.5). The individual components can then be detected on the chromatograms with the Schiff-periodate reagent, or in the case of phosphate esters, by the phosphomolybdate spray.

The technique is still of great value in confirming the identity of known plant lipids, and in establishing the structures of new ones.

D. Isolation of Specific Molecular Species of Individual Lipid Classes

The various biochemical and biophysical functions proposed for acyl lipids in plants imply that these different functions may involve individual lipids of specific fatty acid composition and that the "pure" lipid classes isolated from whole cells may represent a conglomeration of molecular species having different functions. To determine the detailed

10

fatty acid structure of individual molecular entities and to test such hypotheses as to their different roles in plant metabolism, it is clearly necessary to obtain specific fractionations of the individual lipid classes into components of different fatty acid composition.

To date, most such separations have been obtained by the technique of argentation thin layer chromatography in which the lipid is chromatographed on thin layer chromatograms of silicic acid impregnated with silver nitrate. Complex formation between silver ions in the adsorbent and the π-orbitals of the olefinic groups in the acyl chains results in separation of lipid classes containing different overall degrees of unsaturation, those most highly unsaturated complexing most strongly and consequently running more slowly on the chromatogram.

1. Triglycerides

Seed oil triglycerides containing epoxy, hydroxy and other substituent groups can often be substantially separated by conventional TLC on silicic acid according to the type of substituent group and the number of substituted acids per triglyceride molecule (Miller *et al.*, 1965; Evans *et al.*, 1965), and adsorbtion TLC on unmodified layers has also been used to separate the tetra-acid triglycerides of the oils of *Stillingia* (Sprecher *et al.*, 1965) and *Sapium sebiferum* (Christie, 1969) while Morris and Hall (1966) used multiple development TLC for the separation and isolation of the estolides of ricinoleate-containing triglycerides which comprise ergot oils.

Glycerides containing only non-substituted saturated and unsaturated acids cannot normally be subfractionated according to their constituent acids unless argentation chromatography is used, and two of the original papers concerning argentation chromatography described the fractionation of diglycerides and triglycerides on columns (de Vries, 1963) and on thin layers (Barrett *et al.*, 1962) in which the separations were primarily according to the total number of *cis* double bonds in each glyceride although some *cis, trans* isomers were also separated.

It soon became evident that separations of triglyceride species are not simply in the order of the total number of *cis* double bonds in each molecule, but that one linoleic acid residue, for example, has a greater effect on retention than two oleic acid residues. Gunstone and Padley (1965) determined the order of complexing affinities of triglycerides, neglecting effects of positional isomers, to be 333, 332, 331, 330, 322, 321, 320, 311, 222, 310, 221, 300, 220, 211, 210, 111, 200, 110, 100 and 000, where 3 = linolenic, 2 = linoleic, 1 = oleic and 0 = saturated acids. The authors assigned arbitrary values for the complexing effect of each acyl chain, viz. saturated = 0, oleic = 1, linoleic = (2 + a) and linolenic =

(4 + 4a), where a is some fraction less than unity, to permit prediction of the order of migration of polyacyl compounds such as triglycerides. Positional isomerism may result in further fractionation of some of the classes listed.

Argentation column chromatography, which is convenient for relatively large scale separations, does not give very efficient fractionation of glycerides containing more than four double bonds per molecule (Subbaram and Youngs, 1964) but Gunstone and coworkers (1965) partly circumvented this difficulty by preliminary fractionation by low temperature crystallization from solvents containing silver nitrate. Argentation TLC is very effective for the fractionation of the normal range of vegetable triglycerides (Fig. 11.6) but is not suitable for mixtures

FIG. 11.6. Separation of triglycerides according to their overall degree of unsaturation by one-dimensional AgNO₃-TLC (from Morris, 1966). Chromatogram of cottonseed oil triglycerides. The numbers represent total numbers of double bonds in each triglyceride molecule. Solvent: isopropanol–chloroform (1·5:98·5 v/v). Stationary phase: Silica gel impregnated with silver nitrate (5% w/w).

of highly unsaturated triglycerides for which Litchfield (1968) has described an alternative method of fractionation involving the combination of reversed-phase partition TLC and gas chromatography.

Natural glyceride mixtures can seldom be fully resolved into all their individual molecular species by one technique alone and argentation chromatography of these substances reaches its full potential utility only when used in conjunction with other chromatographic methods. For example, argentation TLC was effectively complemented by normal adsorbtion TLC in studies of the composition of seed oil triglycerides containing hydroxy fatty acids which, having both short chain and hydroxy acid constituents as well as the more usual fatty acids, are amenable to preliminary fraction by adsorption chromatography (Gunstone and Qureshi, 1968). More generally, argentation-TLC is combined with reversed-phase partition TLC (e.g. Kaufmann and Wessels, 1964) or GLC of the intact glyceride subfractions so obtained (Litchfield et al., 1964; Jurriens and Kroesen, 1965).

2. Phospholipids and glycolipids

With triglycerides, diglycerides and other comparatively apolar lipids, argentation-TLC of the unmodified lipid normally gives acceptable results. With more polar lipids, such as phospholipids, fractionation of the unaltered material can give the desired separation in some cases, but many authors have preferred to remove the polar phosphate group enzymically and then fractionate the resultant diglyceride mixture. Isolation of phospholipids and glycolipids species according to fatty acid content have so far been largely confined to preparations of animal origin and the only reported studies relating to plant metabolism are those described by Haverkate and van Deenen (1965) for the phosphatidyl glycerol fraction of spinach leaves and by Nichols and Moorhouse (1969) regarding the monogalactosyl diglyceride fraction of *Chlorella vulgaris*. The conditions required for fractionation of lipids into their component fatty acid species naturally depend not only on the nature of the acyl lipid in question but also on the complexity and nature of the component fatty acids. It is therefore not particularly useful to quote here the specific conditions which have been employed for specific separations, but for a detailed consideration of this field the reader is referred to recent reviews by Viswanathan (1968) and Morris and Nichols (1970).

E. Detection of Lipids on Thin Layer Chromatograms

A wide variety of specific and non-specific reagents are now available for the detection and characterization of lipids on thin-layer chromatograms. Some reagents, for example 2′,7′-dichlorofluorescein (Dunphy *et al.*, 1965) and Rhodamine 6G (Rouser *et al.*, 1961) permit detection without chemical alteration of the lipids which may then be recovered by eluting the relevant areas from the chromatograms. Iodine also comes into the category of non-destructive reagents except that it can alter some fatty acids and is therefore not suitable for the location of lipids whose fatty acid compositions are to be subsequently determined (Nichaman *et al.*, 1963). Other reagents, including the ninhydrin reagent for primary amines such as phosphatidyl ethanolamine and phosphatidyl serine (Skidmore and Entenman, 1962), the Schiff-periodate reagent for glycolipids and other diols (Nichols, 1964) and Zinzadze's reagent for phospholipids (Dittmer and Lester, 1964) are more specific and may give some information regarding the structure of the lipids separated, although these reagents are of variable sensitivity and do not permit the recovery of unaltered lipids.

Various spray reagents based on sulphuric acid (e.g. Morris and Nichols, 1970) are non-specific but are extremely sensitive, detecting as little as 1 μg lipid, and may also give some information regarding the nature of the lipids present. For example, when thin layer chromatograms of mixtures of plant lipids are sprayed with 25% sulphuric acid and placed in an oven at 230° then initially certain classes of lipid give typical colour reactions. Thus free and combined sterols give the red-purple colouration typical of the Liebermann-Burchard reaction, glycolipids turn red-brown (galactosyl diglycerides) or bright red (sulpholipid) while phospholipids and neutral lipids are observed as pale brown areas.

F. Stability of Lipids During Chromatography

Many classes of lipid, particularly those containing polyenoic acids are readily oxidized by exogenous oxygen so that during all manipulative procedures every precaution must be taken to avoid deteriorative reactions. Thus, in column chromatography all solvents including those used to slurry adsorbent into the column should be oxygen-free. This condition may be obtained either by bubbling nitrogen through all solvents immediately prior to use, or by distillation of the solvents in an inert atmosphere. Since ultraviolet radiation can catalyse oxidative and other deteriorative reactions in lipids it may also be advisable to shield the chromatographic apparatus from direct sunlight.

Oxidation of lipids is a particular hazard during TLC. There is good evidence that oxidation does not occur during the development stage (Malins and Mangold, 1960) and deterioration is most likely to occur when the material is spread on the dry chromatogram both before and after development, at which time the lipid has a large surface area exposed to the atmosphere. Such a situation also occurs between developments in multiple development or two-dimensional TLC. This problem may be resolved either by maintaining an atmosphere of nitrogen or other inert gas over the chromatogram during the loading and drying operations or by incorporating an antioxidant such as BHT in all storage and chromatographic solvents.

G. Quantitative Determination of Plant Lipids

If a pure lipid is obtained directly from a chromatographic column it may be determined quantitatively by any suitable chemical, gravimetric or spectrophotometric method. This approach was employed by Allen et al. (1966) who isolated the major phospholipids and glycolipids of spinach lamellae and Anacystis nidulans by column chromatography on DEAE cellulose and subsequently quantitated the individual lipids gravimetrically.

If, as is often the case, pure lipids are not recovered they must be quantitated following a further purification step which usually involves TLC. The available methods for the quantitative determination of lipids on thin layer chromatograms may be divided into two groups according to whether they are estimated while still adsorbed to the chromatogram or whether they are first removed from it.

Methods which have received extensive application to the determination of lipids in general are those based on the densitometric analysis of the carbonized areas produced by reacting this layer chromatograms with sulphuric acid-containing reagents. The method, which has been applied to the determination of neutral glycerides (Privett and Blank, 1961; Barrett et al., 1962; and Blank et al., 1964), phospholipids (Rouser et al., 1964; Privett and Blank, 1963; Blank et al., 1964) and brain phospholipids and glycolipids (Payne, 1964) has the limitation of requiring relatively expensive apparatus and samples of pure reference compounds of similar fatty acid composition to those being quantitated.

Alternatively the sectors of adsorbent containing the lipids may be scraped or cut from the chromatogram and after elution from the adsorbent the lipids are quantitated in the eluate by a suitable micro method. When a lipid is eluted from a chromatogram prior to quantitative analysis it must be confirmed in a preliminary study that quantitative recovery can be achieved and this is probably most satisfactorily established by chromatographing radioactively labelled analogues and comparing the activity in the eluate with that retained by the adsorbent.

Where the presence of adsorbent does not affect the accuracy of the analytical method employed, the elution step may be eliminated and the quantitative technique directly applied to the adsorbent-lipid mixture removed from the chromatoplate. This approach has been applied by Roughan and Batt (1968, 1969) for the estimation of the individual phospholipids and glycolipids in a variety of plant tissues, and their general procedure is as follows. Lipid mixtures are bulk-fractionated by column chromatography on DEAE cellulose to give two fractions, one containing "neutral" polar lipids and the other acidic lipids. Both fractions are then separated into their components by preparative TLC, and after location of the individual fractions with iodine vapour the halogen is then allowed to evaporate. The glycolipid sectors of the chromatograms are then scraped into centrifuge tubes where they are reacted with the phenol–sulphuric acid reagent of Dubois *et al.*, as modified by Galanos and Kapoulas (1965a), and following centrifugation the clear supernatants are measured spectrophotometrically at 485 nm. The phenol–sulphuric acid reagent is more sensitive than those based on anthrone and is more reliable for the quantitation of sulphoquinovose. Phospholipids can be determined by the estimation of organic phosphorus in the relevant sectors of the chromatogram employing the method of Rouser *et al.* (1966) which involves digestion of the relevant zones of adsorbent with perchloric acid followed by reaction with molybdate reagent and determination of the colour produced by spectrophotometry at 820 nm.

Wintermans (1960) obtained quantitative data for the lipid composition of a variety of plant tissues by a method based on the fractionation and estimation of the products of deacylation of lipid mixtures. The limitations of this technique lie in the difficulties encountered in the location of these water-soluble derivatives on paper chromatograms by non-destructive means, and in achieving their quantitative desorption from the paper.

3. FRACTIONATION AND ANALYSIS OF COMPONENT FATTY ACIDS

A. Gas Chromatography

The introduction by James and Martin (1956) of gas–liquid chromatographic analysis (GLC) revolutionised the determination of the fatty acid composition of complex lipid fractions. It is probable that today the majority of fatty acid analyses of plant lipid extracts are determined solely by this technique following the conversion of the component acids into their methyl esters by any of a variety of simple methods (e.g. Morrison and Smith, 1964; Kishimoto and Radin, 1965).

In general two standard types of chromatographic stationary phase are used for this purpose, of which the polyester phases such as polyethylene glycol adipate (PEG-A) and diethylene glycol succinate (DEG-S) are the most versatile. Polyester columns resolve fatty acid esters according to both chain length and their degree of unsaturation, increases in chain length and/or number of double bonds giving increasing retention volumes (Fig. 11.7). Because polyester columns resolve methyl esters at least partly according to their polarizability they often give unacceptably low migration rates with compounds containing, for

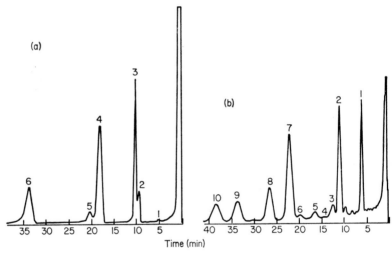

Peak identification is as follows:
  1: methyl myristate
  2: methyl palmitoleate
  3: methyl palmitate
  4: methyl oleate and linoleate
  5: methyl stearate
  6: methyl ricinoleate

Peak identification is as follows:
  1: methyl myristate
  2: methyl palmitate
  3: methyl palmitoleate
  4: methyl ester of a C_{16} dienoic acid
  5: methyl ester of a C_{16} trienoic acid
  6: methyl stearate
  7: methyl oleate
  8: methyl linoleate
  9: methyl linolenate
  10: methyl ester of a C_{20} monoenoic acid

Fig. 11.7. The separation of fatty acid methyl esters by gas–liquid chromatography (a) on a 4 ft SE-30 column at 204°C; and (b) on a 4 ft polyethylene glycol-adipate column at 200°C.

example, free hydroxyl groups. In these circumstances it is advisable to use comparatively non-polar stationary phases such as the silicone greases which separate components more on the basis of their molecular weights than on the nature of their functional groups (Fig. 11.7). Alternatively suitable derivatives (such as acetyl or trimethyl silyl derivatives for hydroxyl groups) of the highly polarizable groups can be made which then permits the methyl esters to be chromatographed acceptably on polyester columns.

When components of methyl ester fractions are unidentified, clues to their structure can be frequently obtained by a comparison of their retention volumes with those of reference substances analysed under the same conditions. However if methyl esters of hitherto unknown acids are encountered, or where GLC cannot give adequate resolution of the material in question then other methods of characterization may be required some of which will be referred to in the following sections.

Many natural lipids contain positionally isomeric unsaturated acids which are frequently either partially or completely resolved by GLC. For example, the methyl esters of the two most abundant natural octa-decatrienoic acids, namely 6,9,12-octadecatrienoic acid (γ-linolenic acid) and 9,12,15-octadecatrienoic acid (α-linolenic acid) are clearly separated on most polyester columns although the natural isomers of eicosatetraenoic acid are normally only partially separated. Ackman (1963) systematically studied the gas chromatographic behaviour of a large number of such isomers and has elaborated a system of separation factors by which they may be identified, and more recently he and Burgher (Ackman and Burgher, 1965) proposed the use of a complex but well authenticated mixture of methyl esters, namely that derived from cod liver oil, as a reference to facilitate identification of polyunsaturated methyl esters.

Many isomeric monoenoic acids are not resolved by GLC using the conventional type of GLC column but Ackman (1966) has demonstrated that the use of open tubular (capillary) columns with polyester coatings has an excellent potential in the detection and identification of such isomers. Figure 11.8 shows the resolution obtained by this worker using this technique for the analysis of rapeseed oil fatty acids.

It must be emphasized that GLC retention characteristics alone, even on several different types of stationary phase, are normally insufficient for the unequivocal identification of unusual unsaturated components. Identification is assisted if a fully hydrogenated sample is chromato-graphed to verify chain-length assignations and Mounts and Dutton (1965) have described a microvapour phase hydrogenation accessory for GLC which facilitates studies of this nature.

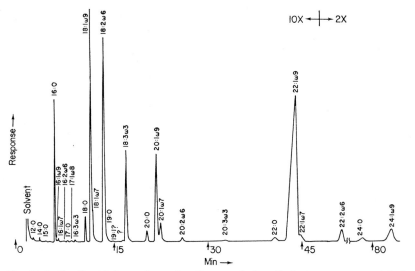

FIG. 11.8. Gas-liquid chromatography of rapeseed oil methyl esters on an open tubular column (Ackman, 1966).

B. Argentation Thin Layer Chromatography

For many problems in fatty acid analysis the use of argentation TLC has become an indispensible tool especially when applied in conjunction with GLC. As normally applied, the technique provides a rapid reliable method for the fractionation of fatty acid methyl ester mixtures into subfractions on the basis of the number of cis and/or trans double bonds of their constituents (Morris, 1966; Morris and Nichols, 1970).

The most common applications of the technique are to obtain simpler defined fractions from a complex mixture as an aid to analysis and characterization by GLC or to isolate pure components for elucidation of structure and/or of extent and position of radioactive labelling.

Argentation chromatography also separates geometric isomers of mono- and poly-unsaturated esters and was applied by Morris and co-workers (1968) to the detection and isolation of trans-3-monoenoic, 3,9-octadecadienoic and 3,9,12-octadecatrienoic acids in aster seed oil, and of geometric isomers of linoleic acid in other seed-oils (Morris and Marshall, 1966b). Similar procedures can be adapted to the unequivocal elucidation of the position and configuration of each individual double bond in a cis,trans-polyunsaturated acid by partial reduction of the acid by hydrazine to give a mixture of cis and trans monoenes which are isolated by argentation TLC and oxidized to reveal the position of the double bonds (Bhatty and Craig, 1966).

C. Determination of Double Bond Position

The position of double bonds in fatty acids is most commonly obtained by chemical oxidation of the double bond followed by analysis, usually employing GLC, of the oxidation products. One of the most commonly applied means of oxidation is reductive ozonolysis in which oxidation of the olefinic group with ozone is followed by reduction of the reaction product with GLC of the aldehydes thus formed. Because ozonolysis of acids containing more than one double bond usually produces mixtures of such complexity that interpretation of the results is a matter of some difficulty, it has become common practice to reduce the acid or its methyl ester with hydrazine to a mixture of monoenoic acids prior to ozonolysis. Aylward and Rao (1956) first demonstrated that this reducing reagent does not result in a positional rearrangement of double bonds during reduction and is comparatively non-specific in its site of attack. Consequently partial reduction of polyunsaturated fatty acids with hydrazine will result in the formation of a series of monoenoic acids the location of the double bonds in which reflects the ethylenic sites in the parent polyunsaturated acid (Roehm and Privett, 1969). Recently, Kleiman and coworkers (1969) have shown that the position of double bonds in unsaturated acids containing up to four double bonds may be obtained by controlled ozonolysis of the fatty acid methyl ester without prior reduction with hydrazine.

Alternative reagents for the cleavage of ethylenic groups include the periodate-permanganate reagent of von Rudloff (1950) which was adapted by Downing and Greene (1968) for the convenient and rapid determination of double bond positions in monoenoic acids, and when combined with partial reduction with hydrazine the technique can also be applied to similar studies on polyunsaturated acids.

More recently proposed methods for the determination of double bond positions in acids are based on the mass spectrographic analysis of volatile derivatives of the polyhydroxy fatty acids derived by mild oxidation of unsaturated fatty acids (McClcoskey and McClelland, 1965; Niehaus and Ryhage, 1967; Argoudelis and Perkins, 1968).

D. Specific Methods of Fatty Acid Analysis

In certain cases standard analytical methods are not applicable and the fatty acids in question must be located and estimated by alternative means. The cyclopropenoic acids are good examples of molecules which in an unmodified form are chemically degraded both during GLC and argentation-TLC, and the deliberate treatment of mixtures containing these acids with silver nitrate, which quantitatively cleaves their ring structures, forms the basis of two GLC procedures for their analysis (Johnson *et al.*, 1967b; Schneider *et al.*, 1968).

4. Fractionation and Analysis of Sphingolipid Bases (Sphingosines)

Sphingosine bases must normally be released by alcoholic acid hydrolysis of the parent lipid. After this, they can be separated from free fatty acids by rendering the alcoholic solution alkaline, diluting with water and extracting the free bases with diethyl ether (Kates, 1964). The conditions used for acid hydrolysis of sphingolipids invariably result in the formation of artefacts which in the case of plant bases are usually the anhydro-bases formed from the corresponding trihydroxyamines (Carter and Koob, 1969). If necessary these can be removed via formation of the DNP-bases (Karlsson, 1965).

Analyses of mixtures of long chain bases can be obtained by TLC on silica gel employing chloroform–methanol–N NH_4OH (40:10:1 v/v) as mobile phase and using ninhydrin for the location of the individual components (Sambasivarao and McCluer, 1963). More recently, the preferred method of analysis has been GLC analysis of the trimethylsilyl ethers of the sphingosine bases. For example, Karlsson (1965) prepared TMS ethers of corn cerebroside bases by refluxing up to 2 mg of sphingosine hydrochlorides for 15 minutes at 120°C with 0·1 ml of freshly prepared silylating reagent. After evaporation of the reagents from the reaction mixture in a stream of nitrogen, the residue was taken up in 5% pyridine in heptane and injected onto a GLC column using 6% silicone as stationary phase and an operating temperature of 201°C.

5. Identification of Isolated Plant Lipids

The chromatographic behaviour displayed by lipids, during their isolation by the methods described in previous sections, provides some information about their chemical structures (Hitchcock, 1967). For instance, tentative identification of fatty esters is possible from their retention times on gas–liquid chromatography (Jamieson, 1970). More detailed data may be obtained by examination of other physical properties of the isolated lipids, including ultraviolet, infrared, mass and nuclear magnetic resonance spectra; these techniques have been reviewed by Chapman (1965). The lipids may also be characterized by chemical reaction followed by identification of the products. Such approaches have already been described in connection with the determination of double bond position in fatty acids (p. 299).

A. Ultraviolet Spectroscopy

Most natural acyl lipids show no distinctive absorption bands in the ultraviolet above 220 nm. Saturated acyl chains absorb hardly any such radiation while non-conjugated unsaturated acids give rise to some featureless end-absorption. However, ethylenic and acetylenic bonds are suitable chromophores when in conjugation; conjugated poly-unsaturated acids (p. 14) therefore provide characteristic spectra which may be used for identification and estimation (Pitt and Morton, 1957). For instance, α-eleostearic acid (18:3, 9c 11t 13t) shows three absorption peaks, the major central one at 270 nm. The spectrum of β-eleostearic acid (18:3, 9t 11t 13t) is similar, but with punicic acid (18:3, 9c 11c 13t) the central peak appears at 275 nm. Conjugated tetraenes give rise to three peaks of higher extinction and at longer wavelength: thus the maximum absorbance of α-parinaric acid (18:4, 9c 11t 13t 15c) occurs at 302 nm. The more usual non-conjugated fatty acids may be similarly estimated after rearrangement to their conjugated analogues by treatment with hot alkali; such "alkali isomerization" techniques are, however, not very accurate, and have been superseded by more direct chromatographic methods. The dinitrophenylhydrazones of keto compounds also give useful characterization spectra in the ultraviolet.

B. Infrared Spectroscopy

The various vibrations of the methylene group are associated with absorption of a number of infrared frequencies, giving rise to characteristic bands in the infrared spectrum of all acyl lipids; they are accompanied by other bands due to methyl and carboxy groups. The presence of cis-ethylenic bonds causes the appearance of extra bands, but these do not help much in the identification of the unsaturated acids present. Trans-ethylenic bonds are associated with bands strong and characteristic enough to diagnose the presence of trans-monoene (10·34 μ) or of conjugated cis,trans-diene (10·17 μ), trans,trans-diene (10·12 μ), cis,-cis,trans-triene (10·11 μ), cis,trans,trans-triene (10·09 μ) or trans,trans,-trans-triene (10·06 μ). Unusual substituents in the acyl chain also give rise to recognizable bands (O'Connor, 1956): examples are free hydroxyl ($2 \cdot 75$–$2 \cdot 80$ μ); cis-epoxide ($12 \cdot 0$ μ) and trans-epoxide ($11 \cdot 2$ μ).

Infrared spectra are obtained from lipids in the vapour, liquid or solid state, or in solution, and can be used to infer certain arrangements of atoms within the molecule. Interpretation of such spectra is complicated by any association between lipid molecules that can occur, such as hydrogen bonding in liquids or solutions. Esters are therefore preferred to acids for such examination. In particular, solid lipids display polymorphism, transitions between different forms causing spectral changes. The arrangement of molecules in these solid states, and in lipid–water and lipid–protein systems, can also be investigated by X-ray diffraction.

C. Mass Spectroscopy

The fragmentation of a lipid vapour in a high vacuum when bombarded with electrons leads to a characteristic mass spectrum, which records the mass/charge ratio and relative abundance of each ion collected. From this and an understanding of the known energetically favoured pathways by which different structures fragment, the structure of the lipid can be deduced (McCloskey, 1970). However, some uncertainty arises due to the possibility during fragmentation of atomic rearrangement, particularly of hydrogen migration. Thus, while polyunsaturated esters give characteristic spectra, the position or geometry of the double bonds cannot be inferred directly. If a peak corresponding to the intact molecular ion is observed, its position gives the molecular weight of the lipid; this alone is of considerable help in identification and immediately distinguishes between fatty esters that are not isomers. The presence and position of keto-hydroxy, methoxy and methyl substituents may usually be determined. The examination of complex lipids is impeded by their high molecular weight and lack of volatility, which makes it difficult to introduce them into the mass spectrometer for study and, once inside, to purge them from it.

D. Nuclear Magnetic Resonance Spectroscopy

Determination of the structure of a complex organic compound with the help of its high-resolution NMR spectrum is a most powerful technique, for it is possible to infer the presence of many definite types of chemical grouping from the experimental data. Normally, 2–3 mg of pure components are required to obtain a satisfactory spectrum, and this limits the usefulness of biological applications; however, with the aid of a spherical microcell (Frost *et al.*, 1967) and a signal accumulator, 200 μg or less of sample can be characterized.

6. DEVELOPMENTS IN LIPID METHODOLOGY

The limitations of some established analytical techniques suggest that there is room for further advances, and here we shall briefly indicate some of these.

A. Gas Chromatography

Although already highly sophisticated and widely applied, gas–liquid chromatography is developing further, particularly in the areas of its ancillary techniques. These include improvements in automatic loading, pre-column chemistry, detector sensitivity and selectivity, and examination of the chemical or physical properties of the eluate (Ettre and McFadden, 1969). For preparative work, components of mixtures may be isolated by trapping them in simple condensers, but these will not be effective if the required component is too dilute or too volatile or forms an aerosol. In such cases it may be expedient to collect the component in a cooled trap packed with supported stationary phase (Howlett and Welti, 1966); chromatography through this is so slow that no material is lost, and it may be recovered by elution with a limited amount of argon at higher temperatures. The vapour is now more concentrated than before, and is quantitatively trapped a second time by condensing the argon at liquid nitrogen temperatures (Fowlis and Welti, 1967). Slow evaporation of the argon leaves behind most of the component for further study; in this way microgram quantities can be handled. The need for preparative chromatography on such a scale can be circumvented by examining the eluate directly.

The radioactivity of the eluate can be continuously measured in a number of ways (James, 1964; Scott, 1967). If the mixture is labelled with ^{14}C, one of the more convenient involves chromatography in argon as moving phase, followed by burning on copper oxide at 600°C to $^{14}CO_2$; the gas stream is now fed with a small proportion of $^{12}CO_2$ which acts as a quencher, and the mixture is passed through a proportional counter at ambient temperatures (Hitchcock *et al.*, 1969). Such an arrangement provides a continuous radiochromatogram which is useful down to activities of one nCi or less per component.

For the estimation of stable isotopic labelling, as well as for identification of unknown unlabelled components, the eluate may be concentrated and bled continuously into a fast-scan mass spectrometer, which continuously sweeps the spectrum: such spectra may be recorded when necessary, and these together with the corresponding chromatographic data may well be enough to characterize most components in a complex volatile mixture. Similar on-line examination of high resolution infra-red spectra involves computerized Fourier transform spectroscopy, since the ordinary spectrometer is too slow. However, conventional infrared spectra may be taken by leading the eluate through a heated vapour cell which is isolated and by-passed while the spectrum is recorded.

The eluate may be continuously monitored by slow-scan mass and infrared spectrometers only if the flow of the moving phase is stopped when spectra are required. This technique of interrupted elution can also be applied to radiochromatography to allow time for an aliquot of the eluate to be accurately counted. The interruptions have very little effect on the efficiency of separation by the chromatographic column, and the process can be made fully automatic (Scott et al., 1966). However, difficulties due to the complexity of such an instrument seem to outweigh its potential utility; in practice manual operations can be more expedient, if less convenient.

B. Liquid Chromatography

One fundamental limitation of gas–liquid chromatography is that it is not applicable to mixtures of involatile or thermally labile components. This difficulty can never be overcome, even though stationary phases which withstand operating temperatures of 400°C are available, and the volatility of some materials can be increased by chemical reactions before chromatography. Indeed, high molecular weight compounds such as triglycerides and trimethylsilylderivatives of saccharides can be chromatographed as vapours. In fact less volatile and less stable mixtures lend themselves more readily to liquid chromatography, a technique that has not advanced as quickly as gas chromatography. There is no theoretical reason why the potential efficiency of a chromatographic column should not be more nearly attained in practice with a liquid moving phase as with a gas. High-pressure liquid chromatographic systems using controlled surface porosity supports are therefore being developed and sensitive non-specific detectors are being designed which will enable mixtures to be fractionated with the efficiency and convenience now associated with gas chromatography (Hitchcock, 1969).

Thin-layer chromatography is already an effective form of liquid chromatography within the limits of size of load and size of plate, and separated lipids can be scraped from the plate and eluted for further investigation. This quick and popular technique will prove even more useful when methods are developed which will facilitate more exact quantitative estimation of each component *in situ*. Spraying followed by densitometry gives data which are useful but not entirely reliable, and even assay of ^{14}C by scanning with a windowless proportional counter is prone to some error. Methods involving the use of well-established sensitive gas-chromatographic detectors may prove more satisfactory: the intact stationary phase may be scanned with a heater burning organic material sequentially to CO_2, which is swept into a detector (Haahti *et al.*, 1970), or the burning may occur in the flame of an ionization detector itself (Padley, 1969).

Future investigations into the structure and metabolism of plant lipids will require more and more highly developed techniques to enable the biochemist to design experiments necessary to understand more exactly the part played by different fatty acids, acyl lipids, lipoproteins, membranes, and subcellular organelles in the functioning of the plant cell. Lipid biochemistry can only advance as fast as its methodology allows.

References

Abdelkader, A. B. and Mazliak, P. (1970). *Eur. J. Biochem.* **15**, 250–262.

Abdelkader, A. B., Mazliak, P. and Catesson, A. M. (1969). *Phytochemistry* **8**, 1121–1133.

Abrahamsson, S., Ställberg-Stenhagen, S. and Stenhagen, E. (1963). *In* "Progress in the Chemistry of Fats and Other Lipids" (R. T. Holman, ed.), Vol. 7 (Part 1), pp. 1–164. Pergamon Press, Oxford.

Acker, L. and Bücking, H. (1957). *Z. Lebensmittelunters. u. Forsch.* **105**, 32–38.

Ackman, R. G. (1963). *J. Am. Oil Chem. Soc.* **40**, 564–567.

Ackman, R. G. (1966). *J. Am. Oil Chem. Soc.* **43**, 483–486.

Ackman, R. G. and Burgher, R. D. (1965). *J. Am. Oil Chem. Soc.* **42**, 38–42.

Ackman, R. G., Tocher, C. S. and McLachlan, J. (1968). *J. Fish Res. Bd. Can.* **25**, 1603–1620.

Ahlers, N. H. E. and Gunstone, F. D. (1954). *Chemy Ind.* 1291–1292.

Akamatsu, Y. and Law, J. H. (1970). *J. biol. Chem.* **245**, 701–708.

Alam, A. V., Couch, J. R. and Creger, C. R. (1968). *Lipids* **3**, 183–184.

Alberts, A. W. and Vagelos, P. R. (1968). *Proc. natn. Acad. Sci. U.S.A.* **59**, 561–568.

Alberts, A. W., Majerus, P. W., Talamo, B. and Vagelos, P. R. (1964). *Biochemistry, Easton* **3**, 1563–1571.

Alberts, A. W., Majerus, P. W. and Vagelos, P. R. (1969a). *In* "Methods in Enzymology" (J. M. Lowenstein, ed.), Vol. XIV, pp. 50–53. Academic Press, New York and London.

Alberts, A. W., Majerus, P. W., and Vagelos, P. R. (1969b). *In* "Methods in Enzymology" (J. M. Lowenstein, ed.), Vol. XIV, pp. 53–56. Academic Press, New York and London.

Alberts, A. W., Majerus, P. W., and Vagelos, P. R. (1969c). *In* "Methods in Enzymology" (J. M. Lowenstein, ed.), Vol. XIV, pp. 57–60. Academic Press, New York and London.

Alberts, A. W., Nervi, A. M. and Vagelos, P. R. (1969d). *Proc. natn. Acad. Sci. U.S.A.* **63**, 1319–1326.

Allen, C. F., Good, P., Davis, H. F. and Fowler, S. D. (1964). *Biochem. biophys. Res. Commun.* **15**, 424–430.

Allen, C. F., Good, P., Davis, H. F., Chisum, P. and Fowler, S. D. (1966). *J. Am. Oil Chem. Soc.* **43**, 223–231.

Altschul, A. M., Ory, R. L. and St. Angelo, A. J. (1963). *Biochem. Prep.* **10**, 93.

Anderson, R. L. (1967). *Biochim. biophys. Acta* **144**, 18–24.

Anderson, R. L. (1968). *Biochim. biophys. Acta* **152**, 531–538.

Anderson, R. L. and Coots, R. H. (1967). *Biochim. biophys. Acta* **144**, 525–531.

Aneja, R., Chadha, J. S. and Knaggs, J. A. (1969). *Biochem. biophys. Res. Commun.* **36**, 401–406.

Antia, N. J., Bilinski, E. and Lau, Y. C. (1970). *Can. J. Biochem.* **48**, 644–648.

Antony, G. J. and Landau, B. R. (1968). *J. Lipid Res.* **9**, 267–269.

Appelman, D., Fulco, A. J. and Sugarman, P. M. (1966). *Pl. Physiol., Lancaster* **41**, 136–142.

Appelqvist, L.-A. (1968). *Physiologia Pl.* **21**, 455–465.

Appelqvist, L.-A. (1969). *Hereditas* **61**, 9–44.

Appelqvist, L.-A., Boynton, J. E., Stumpf, P. K. and von Wettstein, D. (1968a). *J. Lipid Res.* **9**, 425–436.

Appelqvist, L.-A., Stumpf, P. K. and von Wettstein, D. (1968b). *Pl. Physiol., Lancaster* **43**, 163–187.

Argoudelis, C. J. and Perkins, E. J. (1968). *Lipids* **3**, 379–381.

Arnaud, A. (1892). *Compt. rend. Acad. Sci. (Paris)* **114**, 79.

Arvidson, G. A. E. (1968). *Eur. J. Biochem.* **5**, 415–421.

Asselineau, C., Montrozier, H. and Promé, J. C. (1969). *Eur. J. Biochem.* **10**, 580–584.

Asselineau, J. (1966). "The Bacterial Lipids." Hermann, Paris.

Avins, L. R. (1968). *Biochem. biophys. Res. Commun.* **32**, 138–142.

Awasthi, Y. C., Berezney, R., Ruzicka, F. J. and Crane, F. L. (1969). *Biochim. biophys. Acta* **189**, 457–460.

Ayling, J., Fries, R. and Lynen, F. (1969). *Fedn. Proc. Fedn. Am. Socs. exp. Biol.* **28**, 537–537.

Aylward, F. and Rao, C. V. N. (1956). *J. appl. Chem.* **6**, 559–561.

Badami, R. C. and Morris, L. J. (1965). *J. Am. Oil Chem. Soc.* **42**, 1119–1121.

Baddiley, J. (1955). *Adv. Enzymol.* **16**. 1–21.

Bagby, M. O., Smith, C. R., Miwa, T. K., Lohmar, R. L. and Wolff, I. A. (1961). *J. org. Chem.* **26**, 1261–1265.

Bagby, M. O., Siegl, W. O. and Wolff, I. A. (1965a). *J. Am. Oil Chem. Soc.* **42**, 50–53.

Bagby, M. O., Smith, C. R. and Wolff, I. A. (1965b). *J. org. Chem.* **30**, 4227–4229.

Bagby, M. O., Smith, C. R. and Wolff, I. A. (1966). *Lipids* **1**, 263–267.

Bailey, J. L. and Whyborn, A. G. (1963). *Biochim. biophys. Acta* **78**, 163–174.

Baker, E. A. and Martin, J. T. (1963). *Nature, Lond.* **199**, 1268–1270.

Baranska, J. and Wlodawer, P. (1969). *Comp. Biochem. Physiol.* **28**, 553–570.

Barnes, E. M. and Wakil, S. J. (1968). *J. biol. Chem.* **243**, 2955–2962.

Barnes, E. M., Swindell, A. C. and Wakil, S. J. (1970). *J. biol. Chem.* **245**, 3122–3128.

Barrett, C. B., Dallas, M. S. J. and Padley, F. B. (1962). *Chemy. Ind.* 1050–1051.

Barron, E. J. and Mooney, L. A. (1970). *Biochemistry, Easton* **9**, 2143–2152.

Barron, E. J. and Stumpf, P. K. (1962). *Biochim. biophys. Acta* **60**, 329–337.

Barron, E. J., Squires, C. and Stumpf, P. K. (1961). *J. biol. Chem.* **236**, 2610–2614.

Bartels, C. T. (1969). "Metabolism of Plant Lipids in Relation to *Trans*-3-hexadecenoic Acid" (Thesis, University of Utrecht); Drukkerij, Bronder-Offset NV, Rotterdam.

Bartels, C. T. and van Deenen, L. L. M. (1966). *Biochim. biophys. Acta* **125**, 395–397.

Bartels, C. T., James, A. T. and Nichols, B. W. (1967). *Eur. J. Biochem.* **3**, 7–10.

Bartley, J. C., Abraham, S. and Chaikoff, I. L. (1967). *Biochim. biophys. Acta* **144**, 51–60.

Beevers, H. (1961). *Nature, Lond.* **191**, 433–436.

Benson, A. A. (1963a). *Proc. Fifth Int. Cong. Biochem.* **6**, 340–351.

Benson, A. A. (1963b). *Adv. Lipid Res.* **1**, 387–394.

Benson, A. A. (1964). *A. Rev. Pl. Physiol.* **15**, 1–16.

Benson, A. A. (1968). *In* "Membrane Models and the Formation of Biological Membranes" (L. Bolis and B. A. Pethica, eds), pp. 190–201. North-Holland Publ. Co., Amsterdam.

Benson, A. A. and Maruo, B. (1958). *Biochim. biophys. Acta* **27**, 189–195.
Benson, A. A. and Miyano, M. (1961). *Biochem. J.* **81**, 31P–31P.
Benson, A. A. and Strickland, E. H. (1960). *Biochim. biophys. Acta* **44**, 328–333.
Benson, A. A., Wiser, R., Ferrari, R. A. and Miller, J. A. (1958). *J. Am. chem. Soc.* **80**, 4740–4740.
Benson, A. A., Daniel, H. and Wiser, R. (1959a). *Proc. natn. Acad. Sci. U.S.A.* **45**, 1582–1587.
Benson, A. A., Wintermans, J. F. G. M. and Wiser, R. (1959b). *Pl. Physiol., Lancaster* **34**, 315–317.
Bergelson, L. D. (1969). *In* "Progress in the Chemistry of Fats and Other Lipids" (R. T. Holman, ed.), Vol. 10 (Part 3), pp. 241–286. Pergamon Press, Oxford.
Bergelson, L. D., Vaver, V. A. and Prokazova, N. V. (1964). *Doklady Akad. Nauk. SSSR* **157**, 122–124.
Berner, D. L. and Hammond, E. G. (1970). *Lipids* **5**, 572–573.
Bhatty, M. K. and Craig, B. M. (1966). *Can. J. Biochem.* **44**, 311–318.
Binder, R. G. and Lee, A. (1966). *J. org. Chem.* **31**, 1477–1479.
Binder, R. G., Applewhite, T. H., Diamond, M. J. and Goldblatt, L. A. (1964). *J. Am. Oil Chem. Soc.* **41**, 108–111.
Birge, C. H., Silbert, D. F. and Vagelos, P. R. (1967). *Biochem. biophys. Res. Commun.* **29**, 808–814.
Bishop, N. I. (1962). *Nature, Lond.* **195**, 55–57.
Blank, M. L., Schmidt, J. A. and Privett, O. S. (1964). *J. Am. Oil Chem. Soc.* **41**, 371–376.
Blass, J. P., Avigan, J. and Steinberg, D. (1969). *Biochim. biophys. Acta* **187**, 36–41.
Bloch, K. (1969). *Acc. Chem. Res.* **2**, 193–202.
Bloch, K. and Chang, S. B. (1964). *Science, N.Y.* **144**, 560–560.
Bloch, K., Constantopoulos, G., Kenyon, C. and Nagai, J. (1967). *In* "Biochemistry of Chloroplasts" (T. W. Goodwin, ed.), Vol. II, pp. 197–211, Academic Press, London and New York.
Bloomfield, D. K. and Bloch, K. (1960). *J. biol. Chem.* **235**, 337–345.
Boatman, S. G. and Crombie, W. M. (1958). *J. exp. Bot.* **9**, 52–74.
Bohlmann, F. (1967). *Fortschr. Chem. org. Naturst.* **25**, 1.
Bohlmann, F. and Schulz, H. (1968a). *Tetrahedron Lett.* 1801–1803.
Bohlmann, F. and Schulz, H. (1968b). *Tetrahedron Lett.* 4795–4798.
Bomstein, R. A. (1965). *Biochem. biophys. Res. Commun.* **21**, 49–54.
Bonner, W. D. (1965). *In* "Plant Biochemistry" (J. Bonner and J. E. Varner, eds.), pp. 89–123. Academic Press, New York and London.
Bonsen, P. P. M., de Haas, G. H. and van Deenen, L. L. M. (1965). *Biochim. biophys. Acta* **106**, 93–105.
Booth, V. H. (1964). *Phytochemistry* **3**, 229–234.
Borgström, B. and Ory, R. L. (1970). *Biochim. biophys. Acta* **212**, 521–523.
Borkenhagen, L. F., Kennedy, E. P. and Fielding, L. (1961). *J. biol. Chem.* **236**, PC 28–PC 30.
Bowen, D. M. and Radin, N. S. (1968). *Adv. Lipid Res.* **6**, 255–272.
Bradbeer, C. and Stumpf, P. K. (1960). *J. Lipid Res.* **1**, 214–220.
Brady, R. O. (1958). *Proc. natn. Acad. Sci. U.S.A.* **44**, 993–998.
Branton, D. and Park, R. B. (1967). *J. Ultrastructure Res.* **19**, 283–303.
Breidenbach, R. W., Kahn, A. and Beevers, H. (1968). *Pl. Physiol., Lancaster* **43**, 705–713.

Bressler, R. and Wakil, S. J. (1962). *J. biol. Chem.* **237**, 1441–1448.

Brett, D., Howling, D. and James, A. T. (1971). Unpublished data.

Brindley, D. N., Matsumura, S. and Bloch, K. (1969). *Nature, Lond.* **224**, 666–669.

Brock, D. J. H., Kass, L. R. and Bloch, K. (1967). *J. biol. Chem.* **242**, 4432–4440.

Brockerhoff, H. (1965a). *J. Lipid Res.* **6**, 10–15.

Brockerhoff, H. (1965b). *Archs Biochem. Biophys.* **110**, 586–592.

Brockerhoff, H. (1967). *J. Lipid Res.* **8**, 167–169.

Brockerhoff, H. and Yurkowski, M. (1966). *J. Lipid Res.* **7**, 62–64.

Brockerhoff, H., Hoyle, R. J. and Wolmark, N. (1965). *Fedn Proc. Fedn Am. Socs exp. Biol.* **24**, 662–662.

Brodie, J. D., Wasson, G. and Porter, J. W. (1964). *J. biol. Chem.* **239**, 1346–1356.

Brodnitz, M. H., Nawar, W. W. and Fagerson, I. S. (1968a). *Lipids* **3**, 59–64.

Brodnitz, M. H., Nawar, W. W. and Fagerson, I. S. (1968b). *Lipids* **3**, 65–71.

Brody, S. S., Brody, M. and Döring, G. (1970). *Z. Naturf.* **25b**, 367–372.

Brooks, J. L. and Stumpf, P. K. (1966). *Archs Biochem. Biophys.* **116**, 108–116.

Brown, C. M. and Rose, A. H. (1969). *J. Bact.* **99**, 371–378.

Brown, J. L. and Johnston, J. M. (1962). *J. Lipid Res.* **3**, 480–481.

Bu'Lock, J. D. (1964). *In* "Progress in Organic Chemistry" (Sir James Cook and W. Carruthers, eds), p. 86. Butterworth.

Bu'Lock, J. D. (1966). *In* "Comparative Phytochemistry" (T. Swain, ed.), pp. 79–95. Academic Press, London and New York.

Bu'Lock, J. D. (1967). "Essays in Biosynthesis and Microbial Development." John Wiley & Sons, Chichester.

Bu'Lock J. D. and Smalley, H. M. (1962). *J. chem. Soc.* 4662–4664.

Bu'Lock, J. D. and Smith, G. N. (1963). *Phytochemistry* **2**, 289–296.

Bu'Lock, J. D. and Smith, G. N. (1967). *J. chem. Soc.* (*C*) 332–336.

Burkhardt, H. J. (1970). *J. Am. Oil Chem. Soc.* **47**, 69–72.

Burton, D. and Stumpf, P. K. (1966). *Archs Biochem. Biophys.* **117**, 604–614.

Butler, R. D. (1967). *J. exp. Bot.* **18**, 535–543.

Callely, A. G. and Lloyd, D. (1964). *Biochem. J.* **92**, 338–345.

Canvin, D. T. (1963). *Can. J. Biochem. Physiol.* **41**, 1879–1885.

Canvin, D. T. (1965a). *Can. J. Bot.* **43**, 49–62.

Canvin, D. T. (1965b). *Can. J. Bot.* **43**, 63–69.

Canvin, D. T. and Beevers, H. (1961). *J. biol. Chem.* **236**, 988–995.

Carroll, K. K. (1961). *J. Lipid Res.* **2**, 135–141.

Carroll, K. K. (1963). *J. Am. Oil Chem. Soc.* **40**, 413–419.

Carter, H. E. and Greenwood, F. L. (1952). *J. biol. Chem.* **199**, 283–288.

Carter, H. E. and Hendrickson, H. S. (1963). *Biochemistry, Easton* **2**, 389–393.

Carter, H. E. and Kisic, A. (1969). *J. Lipid Res.* **10**, 356–362.

Carter, H. E. and Koob, J. L. (1969). *J. Lipid Res.* **10**, 363–369.

Carter, H. E., Gigg, R. H., Law, J. H., Nakayama, T. and Weber, E. (1958). *J. biol. Chem.* **233**, 1309–1314.

Carter, H. E., Ohno, K., Nojima, S., Tipton, C. L. and Stanacev, N. Z. (1961a). *J. Lipid Res.* **2**, 215–222.

Carter, H. E., Hendry, R. A. and Stanacev, N. Z. (1961b). *J. Lipid Res.* **2**, 223–227.

Carter, H. E., Hendry, R. A., Nojima, S., Stanacev, N. Z. and Ohno, K. (1961c). *J. biol. Chem.* **236**, 1912–1916.

Carter, H. E., Galanos, D. S., Hendrickson, H. S., Jann, B., Nakayama, T., Nakazawa, Y. and Nichols, B. (1962). *J. Am. Oil Chem. Soc.* **39**, 107–115.

Carter, H. E., Betts, B. E. and Strobach, D. R. (1964a). *Biochemistry, Easton* **3**, 1103–1107.

Carter, H. E., Brooks, S., Gigg, R. H., Strobach, D. R. and Suami, T. (1964b). *J. biol. Chem.* **239**, 743–746.

Carter, H. E., Strobach, D. R. and Hawthorne, J. N. (1969). *Biochemistry, Easton* **8**, 383–388.

Castelfranco, P., Stumpf, P. K. and Contopoulou, R. (1955). *J. biol. Chem.* **214**, 567–577.

Cerletti, P., Giovenco, M. A., Giordano, M. G., Giovenco, S. and Strom, R. (1967). *Biochim. biophys. Acta* **146**, 380–396.

Chang, S. B. and Kulkarni, N. D. (1970). *Phytochemistry* **9**, 927–934.

Chang, S. B. and Lundin, K. (1965). *Biochem. biophys. Res. Commun.* **21**, 424–430.

Chang, Y. Y. and Kennedy, E. P. (1967). *J. biol. Chem.* **242**, 516–519.

Channon, H. J. and Chibnall, A. C. (1929). *Biochem. J.* **23**, 168–175.

Chapman, D. (1965). "The Structure of Lipids." Methuen.

Cheniae, G. M. (1964). *Archs Biochem. Biophys.* **105**, 163–169.

Cheniae, G. M. (1965). *Pl. Physiol., Lancaster* **40**, 235–243.

Cheniae, G. M. and Kerr, P. C. (1965). *Pl. Physiol., Lancaster* **40**, 452–457.

Cherry, J. H. (1963). *Pl. Physiol., Lancaster* **38**, 440–446.

Ching, T. M. (1968). *Lipids* **3**, 482–488.

Chisholm, M. J. and Hopkins, C. Y. (1958). *Can. J. Chem.* **36**, 1537–1540.

Chisholm, M. J. and Hopkins, C. Y. (1959). *Chemy Ind.* 1154–1155.

Chisholm, M. J. and Hopkins, C. Y. (1960a) *Can. J. Chem.* **38**, 805–812.

Chisholm, M. J. and Hopkins, C. Y. (1960b). *Can. J. Chem.* **38**, 2500–2507.

Chisholm, M. J. and Hopkins, C. Y. (1962). *J. org. Chem.* **27**, 3137–3139.

Chisholm, M. J. and Hopkins, C. Y. (1963). *Can. J. Chem.* **41**, 1888–1892.

Chisholm, M. J. and Hopkins, C. Y. (1965). *J. Am. Oil Chem. Soc.* **42**, 49–50.

Christiansen, K., Marcel, Y., Gan, M. V., Mohrhauer, H. and Holman, R. T. (1968). *J. biol. Chem.* **243**, 2969–2974.

Christie, W. W. (1969). *Biochem. biophys. Acta* **187**, 1–5.

Christopher, J., Pistorius, E. and Axelrod, B. (1970). *Biochim. biophys. Acta* **198**, 12–19.

Chuecas, L. and Riley, J. P. (1969). *J. mar. biol. Ass.* (*UK*) **49**, 97–116.

Chung, A. E. and Law, J. H. (1964). *Biochemistry, Easton* **3**, 967–974.

Clayton, T. A., MacMurray, T. A. and Morrison, W. R. (1970). *J. Chromatog.* **47**, 277–281.

Clermont, H. and Douce, R. (1970). *FEBS Lett.* **9**, 284–286.

Click, R. E. and Hackett, D. P. (1963). *Proc. natn. Acad. Sci. U.S.A.* **50**, 243–250.

Cobern, D., Hobbs, J. S., Lucas, R. A. and Mackenzie, D. J. (1966). *J. chem. Soc.* (*C*) 1897–1902.

Cole, H. I. and Cardoso, H. T. (1939). *J. Am. Chem. Soc.* **61**, 2351–2353.

Coleman, M. H. (1963). *Adv. Lipid Res.* **1**, 1–64.

Colli, W., Hinkle, P. C. and Pullman, M. E. (1969). *J. biol. Chem.* **244**, 6432–6443.

Collins, F. D. (1959). *Biochem. J.* **72**, 532–537.

Conacher, H. B. S. and Gunstone, F. D. (1970). *Lipids* **5**, 137–141.

Constantopoulos, G. (1970). *Pl. Physiol., Lancaster* **45**, 76–80.

Constantopoulos, G. and Bloch, K. (1967). *J. biol. Chem.* **242**, 3538–3542.

Cooper, T. G. and Beevers, H. (1969a). *J. biol. Chem.* **244**, 3507–3513.

Cooper, T. G. and Beevers, H. (1969b). *J. biol. Chem.* **244**, 3514–3520.

Coppens, N. (1956). *Nature, Lond.* **177**, 279–279.

Coulon-Morelec, M. J. and Douce, R. (1968). *Bull. Soc. Chim. biol.* **50**, 1547–1559.
Craig, B. M. and Murti, N. L. (1959). *J. Am. Oil Chem. Soc.* **36**, 549–552.
Crombie, L. and Jacklin, A. G. (1957). *J. chem. Soc.* 1632–1646.
Crombie, W. M. (1956). *J. exp. Bot.* **7**, 181–193.
Crombie, W. M. (1958). *J. exp. Bot.* **9**, 254–261.
Crombie, W. M. and Comber, R. (1956). *J. exp. Bot.* **7**, 166–180.
Crombie, W. M. and Hardman, E. E. (1958). *J. exp. Bot.* **9**, 247–253.
Cronan, J. E. (1967). *Biochim. biophys. Acta* **144**, 695–697.
Crossley, A. and Hilditch, T. P. (1949). *J. chem. Soc.* 3353–3357.
Dahlen, J. V. and Porter, J. W. (1968). *Archs Biochem. Biophys.* **127**, 207–223.
Dalgarno, M. and Birt, L. M. (1963). *Biochem. J.* **87**, 586–596.
Daniel, H., Miyano, M., Mumma, R. O., Yagi, T., Lepage, M., Shibuya, I. and Benson, A. A. (1961). *J. Am. chem. Soc.* **83**, 1765–1766.
Das, M. L., Orrenius, S. and Ernster, L. (1968). *Eur. J. Biochem.* **4**, 519–523.
Davenport, J. B. (1966). *Nature, Lond.* **210**, 198–198.
Davidoff, F. and Korn, E. D. (1963). *J. biol. Chem.* **238**, 3199–3209.
Davidoff, F. and Korn, E. D. (1964). *J. biol. Chem.* **239**, 2496–2506.
Davidson, F. M. and Long, C. (1958). *Biochem. J.* **69**, 458–466.
Davies, W. E. and Radin, N. S. (1966). Unpublished data. Quoted in Bowen, D. M. and Radin, N. S. (1968).
Davies, W. E., Hajra, A. K., Parmar, S. S., Radin, N. S. and Mead, J. F. (1966). *J. Lipid Res.* **7**, 270–275.
Davies, W. H., Mercer, E. I. and Goodwin, T. W. (1966). *Biochem. J.* **98**, 369–373.
Davis, E. N., Wallen, L. L., Goodwin, J. C., Rohwedder, W. K. and Rhodes, R. A. (1969). *Lipids* **4**, 356–362.
Dawson, R. M. C. (1967). *Biochem. J.* **102**, 205–210.
Dawson, R. M. C. and Hemington, N. (1967). *Biochem. J.* **102**, 76–86.
Dawson, R. M. C., Clarke, N. and Quarles, R. H. (1969). *Biochem. J.* **114**, 265–270.
Debuch, H. (1961). *Z. Naturf.* **16b**, 561–567.
Debuch, H. and Rotsch, E. (1966). *Z. physiol. Chem.* **346**, 79–86.
De Gier, J., Mandersloot, J. G. and van Deenen, L. L. M. (1968). *Biochim. biophys. Acta* **150**, 666–675.
de Haas, G. H. and van Deenen, L. L. M. (1964). *Biochim. biophys. Acta* **84**, 469–471.
de Haas, G. H., Bonsen, P. P. M. and van Deenen, L. L. M. (1966). *Biochim. biophys. Acta* **116**, 114–124.
Devine, J. (1950). *J. Sci. Food. Agric.* **1**, 88–92.
Devor, K. A. and Mudd, J. B. (1968). *Pl. Physiol., Lancaster* **43**, 853–858.
de Vries, B. (1963). *J. Am. Oil Chem. Soc.* **40**, 184–186.
Dickson, L. G., Galloway, R. A. and Patterson, G. W. (1969). *Pl. Physiol., Lancaster* **44**, 1413–1416.
Dimick, P. S., Walker, N. J. and Patton, S. (1969). *Biochem. J.* **111**, 395–399.
Dittmer, J. C. and Lester, R. L. (1964). *J. Lipid Res.* **5**, 126–127.
Dolev, A., Rohwedder, W. K. and Dutton, H. J. (1967a). *Lipids* **2**, 28–32.
Dolev, A., Rohwedder, W. K., Mounts, T. L. and Dutton, H. J. (1967b). *Lipids* **2**, 33–36.
Donaldson, W. E. (1967). *Biochem. biophys. Res. Commun.* **27**, 681–685.
Dorsey, J. A. and Porter, J. W. (1968). *J. biol. Chem.* **243**, 3512–3516.
Douce, R. (1968). *Compt. rend. Acad. Sci. (Paris)* **267**, 534–537.
Douce, R. and Dupont, J. (1969). *Compt. rend. Acad. Sci. (Paris)* **268**, 1657–1660.

Douce, R., Faure, M. and Marechal, J. (1966). *Compt. rend. Acad. Sci. (Paris)* **262**, 1549–1552.

Downey, R. K. (1964). *Can. J. Plant. Sci.* **44**, 295–295.

Downey, R. K. and Craig, B. M. (1964). *J. Am. Oil Chem. Soc.* **41**, 475–478.

Downey, R. K. and Harvey, B. L. (1963). *Can. J. Plant Sci.* **43**, 271–275.

Downing, D. T. and Greene, R. S. (1968). *Lipids* **3**, 96–100.

Draper, S. R. (1969). *Phytochemistry* **8**, 1641–1647.

Drapron, R., Anh, N'G. X., Launay, B. and Guilbot, A. (1969). *Cereal Chem.* **46**, 647–655.

Drennan, C. H. and Canvin, D. T. (1969). *Biochim. biophys. Acta* **187**, 193–200.

Drysdale, G. R. and Lardy, H. A. (1953). *J. biol. Chem.* **202**, 119–136.

Dunphy, P. J. and Allcock, C. (1971). In preparation.

Dunphy, P. J., Whittle, K. J. and Pennock, J. F. (1965). *Chemy Ind.* 1217–1218.

Dupont, J. and Mathias, M. M. (1969). *Lipids* **4**, 478–488.

Duttera, S. M., Byrne, W. L. and Ganoza, M. C. (1968). *J. biol. Chem.* **243**, 2216–2228.

Earle, F. R., Barclay, A. S. and Wolff, I. A. (1966). *Lipids* **1**, 325–327.

Easter, D. J. and Dils, R. (1968). *Biochim. biophys. Acta* **152**, 653–668.

Eberhardt, F. M. and Kates, M. (1957). *Can. J. Bot.* **35**, 907–921.

Echlin, P. and Morris, I. (1965). *Biol. Rev.* **40**, 143–187.

Eglinton, G. and Hamilton, R. J. (1967). *Science, N.Y.* **156**, 1322–1335.

Eichenberger, W. and Newman, D. W. (1968). *Biochem. biophys. Res. Commun.* **32**, 366–374.

Elix, J. A. and Sargent, M. V. (1968). *J. chem. Soc. (C)* 595–596.

Elovson, J. (1964). *Biochim. biophys. Acta* **84**, 275–293.

Elovson, J. and Vagelos, P. R. (1968). *J. biol. Chem.* **243**, 3603–3611.

Endo, K., Helmkamp, G. M. and Bloch, K. (1970). *J. biol. Chem.* **245**, 4293–4296.

English, J., Bonner, J. and Haagen-Smit, A. J. (1939). *J. Am. chem. Soc.* **61**, 3434–3436.

Eriksson, C. E. and Svensson, S. G. (1970). *Biochim. biophys. Acta* **198**, 449–459.

Erwin, J. and Bloch, K. (1964). *Science, N.Y.* **143**, 1006–1012.

Erwin, J., Hulanicka, D. and Bloch, K. (1964). *Comp. Biochem. Physiol.* **12**, 191–207.

Ettre, L. S., and McFadden, W. H. (1969). "Ancillary Techniques of Gas Chromatography." Wiley-Interscience, New York.

Evans, C. D., McConnell, D. G., Hoffman, R. L. and Peters, H. M. (1965). *J. Am. Oil Chem. Soc.* **42**, 462A–462A.

Evans, C. D., McConnell, D. G., List, G. R. and Scholfield, C. R. (1969). *J. Am. Oil Chem. Soc.* **46**, 421–424.

Faulkner, R. N. (1958). *J. appl. Chem.* **8**, 448–458.

Fawcett, C. H., Ingram, J. M. A. and Wain, R. L. (1954). *Proc. Roy. Soc. Ser. B* **142**, 60–72.

Fawcett, C. H., Spencer, D. M., Wain, R. L., Fallis, A. G., Jones, E. R. H., LeQuan, M., Page, C. B., Thaller, V., Schubrook, D. C. and Witham, P. M. (1968). *J. chem. Soc. (C)* 2455–2462.

Ferrari, R. A. and Benson, A. A. (1961). *Archs Biochem. Biophys.* **93**, 185–192.

Fisher, N. (1962). *In* "Recent Advances in Food Science" (J. Hawthorn and J. M. Leitch, eds.), Vol. I, pp. 226–245. Butterworths, London.

Fisher, N. and Broughton, M. E. (1960). *Chemy Ind.* 869–870.

Flavin, M. and Ochoa, S. (1957). *J. biol. Chem.* **229**, 965–979.

Fleischer, S., Brierley, G., Klouwen, H. and Slautterback, D. B. (1962). *J. biol. Chem.* **237**, 3264–3272.

Fleischer, S., Fleischer, B. and Stockenius, W. (1967). *J. Cell. Biol.* **32**, 193–208.

Fowlis, I. A. and Welti, D. (1967). *Analyst* **92**, 639–641.

Frankel, E. N., Evans, C. D., McConnel, D. G., Selke, E. and Dutton, H. J. (1961). *J. org. Chem.* **26**, 4663–4669.

French, R. B. (1962). *J. Am. Oil Chem. Soc.* **39**, 176–178.

Frost, D. J., Hall, G. E., Green, M. J. and Leane, J. B. (1967). *Chemy Ind.* 116–117.

Fulco, A. J. (1965). *Biochim. biophys. Acta* **106**, 211–212.

Fulco, A. J. (1967). *J. biol. Chem.* **242**, 3608–3613.

Fulco, A. J. (1969a). *Biochim. biophys. Acta* **187**, 169–171.

Fulco, A. J. (1969b). *J. biol. Chem.* **244**, 889–895.

Fulco, A. J. (1970). *J. biol. Chem.* **245**, 2985–2990.

Fulco, A. J. and Bloch, K. (1964). *J. biol. Chem.* **239**, 993–997.

Fulco, A. J., Levy, R. and Bloch, K. (1964). *J. biol. Chem.* **239**, 998–1003.

Galanos, D. S. and Kapoulas, V. M. (1965a). *Biochim. biophys. Acta* **98**, 278–312.

Galanos, D. S. and Kapoulas, V. M. (1965b). *Biochim. biophys. Acta* **98**, 313–328.

Galliard, T. (1968a). *Phytochemistry* **7**, 1907–1914.

Galliard, T. (1968b). *Phytochemistry* **7**, 1915–1922.

Galliard, T. (1969). *Biochem. J.* **115**, 335–339.

Galliard, T. (1970). *Phytochemistry* **9**, 1725–1734.

Galliard, T. and Stumpf, P. K. (1966). *J. biol. Chem.* **241**, 5806–5812.

Galliard, T., Rhodes, M. J. C., Wooltorton, L. S. C. and Hulme, A. C. (1968a). *FEBS Lett.* **1**, 283–286.

Galliard, T., Rhodes, M. J. C., Wooltorton, L. S. C. and Hulme, A. C. (1968b). *Phytochemistry* **7**, 1453–1463.

Galzigna, L., Rossi, C. R., Sartorelli, L. and Gibson, D. M. (1967). *J. biol. Chem.* **242**, 2111–2115.

Gardner, H. W. (1970). *J. Lipid Res.* **11**, 311–321.

Gardner, H. W. and Weisleder, D. (1970). *Lipids* **5**, 678–683.

Garton, G. A. (1960). *Nature, Lond.* **187**, 511–512.

Gastambide-Odier, M. and Lederer, E. (1959). *Nature, Lond.* **184**, 1563–1564.

Gastambide-Odier, M., Delaumeny, J.-M. and Lederer, E. (1963). *Biochim. biophys. Acta* **70**, 670–678.

Gatt, S. and Barenholz, Y. (1968). *Biochem. biophys. Res. Commun.* **32**, 588–594.

Gellerman, J. L. and Schlenk, H. (1964). *Experientia*, **20**, 426–427.

Gerloff, E. D., Richardson T. and Stahmann, M.A. (1966). *Pl. Physiol., Lancaster* **41**, 1280–1284.

Gibble, W. P. and Kurtz, E. B. (1956). *Archs Biochem. Biophys.* **64**, 1–5.

Gibson, D. M. *In* "Progress in the Chemistry of Fats and Other Lipids" (R. T. Holman, W. O. Lundberg and T. Malkin, eds), Vol. 6, pp. 117–136. Pergamon Press, Oxford.

Gibson, D. M., Titchener, E. B. and Wakil, S. J. (1958a). *J. Am. chem. Soc.* **80**, 2908–2908.

Gibson, D. M., Titchener, E. B. and Wakil, S. J. (1958b). *Biochim. biophys. Acta* **30**, 376–383.

Gini, B. and Koch, R. B. (1961). *J. Food. Sci.* **26**, 359–364.

Giovanelli, J. and Stumpf, P. K. (1957). *J. Am. chem. Soc.* **79**, 2652–2653.

Giovanelli. J. and Stumpf, P. K. (1958). *J. biol. Chem.* **231**, 411–426.

Goldfine, H. and Bloch, K. (1961). *J. biol. Chem.* **236**, 2596–2601.

Grace, N. H. (1939). *Can. J. Research* **17**, 247–255.

Graham, A. B. and Wood, G. C. (1969). *Biochem. biophys. Res. Commun.* **37**, 567–575.

Green, D. E. (1954). *Biol. Rev.* **29**, 330–366.

Green, D. E. and Allman, D. W. (1968a). *In* "Metabolic Pathways" (D. M. Greenberg, ed.), Vol. II (3rd Ed.), pp. 1–36. Academic Press, New York and London.

Green, D. E. and Allman, D. W. (1968b). *In* "Metabolic Pathways" (D. M. Greenberg, ed.), Vol. II (3rd Ed.), pp. 37–67. Academic Press, New York and London.

Green, D. E. and Fleischer, S. (1963). *Biochim. biophys. Acta* **70**, 554–582.

Green, D. E. and Fleischer, S. (1964). *In* "Metabolism and Physiological Significance of Lipids" (R. M. C. Dawson and D. N. Rhodes, eds), pp. 581–618. Wiley and Sons, New York.

Green, D. E., Asai, J., Harris, R. A. and Penniston, J. T. (1968). *Archs Biochem. Biophys.* **125**, 684–705.

Green, T. G., Hilditch, T. P. and Stainsby, W. J. (1936). *J. chem. Soc.* 1750–1755.

Greenspan, M. D., Alberts, A. W. and Vagelos, P. R. (1969). *J. biol. Chem.* **244**, 6477–6485.

Grob, E. C. and Csupor, L. (1967). *Experientia* **23**, 1004–1005.

Guchhait, R. B., Putz, G. R. and Porter, J. W. (1966). *Archs Biochem. Biophys.* **117**, 541–549.

Guinand, M. and Michel, G. (1966). *Biochim. biophys. Acta* **125**, 75–91.

Gunstone, F. D. (1952). *J. chem. Soc.* 1274–1278.

Gunstone, F. D. (1954). *J. chem. Soc.* 1611–1616.

Gunstone, F. D. (1962). *Chemy Ind.* 1214–1223.

Gunstone, F. D. (1965). *Chemy Ind.* 1033–1034.

Gunstone, F. D. (1966). *Chemy Ind.* 1551–1554.

Gunstone, F. D. and Morris, L. J. (1959). *J. chem. Soc.* 2127–2132.

Gunstone, F. D. and Padley, F. B. (1965). *J. Am. Oil Chem. Soc.* **42**, 957–961.

Gunstone, F. D. and Qureshi, M. I. (1968). *J. Sci. Food Agric.* **19**, 386–388.

Gunstone, F. D. and Sealy, A. J. (1963). *J. chem. Soc.* 5772–5778.

Gunstone, F. D. and Subbarao, R. (1967). *Chem. Phys. Lip.* **1**, 349–359.

Gunstone, F. D., Hamilton, R. J. and Qureshi, M. I. (1965). *J. chem. Soc.* 319–325.

Gunstone, F. D., Kilgast, D., Powell, R. G. and Taylor, G. M. (1967). *Chem. Commun.* 295–296.

Gurr, M. I. and Bloch, K. (1966). *Biochem. J.* **99**, 16C–18C.

Gurr, M. I., Davey, K. W. and James, A. T. (1968). *FEBS Lett.* **1**, 320–322.

Gurr, M. I. and Robinson, M. P. (1970). *Europ. J. Biochem.* **15**, 335–341.

Gurr, M. I., Robinson, M. P. and James, A. T. (1969). *Eur. J. Biochem.* **9**, 70–78.

Guss, P. L., Richardson, T. and Stahmann, M. A. (1968). *J. Am. Oil Chem. Soc.* **45**, 272–276.

Haahti, E. O. A. and Fales, H. M. (1967). *J. Lipid Res.* **8**, 131–137.

Haahti, E., Vihko, R., Jaakonmäki, I. and Evans, R. S. (1970). *J. Chrom. Sci.* **8**, 370–374.

Haberlandt, G. (1922). *Biol. Zbl.* **42**, 145–171.

Hack, M. H. and Ferrans, V. J. (1959). *Hoppe Seyler's Z. physiol. Chem.* **315**, 157–162.

Hacker, M. and Stohr, H. (1966). *Planta* **68**, 215.

Haeffner, E. W. (1970). *Lipids* **5**, 430–433.

Haest, C. W. M., De Gier, J. and van Deenen, L. L. M. (1969). *Chem. Phys. Lip.* **3**, 413–416.

Haigh, W. G., Morris, L. J. and James, A. T. (1968). *Lipids* **3**, 307–312.

Haigh, W. G., Safford, R. and James, A. T. (1969). *Biochim. biophys. Acta* **176**, 647–650.

Hajra, A. K. and Agranoff, B. W. (1967). *J. biol. Chem.* **242**, 1074–1075.

Hajra, A. K. and Agranoff, B. W. (1968a). *J. biol. Chem.* **243**, 1617–1622.

Hajra, A. K. and Agranoff, B. W. (1968b). *J. biol. Chem.* **243**, 3542–3543.

Hale, S. A., Richardson, T., von Elke, J. H. and Hagedorn, D. J. (1969). *Lipids* **4**, 209–215.

Hall, D. M. and Jones, R. L. (1961). *Nature, Lond.* **191**, 95–96.

Hall, S. W. and Morris, L. J. (1970). Unpublished data.

Hamberg, M. and Samuelsson, B. (1967a). *J. biol. Chem.* **242**, 5329–5335.

Hamberg, M. and Samuelsson, B. (1967b). *J. biol. Chem.* **242**, 5344–5354.

Hamilton, R. J. and Power, D. M. (1969). *Phytochemistry* **8**, 1771–1775.

Hardman, E. E. and Crombie, W. M. (1958). *J. exp. Bot.* **9**, 239–246.

Harlan, W. R. and Wakil, S. J. (1963). *J. biol. Chem.* **238**, 3216–3223.

Harris, P. and James, A. T. (1969a). *Biochem. J.* **112**, 325–330.

Harris, P. and James, A. T. (1969b). *Biochim. biophys. Acta* **187**, 13–18.

Harris, R. V. and James, A. T. (1965). *Biochim. biophys. Acta* **106**, 456–464.

Harris, R. V., Harris, P. and James, A. T. (1965). *Biochim. biophys. Acta* **106**, 465–473.

Harris, R. V., James, A. T. and Harris, P. (1967). *In* "Biochemistry of Chloroplasts" (T. W. Goodwin, ed.), Vol. 2, pp. 241–253. Academic Press, London and New York.

Hartmann, G. R. and Frear, D. S. (1963). *Biochem. biophys. Res. Commun.* **10**, 366–372.

Hatch, M. D. and Stumpf, P. K. (1961). *J. biol. Chem.* **236**, 2879–2885.

Hatch, M. D. and Stumpf, P. K. (1962a). *Archs Biochem. Biophys.* **96**, 193–198.

Hatch, M. D. and Stumpf, P. K. (1962b). *Pl. Physiol., Lancaster* **37**, 121–126.

Hatt, H. H., Triffett, A. C. K. and Wailes, P. C. (1959). *Aust. J. Chem.* **12**, 190–195.

Hatt, H. H., Triffett, A. C. K. and Wailes, P. C. (1960). *Aust. J. Chem.* **13**, 488.

Haverkate, F. (1965). "Phosphatidyl glycerol from Photosynthetic Tissues." Thesis, University of Utrecht.

Haverkate, F. and van Deenen, L. L. M. (1964). *Biochim. biophys. Acta* **84**, 106–108.

Haverkate, F. and van Deenen, L. L. M. (1965). *Biochim. biophys. Acta* **106**, 78–92.

Haverkate, F., de Gier, J. and van Deenen, L. L. M. (1964). *Experientia* **20**, 511–512.

Hawke, J. C. (1963). *J. Dairy Res.* **30**, 67–75.

Hawke, J. C. and Stumpf, P. K. (1965a). *Pl. Physiol., Lancaster* **40**, 1023–1032.

Hawke, J. C. and Stumpf, P. K. (1965b). *J. biol. Chem.* **240**, 4746–4752.

Heinen, W. and Brand, I.v.d. (1963). *Z. Naturf.* **18b**, 67–79.

Heinstein, P. F. and Stumpf, P. K. (1969). *J. biol. Chem.* **244**, 5374–5381.

Heinz, E. (1967a). *Biochim. biophys. Acta* **144**, 321–332.

Heinz, E. (1967b). *Biochim. biophys. Acta* **144**, 333–343.

Heinz, D. E. and Jennings, W. G. (1966). *J. Food Sci.* **31**, 69–80.

Heinz, E. and Tulloch, A. P. (1969). *Hoppe-Seyler's Z. physiol. Chem.* **350**, 493–498.

Heinz, E., Tulloch, A. P. and Spencer, J. F. T. (1969). *J. biol. Chem.* **244**, 882–888.

Heinz, E., Tullock, A. P. and Spencer, J. F. T. (1270). *Biochim. biophys. Acta.* **202**, 49–55.

Heller, M., Aladjeni, E. and Shapiro, B. (1965). *Bull. Soc. Chim. biol.* **50**, 1395–1408.

Helmkamp, G. M. and Bloch, K. (1969). *J. biol. Chem.* **244**, 6014–6022.

Helmkamp, G. M., Rando, R. R., Brock, D. J. H. and Bloch, K. (1968). *J. biol. Chem.* **243**, 3229–3231.

Heyes, J. K. and Shorland, F. B. (1951). *Biochem. J.* **49**, 503–506.

Hilditch, T. P. and Meara, M. L. (1944). *J. Soc. Chem. Ind.* (Lond.) **63**, 112–114.

Hilditch, T. P. and Mendelowitz, A. (1951). *J. Sci. Food. Agric.* **2**, 548–556.

Hilditch, T. P. and Williams, P. N. (1964). "The Chemical Constitution of the Natural Fats", 4th ed. Chapman and Hall, London.

Hirayama, O. and Hujii, K. (1965). *Agr. biol. Chem. Tokyo* **29**, 1–6.

Hirschmann, H. (1960). *J. biol. Chem.* **235**, 2762–2767.

Hitchcock, C. (1967). *Process Biochem.* **2** (May), 13–22.

Hitchcock, C. (1969). *Process Biochem.* **4** (March), 23–32.

Hitchcock, C. (1971). Unpublished data.

Hitchcock, C. and James, A. T. (1964). *J. Lipid Res.* **5**, 593–599.

Hitchcock, C. and James, A. T. (1966). *Biochim. biophys. Acta* **116**, 413–424.

Hitchcock, C. and Morris, L. J. (1970). *Eur. J. Biochem.* **17**, 39–42.

Hitchcock, C., Morris, L. J. and James, A. T. (1968a). *Eur. J. Biochem.* **3**, 419–423.

Hitchcock, C., Morris, L. J. and James, A. T. (1968b). *Eur. J. Biochem.* **3**, 473–475.

Hitchcock, C., James, A. T. and Carr, A. H. (1969). *In* "Radioactive Isotopes in Pharmacology" (P. G. Waser and B. Glasson, eds), pp. 113–119. Wiley–Interscience, New York.

Holden, M. (1970). *Phytochemistry* **9**, 507–512.

Holloway, P. W. and Wakil, S. J. (1964). *J. biol. Chem.* **239**, 2489–2495.

Holman, R. T. and Hanks, D. P. (1955). *J. Am. Oil Chem. Soc.* **32**, 356–357.

Holman, R. T., Egwim, P. O. and Christie, W. W. (1969). *J. biol. Chem.* **244**, 1149–1151.

Holton, R. W., Blecker, H. H. and Onore, M. (1964). *Phytochemistry* **3**, 595–602.

Hooper, N. K. and Law, J. H. (1965). *Biochem. biophys. Res. Commun.* **18**, 426–429.

Hopfer, U., Lehninger, A. L. and Lennarz, W. J. (1970). *J. Membrane Biol.* **2**, 41–58.

Hopkins, C. Y. and Chisholm, M. J. (1960). *J. Am. Oil Chem. Soc.* **37**, 682–684.

Hopkins, C. Y. and Chisholm, M. J. (1962a). *J. chem. Soc.* 573–575.

Hopkins, C. Y. and Chisholm, M. J. (1962b). *Chemy Ind.* 2064–2064.

Hopkins, C. Y. and Chisholm, M. J. (1964a). *J. Am. Oil Chem. Soc.* **41**, 42–44.

Hopkins, C. Y. and Chisholm, M. J. (1964b). *Tetrahedron Lett.* 3011–3013.

Hopkins, C. Y. and Chisholm, M. J. (1964c). *Can. J. Chem.* **42**, 2224–2227.

Hopkins, C. Y. and Chisholm, M. J. (1965a). *J. chem. Soc.* 907–910.

Hopkins, C. Y. and Chisholm, M. J. (1965b). *Can. J. Chem.* **43**, 3160–3164.

Hopkins, C. Y. and Chisholm, M. J. (1966). *Chemy Ind.* 1533–1534.

Hopkins, C. Y. and Chisholm, M. J. (1968). *J. Am. Oil Chem. Soc.* **45**, 176–182.

Hopkins, C. Y., Chisholm, M. J. and Prince, L. (1966). *Lipids* **1**, 118–122.

Hopkins, C. Y., Jevans, A. W. and Chisholm, M. J. (1967). *Chemy Ind.* 998–999.

Hopkins, C. Y., Ewing, D. F. and Chisholm, M. J. (1968a). *Phytochemistry* **7**, 619–624.

Hopkins, C. Y., Jevans, A. W. and Chisholm, M. J. (1968b). *J. chem. Soc.* (*C*) 2462–2465.

Hopkins, C. Y., Chisholm, M. J. and Ogrodnik, J. A. (1969a). *Lipids* **4**, 89–92.

Hopkins, C. Y., Jevans, A. W. and Boch, R. (1969b). *Can. J. Biochem.* **47**, 433–436.

Hörhammer, L., Wagner, H. and Richter, G. (1959). *Biochem. Z.* **331**, 155–161.

Horowitz, B. and Winter, G. (1957). *Nature, Lond.* **179**, 582–583.

Hou, C. T., Umemura, Y., Nakamura, M. and Funahashi, S. (1968). *J. Biochem., Tokyo* **63**, 351–360.

Howard, C. F. (1970). *J. biol. Chem.* **245**, 462–468.
Howell, R. W. and Collins, F. I. (1957). *Agron. J.* **49**, 593–597.
Howlett, M. D. D. and Welti, D. (1966). *Analyst* **91**, 291–293.
Howling, D., Morris, L. J. and James, A. T. (1968). *Biochim. biophys. Acta* **152**, 224–226.
Hsu, R. Y., Butterworth, P. H. W. and Porter, J. W. (1969). *In* "Methods in Enzymology" (J. M. Lowenstein, ed.), Vol. XIV, pp. 33–39. Academic Press, New York and London.
Huang, K. P. and Stumpf, P. K. (1970). *Archs Biochem. Biophys.* **140**, 158–173.
Huber, R. E. and Zalik, S. (1963). *Can. J. Biochem. Phyisol.* **41**, 745–754.
Hulanicka, D., Erwin, J. and Bloch, K. (1964). *J. biol. Chem.* **239**, 2778–2787.
Humphreys, T. E. and Stumpf, P. K. (1955). *J. biol. Chem.* **213**, 941–949.
Humphreys, T. E., Newcomb, E. H., Bokman, A. H. and Stumpf, P. K. (1954). *J. biol. Chem.* **210**, 941–948.
Hutt, H. H., Malkin, T., Poole, A. G. and Watt, P. R. (1950). *Nature, Lond.* **165**, 314–315.
Hutton, D. and Stumpf, P. K. (1969). *Pl. Physiol., Lancaster* **44**, 508–516.
Ichihara, K., Kusunose, E. and Kusunose, M. (1969). *Biochim. biophys. Acta* **176**, 704–712.
Inkpen, J. A. and Quackenbush, F. W. (1969a). *Lipids* **4**, 539–543.
Inkpen, J. A. and Quackenbush, F. W. (1969b). *Cereal Chem.* **46**, 580–587.
Jackman, M. E. and Van Steveninck, R. F. M. (1967). *Aust. J. biol. Sci.* **20**, 1063–1068.
Jacks, T. J., Yatsu, L. Y. and Altschul, A. M. (1967). *Pl. Physiol., Lancaster* **42**, 585–597.
Jackson, L. L. and Frear, D. S. (1967). *Can. J. Biochem.* **45**, 1309–1315.
James, A. T. (1962a). *Biochim. biophys. Acta* **57**, 167–169.
James, A. T. (1962b). *Bull. Soc. Chim. biol.* **44**, 951–963.
James, A. T. (1963). *Biochim. biophys. Acta* **70**, 9–19.
James, A. T. (1964). *In* "New Biochemical Separations" (A. T. James and L. J. Morris, eds), pp. 1–24. Van Nostrand, Princeton, New Jersey.
James, A. T. (1968). *Chem. Brit.* **4**, 484–488.
James, A. T. and Martin, A. J. P. (1956). *Biochem. J.* **63**, 144–152.
James, A. T. and Nichols, B. W. (1966). *Nature, Lond.* **210**, 372–375.
James, A. T., Hadaway, H. C. and Webb, J. P. W. (1965). *Biochem. J.* **95**, 448–452.
James, A. T., Harris, P. and Bezard, J. (1968). *Eur. J. Biochem.* **3**, 318–325.
Jamieson, G. R. (1970). *In* "Topics in Lipid Chemistry" (F. D. Gunstone, ed.), pp. 107–159. Logos Press, London.
Jamieson, G. R. and Reid, E. H. (1968). *J. Sci. Food Agric.* **19**, 628–631.
Jamieson, G. R. and Reid, E. H. (1969). *Phytochemistry* **8**, 1489–1494.
Jauréguiberry, G., Law, J. H., McCloskey, J. A. and Lederer, E. (1965). *Biochemistry, Easton* **4**, 347–353.
Jauréguiberry, G., Lenfant, M., Toubiana, R., Azerad, R. and Lederer, E. (1966). *Chem. Commun.* 855–857.
Jennings, W. G., Creveling, R. K. and Heinz, D. E. (1964). *J. Food Sci.* **29**, 730–734.
Jevans, A. W. and Hopkins, C. Y. (1968). *Tetrahedron Lett.* 2167–2170.
Ji, T. H. and Benson, A. A. (1968). *Biochim. biophys. Acta* **150**, 686–693.
Johnson, A. R., Pearson, J. A., Shenstone, F. S., Fogerty, A. C. and Giovanelli, J. (1967a). *Lipids* **2**, 308–315.

Johnson, A. R., Murray, K. E., Fogerty, A. C., Kennet, B. H., Pearson, J. A. and Shenstone, F. S. (1967b). *Lipids* **2**, 316–322.

Johnson, A. R., Pearson, J. A., Shenstone, F. S. and Fogerty, A. C. (1967c). *Nature, Lond.* **214**, 1244–1245.

Johnson, A. R., Fogerty, A. C., Pearson, J. A., Shenstone, F. S. and Bersten, A. M. (1969). *Lipids* **4**, 265–269.

Johnson, A. W. (1965). *Endeaver* **24**, 126–130.

Jones, D. F. (1968). *J. chem. Soc.* (*C*) 2827–2833.

Jones, D. F. and Howe, R. (1968). *J. chem. Soc.* (*C*) 2801–2808.

Jones, E. R. H. (1966). *Chem. Brit.* **2**, 6–13.

Jones, P. D. and Wakil, S. J. (1967). *J. biol. Chem.* **242**, 5267–5273.

Jones, P. D., Holloway, P. W., Peluffo, R. O. and Wakil, S. J. (1969). *J. biol. Chem.* **244**, 744–754.

Joshi, V. C., Plate, C. A. and Wakil, S. J. (1970). *J. biol. Chem.* **245**, 2857–2867.

Jurriens, G. and Kroesen, A. C. J. (1965). *J. Am. Oil Chem. Soc.* **42**, 9–14.

Jurtshuk, P., Sekuzu, I. and Green, D. E. (1961). *Biochem. biophys. Res. Commun.* **6**, 76–80.

Kaneda, T. (1963a). *J. biol. Chem.* **238**, 1222–1228.

Kaneda, T. (1963b). *J. biol. Chem.* **238**, 1229–1235.

Kanemasa, Y. and Goldman, D. S. (1965). *Biochim. biophys. Acta* **98**, 476–485.

Kaneshiro, T. and Law, J. H. (1964). *J. biol. Chem.* **239**, 1705–1713.

Karlsson, K.-A. (1965). *Acta chem. scand.* **19**, 2425–2427.

Karlsson, K.-A. and Holm, G. A. L. (1965). *Acta chem. scand.* **19**, 2423–2425.

Kartha, A. R. S. and Khan, R. A. (1969). *Chemy Ind.* 1869–1870.

Kartha, A. R. S. and Selvaraj, Y. (1970). *Chemy Ind.* 831–832.

Kartha, A. R. S. and Singh, S. P. (1969). *Chemy Ind.* 1342-1343.

Kasbekar, M. G. and Bringi, N. V. (1969). *J. Am. Oil Chem. Soc.* **46**, 183–183.

Kass, L. R. (1969). *In* "Methods in Enzymology" (J. M. Lowenstein, ed.), Vol. XIV, pp. 73–80. Academic Press, New York and London.

Kass, L. R., Brock, D. J. H. and Bloch, K. (1967). *J. biol. Chem.* **242**, 4418–4431.

Katayama, M. and Funahashi, S. (1969). *J. Biochem., Tokyo* **66**, 479–485.

Kates, M. (1954). *Can. J. Biochem. Physiol.* **32**, 571–583.

Kates, M. (1956). *Can. J. Biochem. Physiol.* **34**, 967–980.

Kates, M. (1964). *J. Lipid Res.* **5**, 132–135.

Kates, M. and Sastry, P. S. (1970). *In* "Methods in Enzymology" (J. M. Lowenstein, ed.), Vol. XIV, pp. 197–203. Academic Press, New York and London.

Kates, M. and Volcani, B. E. (1966). *Biochim. biophys. Acta* **116**, 264–278.

Kaufmann, H. P. and Wessels, H. (1964). *Fette, Seifen, Anstrichmittel,* **66**, 81–86.

Kaufmann, H. P., Hamza, Y. and Mangold, H. K. (1970). *J. Am. Oil Chem. Soc.* **47**, Abs. No. 174.

Kauss, H. (1968). *Z. Naturf.* **23b**, 1522–1526.

Kawanami, J., Kimura, A., Nakagawa, Y. and Otsuka, H. (1969). *Chem. Phys. Lip.* **3**, 29–38.

Kenyon, C. N. and Stanier, R. Y. (1970). *Nature, Lond.* **227**, 1164–1166.

Kies, M. W., Haining, J. L., Pistorius, E., Schroeder, D. H. and Axelrod, B. (1969). *Biochem. biophys. Res. Commun.* **36**, 312–315.

King, G. (1942). *J. chem. Soc.* 387–391.

Kircher, H. W. and Heywang, B. W. (1966). *Poultry Sci.* **45**, 1432–1434.

Kishimoto, Y. and Radin, N. S. (1965). *J. Lipid Res.* **6**, 435–436.

Kleiman, R., Earle, F. R., Wolff, I. A. and Jones, Q. (1964). *J. Am. Oil Chem. Soc.* **41**, 459–460.

Kleiman, R., Earle, F. R. and Wolff, I. A. (1966). *Lipids* **1**, 301–304.
Kleiman, R., Davison, V. L., Earle, F. R. and Dutton, H. J. (1967a). *Lipids* **2**, 339–341.
Kleiman, R., Spencer, G. F., Earle, F. R. and Wolff, I. A. (1967b). *Chemy Ind.* 1326–1327.
Kleiman, R., Spencer, G. F., Earle, F. R. and Wolff, I. A. (1969a). *Lipids* **4**, 135–141.
Kleiman, R., Earle, F. R. and Wolff, I. A. (1969b). *Lipids* **4**, 317–320.
Kleiman, R., Earle, F. R. and Wolff, I. A. (1969c). *J. Am. Oil Chem. Soc.* **46**, 505–505.
Kleinig, H. and Czygan, F. (1969). *Z. Naturf.* **24b**, 927–930.
Kleinschmidt, A. K., Moss, J. and Lane, M. D. (1969). *Science, N.Y.* **166**, 1276–1278.
Klenk, E., Knipprath, W., Eberhagen, D. and Koof, H. P. (1963). *Hoppe-Seyler's Z. physiol. Chem.* **334**, 44–59.
Klopfenstein, W. E. and Shigley, J. W. (1967). *J. Lipid Res.* **8**, 350–351.
Knipprath, W. G. and Mead, J. F. (1966). *Lipids* **1**, 113–117.
Knipprath, W. G. and Mead, J. F. (1968). *Lipids* **3**, 121–128.
Knoche, H. W. (1968). *Lipids* **3**, 163–169.
Knoche, H. W. and Shively, J. M. (1969). *J. biol. Chem.* **244**, 4773–4778.
Knoop, F. (1904). *Beitr. chem. Physiol. Path.* **6**, 150.
Knowles, P. F. (1969). *J. Am. Oil Chem. Soc.* **46**, 130–132.
Knowles, P. F. and Mutwakil, A. (1963). *Econ. Bot.* **17**, 139–145.
Koch, R. B. (1968). *Archs. Biochem. Biophys.* **125**, 303–307.
Koike, M., Reed, L. J. and Carroll, W. R. (1963). *J. biol. Chem.* **238**, 30–39.
Kolattukudy, P. E. (1965). *Biochemistry, Easton* **4**, 1844–1855.
Kolattukudy, P. E. (1966). *Biochemistry, Easton* **5**, 2265–2275.
Kolattukudy, P. E. (1967a). *Phytochemistry* **6**, 963–975.
Kolattukudy, P. E. (1967b). *Biochemistry, Easton* **6**, 2705–2717.
Kolattukudy, P. E. (1968). *Science, N.Y.* **159**, 498–505.
Kolattukudy, P. E. (1969). *Pl. Physiol., Lancaster* **44**, 315–317.
Kolattukudy, P. E. (1970). *Lipids* **5**, 259–275.
Korn, E. D. (1964a). *J. Lipid Res.* **5**, 352–362.
Korn, E. D. (1964b). *J. biol. Chem.* **239**, 396–400.
Korn, E. D. (1968). *In* "Biological Interfaces: Flows and Exchanges", p. 257. Little, Brown and Co., Boston.
Kreger, D. R. (1948). *Rec. Trav. Bot. Neer.* **41**, 603.
Krzymanski, J. and Downey, R. K. (1969). *Can. J. Plant Sci.* **49**, 313–319.
Kuhn, R., Winterstein, A. and Lederer, E. (1931). *Hoppe-Seyler's Z. physiol. Chem* **197**, 141.
Kusunose, M., Kusunose, E. and Coon, M. J. (1964a). *J. biol. Chem.* **239**, 1374–1380.
Kusunose, M., Kusunose, E. and Coon, M. J. (1964b). *J. biol. Chem.* **239**, 2135–2139.
Lands, W. E. M. and Hart, P. (1965). *Biochim. biophys. Acta* **98**, 532–538.
Lands, W. E. M., Pieringer, R. A., Slakey, P. M. and Zschocke, A. (1966). *Lipids* **1**, 444–448.
Larrabee, A. R., McDaniel, E. G., Bakerman, H. A. and Vagelos, P. R. (1965). *Proc. natn. Acad. Sci. U.S.A.* **54**, 267–273.
Laties, G. G. and Hoelle, C. (1967). *Phytochemistry* **6**, 49–57.
Lawrence, R. C. and Hawke, J. C. (1966). *Biochem. J.* **98**, 25–29.
Lee, B. and Priestley, J. H. (1924). *Ann. Bot.* **38**, 525–545.

Lee, S. G. and Chasson, R. M. (1966). *Physiologia Pl.* **19**, 199–206.
Legler, G. (1965). *Phytochemistry* **4**, 29.
Lehninger, A. L. and Greville, G. D. (1953). *Biochim. biophys. Acta* **12**, 188–202.
Lemieux, R. U. (1953). *Can. J. Chem.* **31**, 396–417.
Lennarz, W. J. (1961). *Biochem. biophys. Res. Commun.* **6**, 112–116.
Lennarz, W. J. (1966). *Adv. Lipid Res.* **4**, 175–225.
Lennarz, W. J., Scheuerbrandt, G. and Bloch, K. (1962). *J. biol. Chem.* **237**, 664–671.
Lepage, M. (1964a). *J. Chromat.* **13**, 99–103.
Lepage, M. (1964b). *J. Lipid Res.* **5**, 587–592.
Lepage, M. (1967). *Lipids* **2**, 244–250.
Lepage, M. (1968). *Lipids* **3**, 477–481.
Lepage, M., Mumma, R. O. and Benson, A. A. (1960). *J. Am. chem. Soc.* **82**, 3713–3715.
Levis, G. M. (1970). *Biochem. biophys. Res. Commun.* **38**, 470–477.
Levis, G. M. and Mead, J. F. (1964). *J. biol. Chem.* **239**, 77–80.
Lewis, H. L. and Johnson, G. T. (1966). *Mycologia* **58**, 136–147.
Light, R. J., Lennarz, W. J. and Bloch, K. (1962). *J. biol. Chem.* **237**, 1793–1800.
Ligthelm, S. P. (1954). *Chemy. Ind.* 249–250.
Ligthelm, S. P., Horn, D. H. S., Schwartz, H. M. and von Holdt, M. M. (1954). *J. Sci. Food Agric.* **5**, 281–288.
Lippel, K. and Mead, J. F. (1968). *Biochim. biophys. Acta* **152**, 669–680.
Lippel, K. and Mead, J. F. (1969). *Lipids* **4**, 129–134.
Litchfield, C. (1968). *Lipids* **3**, 170–177.
Litchfield, C. (1970). *Lipids* **5**, 144–145.
Litchfield, C., Farquhar, M. and Reiser, R. (1964). *J. Am. Oil Chem. Soc.* **41**, 588–592.
Lloyd, D. and Venables, S. E. (1967). *Biochem. J.* **104**, 639–646.
Long, C., Odavic, R. and Sargent, E. J. (1963). *Biochem. J.* **87**, 13P–13P.
Lu, A. Y. H., Junk, K. W. and Coon, M. J. (1969). *J. biol. Chem.* **244**, 3714–3721.
Lust, G. and Lynen, F. (1968). *Eur. J. Biochem.* **7**, 68–72.
Lynen, F. (1953). *Fedn Proc. Fedn Am. Socs exp. Biol.* **12**, 683–691.
Lynen, F. (1955). *A. Rev. Biochem.* **24**, 653–688.
Lynen, F. (1959). *J. Cell Comp. Physiol.* **54** (Suppl. 1), 33–49.
Lynen, F. (1961). *Fedn Proc. Fedn Am. Socs exp. Biol.* **20**, 941–951.
Lynen, F. (1967). *Biochem. J.* **102**, 381–400.
Lynen, F. (1969). *In* "Methods in Enzymology" (J. M. Lowenstein, ed.), Vol. XIV, pp. 17–33. Academic Press, New York and London.
Lynen, F. and Ochoa, S. (1953). *Biochim. biophys. Acta* **12**, 299–314.
McCloskey, J. A. (1970). *In* "Topics in Lipid Chemistry" (F. D. Gunstone, ed.), pp. 369–440. Logos Press, London.
McCloskey, J. A. and McClelland, M. J. (1965). *J. Am. chem. Soc.* **87**, 5090–5093.
McConnell, W. B. and Finlayson, A. J. (1964). *Can. J. Biochem.* **42**, 187–193.
MacDonald, R. C. and Mead, J. F. (1968). *Lipids* **3**, 275–283.
McElroy, F. A., and Stewart, H. B. (1968). *Can. J. Biochem.* **46**, 303–313.
Macey, M. J. K. and Barber, H. N. (1970). *Phytochemistry* **9**, 5–12.
Macfarlane, M. G. (1958). *Nature, Lond.* **182**, 946–946.
McKenna, E. J. and Coon, M. J. (1970). *J. biol. Chem.* **245**, 3882–3889.
McKillican, M. E. (1966). *J. Am. Oil Chem. Soc.* **43**, 461–465.
McKillican, M. E. (1967). *J. Am. Oil chem. Soc.* **44**, 200–201.
McLean, J. and Clark, A. H. (1956). *J. chem. Soc.* 777–778.

McMahon, V. and Stumpf, P. K. (1964). *Biochim. biophys. Acta* **84**, 359–361.

McMahon, V. and Stumpf, P. K. (1966). *Pl. Physiol., Lancaster* **41**, 148–156.

Mahler, H. R. (1964). *In* "Fatty Acids" (K. S. Markley, ed.), 2nd Ed., Pt. 3, pp. 1487–1550. Interscience, New York.

Mahler, H. R. and Cordes, E. H. (1966). "Biological Chemistry." Harper and Row (New York, Evanston and London)/John Weatherhill, Inc. (Tokyo).

Maier, R. and Holman, R. T. (1964). *Biochemistry, Easton* **3**, 270–274.

Majerus, P. W. (1967). *J. biol. Chem.* **242**, 2325–2332.

Majerus, P. W. and Vagelos, P. R. (1967). *Adv. Lipid Res.* **5**, 2.

Majerus, P. W., Alberts, A. W. and Vagelos, P. R. (1964). *Proc. natn. Acad. Sci. U.S.A.* **51**, 1231–1238.

Majerus, P. W., Alberts, A. W. and Vagelos, P. R. (1965a). *J. biol. Chem.* **240**, 618–621.

Majerus, P. W., Alberts, A. W. and Vagelos, P. R. (1965b). *J. biol. Chem.* **240**, 4723–4726.

Majerus, P. W., Alberts, A. W. and Vagelos, P. R. (1969a). *In* "Methods in Enzymology" (J. M. Lowenstein, ed.), Vol. XIV, pp. 43–50. Academic Press, New York and London.

Majerus, P. W., Alberts, A. W. and Vagelos, P. R. (1969b). *In* "Methods in Enzymology" (J. M. Lowenstein, ed.), Vol. XIV, pp. 64–66. Academic Press, New York and London.

Malins, D. C. and Mangold, H. K. (1960). *J. Am. Oil Chem. Soc.* **37**, 576–578.

Malkin, T. and Poole, A. G. (1953). *J. chem. Soc.* 3470–3478.

Mapson, L. W. and Moustafa, E. M. (1955). *Biochem. J.* **60**, 71–80.

Mapson, L. W. and Wardale, D. A. (1967). *Biochem. J.* **102**, 574–585.

Mapson, L. W. and Wardale, D. A. (1968). *Biochem. J.* **107**, 433–442.

Mapson, L. W., March, J. F., Rhodes, M. J. C. and Wooltorton, L. S. C. (1970). *Biochem. J.* **117**, 473–479.

Marcus, A. and Velasco, J. (1960). *J. biol. Chem.* **235**, 563–567.

Markley, K. S. (1960). *In* "Fatty Acids" (K. S. Markley, ed.), 2nd Ed., Pt. I, pp. 23–249. Interscience, New York.

Markley, K. S. (1968). *In* "Fatty Acids" (K. S. Markley, ed.), 2nd Ed., Pt. V, pp. 3133–3285. Interscience, New York.

Marsh, J. B. and James, A. T. (1962). *Biochim. biophys. Acta* **60**, 320–328.

Martin, J. T. (1964). *A. Rev. Phytopathol.* **2**, 81.

Martin, R. O. and Stumpf, P. K. (1959). *J. biol. Chem.* **234**, 2548–2554.

Martinosi, A., Donley, J. and Halpin, R. A. (1968). *J. biol. Chem.* **243**, 61–70.

Matic, M. (1956). *Biochem. J.* **63**, 168–176.

Matsuhashi, M. *In* "Methods in Enzymology" (J. M. Lowenstein, ed.), Vol. XIV, pp. 3–8. Academic Press, New York and London.

Matsuhashi, M., Matsuhashi, S. and Lynen, F. (1964). *Biochem. Z.* **340**, 263–289.

Matsumura, S. and Stumpf, P. K. (1968). *Archs Biochem. Biophys.* **125**, 932–941.

Matsumura, S., Brindley, D. N. and Bloch, K. (1970). *Biochem. biophys. Res. Commun.* **38**, 369–377.

Mattson, F. H. and Volpenhein, R. A. (1961). *J. biol. Chem.* **236**, 1891–1894.

Mattson, F. H. and Volpenhein, R. A. (1963). *J. Lipid Res.* **4**, 392–396.

Mattson, F. H. and Volpenhein, R. A. (1966). *J. Lipid Res.* **7**, 536–543.

Mattson, F. H. and Volpenhein, R. A. (1968). *J. Lipid Res.* **9**, 79–84.

Mazliak, P. (1967). *Phytochemistry* **6**, 687–702.

11

Mazliak, P. (1968). *In* "Progress in Phytochemistry" (L. Reinhold and Y. Liwschitz, eds), Vol. 1, pp. 49–111. Interscience, New York.

Mead, J. F. and Howton, D. R. (1957). *J. biol. Chem.* **229**, 575–582.

Mecham, D. K. and Mohammad, A. (1955). *Cereal Chem.* **32**, 405–415.

Meyer, F. and Bloch, K. (1963). *Biochim. biophys. Acta* **77**, 671–673.

Mikolajczak, K. L. and Smith, C. R. (1967). *Lipids* **2**, 261–265.

Mikolajczak, K. L., Smith, C. R., Bagby, M. O. and Wolff, I. A. (1964). *J. org. Chem.* **29**, 318–322.

Mikolajczak, K. L., Smith, C. R. and Wolff, I. A. (1965). *J. Am. Oil Chem. Soc.* **42**, 939–941.

Mikolajczak, K. L., Rogers, M. F., Smith, C. R. and Wolff, I. A. (1967). *Biochem. J.* **105**, 1245–1249.

Mikolajczak, K. L., Freidinger, R. M., Smith, C. R. and Wolff, I. A. (1968a). *Lipids* **3**, 489–494.

Mikolajczak, K. L., Smith, C. R. and Wolff, I. A. (1968b). *Lipids* **3**, 215–220.

Mikolajczak, K. L., Seigler, D. S., Smith, C. R., Wolff, I. A. and Bates, R. B. (1969). *Lipids* **4**, 617–619.

Mikolajczak, K. L., Smith, C. R. and Tjarks, L. W. (1970). *Biochim. biophys. Acta* **210**, 306–314.

Miller, A. L. and Levy, H. R. (1969). *J. biol. Chem.* **244**, 2334–2342.

Miller, R. W., Earle, F. R., Wolff, I. A. and Jones, Q. (1965). *J. Am. Oil Chem. Soc.* **42**, 817–821.

Miller, R. W., Earle, F. R. and Wolff, I. A. (1968). *Lipids* **3**, 43–45.

Mitsuda, H., Yasumoto, K. and Yamamoto, A. (1967). *Archs Biochem. Biophys.* **118**, 664–669.

Miwa, T. K., Earle, F. R., Miwa, G. C. and Wolff, I. A. (1963). *J. Am. Oil Chem. Soc.* **40**, 225–229.

Miyachi, S. and Miyachi, S. (1966). *Pl. Physiol., Lancaster* **41**, 479–486.

Miyachi, S., Miyachi, S. and Benson, A. A. (1965). *Plant and Cell Physiol., Tokyo* **6**, 789–792.

Mize, C. E., Avigan, J., Steinberg, D., Pittman, R. C., Fales, H. M. and Milne, G. W. A. (1969a). *Biochim. biophys. Acta* **176**, 720–739.

Mize, C. E., Herndon, J. H., Blass, J. P., Milne, G. W. A., Follansbee, C., Laudat, P. and Steinberg, D. (1969b). *J. Clin. Invest.* **48**, 1033.

Mizugaki, M., Weeks, G., Toomey, R. E. and Wakil, S. J. (1968a). *J. biol. Chem.* **243**, 3661–3670.

Mizugaki, M., Swindell, A. C. and Wakil, S. J. (1968b). *Biochem. biophys. Res. Commun.* **33**, 520–527.

Mooney, L. A. and Barron, E. J. (1970). *Biochemistry, Easton* **9**, 2138–2143.

Morré, D. J., Nyquist, S. and Rivera, E. (1970). *Pl. Physiol., Lancaster* **45**, 800–804.

Morris, L. J. (1963). *J. chem. Soc.* 5779–5781.

Morris, L. J. (1965a). *Biochem. biophys. Res. Commun.* **18**, 495–500.

Morris, L. J. (1965b). *Biochem. biophys. Res. Commun.* **20**, 340–345.

Morris, L. J. (1966). *J. Lipid Res.* **7**, 717–732.

Morris, L. J. (1967). *Biochem. biophys. Res. Commun.* **29**, 311–315.

Morris, L. J. (1970). *Biochem. J.* **118**, 681–693.

Morris, L. J. and Crouchman, M. L. (1969). *Lipids* **4**, 50–54.

Morris, L. J. and Hall, S. W. (1966). *Lipids* **1**, 188–196.

Morris, L. J. and Hall, S. W. (1967). *Chemy. Ind.* 32–34.

Morris, L. J. and Hitchcock, C. (1968). *Eur. J. Biochem.* **4**, 146–148.

Morris, L. J. and Marshall, M. O. (1966a). *Chemy. Ind.* 460–461.

Morris, L. J. and Marshall, M. O. (1966b). *Chemy. Ind.* 1493–1494.

Morris, L. J. and Nichols, B. W. (1970). *In* "Progress in Thin Layer Chromatography and Related Methods" (A. Niederwieser and G. Pataki, eds), Vol. I, pp. 74–93. Ann Arbor Science Publishers Inc.

Morris, L. J. and Wharry, D. M. (1966). *Lipids* 1, 41–46.

Morris, L. J., Holman, R. T. and Fontell, K. (1960). *J. Am. Oil Chem. Soc.* 37, 323–327.

Morris, L. J., Hall, S. W. and James, A. T. (1966a). *Biochem. J.* 100, 29c–30c.

Morris, L. J., Marshall, M. O. and Kelly, W. (1966b). *Tetrahedron Lett.* 4249.

Morris, L. J., Harris, R. V., Kelly, W. and James, A. T. (1967). *Biochem. biophys. Res. Commun.* 28, 904–908.

Morris, L. J., Marshall, M. O. and Hammond, E. W. (1968). *Lipids* 3, 91–95.

Morrison, W. R. and Smith, L. M. (1964). *J. Lipid Res.* 5, 600–608.

Mounts, T. L. and Dutton, H. J. (1965). *Anal. Chem.* 37, 641–644.

Mudd, J. B. (1967). *A. Rev. Pl. Physiol.* 18, 229–252.

Mudd, J. B. and McManus, T. T. (1962). *J. biol. Chem.* 237, 2057–2063.

Mudd, J. B. and Stumpf, P. K. (1961). *J. biol. Chem.* 236, 2602–2609.

Mudd, J. B., van Vliet, H. H. D. M. and van Deenen, L. L. M. (1969). *J. Lipid Res.* 10, 623–630.

Myhre, D. V. (1968). *Can. J. Chem.* 46, 3071–3077.

Nagai, J. and Bloch, K. (1965). *J. biol. Chem.* 240, 3702–3703.

Nagai, J. and Bloch, K. (1966). *J. biol. Chem.* 241, 1925–1927.

Nagai, J. and Bloch, K. (1967). *J. biol. Chem.* 242, 357–362.

Nagai, J. and Bloch, K. (1968). *J. biol. Chem.* 243, 4626–4633.

Nandedkar, A. K. N. and Kumar, S. (1969). *Archs Biochem. Biophys.* 134, 563–571.

Nandedkar, A. K. N., Schirmer, E. W., Pynadath, T. I. and Kumar, S. (1969). *Archs Biochem. Biophys.* 134, 554–562.

Neufeld, E. F. and Hall, C. W. (1964). *Biochem. biophys. Res. Commun.* 14, 503–508.

Newcomb, E. H. and Stumpf, P. K. (1952). *In* "Phosphorus Metabolism" (W. D. McElroy and B. Glass, eds), Vol. II, pp. 291–300. Johns Hopkins Press, Baltimore.

Newman, D. W. (1962). *Biochem. biophys. Res. Commun.* 9, 179–183.

Newman, D. W. (1964). *J. exp. Bot.* 15, 525–529.

Newman, D. W. (1966). *Pl. Physiol., Lancaster* 41, 328–334.

Nichaman, M. Z., Sweeley, C. C., Oldham, N. M. and Olson, R. E. (1963). *J. Lipid Res.* 4, 484–485.

Nichols, B. W. (1963). *Biochim. biophys. Acta* 70, 417–422.

Nichols, B. W. (1964a). *Lab. Pract.* 13, 299–305.

Nichols, B. W. (1964b). *In* "New Biochemical Separations" (L. J. Morris and A. T. James, eds), pp. 321–327. Van Nostrand, London and Princeton, New Jersey.

Nichols, B. W. (1965a). *Phytochemistry* 4, 769–772.

Nichols, B. W. (1965b). *Biochim. biophys. Acta* 106, 274–279.

Nichols, B. W. (1968). *Lipids* 3, 354–360.

Nichols, B. W. (1970). *In* "Phytochemical Phylogeny" (J. B. Harborne, ed.), pp. 105–117. Academic Press, London and New York.

Nichols, B. W. and Appleby, R. S. (1969), *Phytochemistry* 8, 1907–1915.

Nichols, B. W. and James, A. T. (1964). *Fette Seifen, Anstrichmittel* 66, 1003–1006.

Nichols, B. W. and James, A. T. (1968). *In* "Progress in Phytochemistry" (L. Reinhold and Y. Liwschitz, eds), Vol. I, pp. 1–48. Interscience Publishers, London and New York.

Nichols, B. W. and Moorhouse, R. (1969). *Lipids* **4**, 311–316.

Nichols, B. W. and Wood, B. J. B. (1968). *Lipids* **3**, 46–50.

Nichols, B. W., Harris, R. V. and James, A. T. (1965a). *Biochem. biophys. Res. Commun.* **20**, 256–262.

Nichols, B. W., Harris, P. and James, A. T. (1965b). *Biochem. biophys. Res. Commun.* **21**, 473–479.

Nichols, B. W., Stubbs, J. M. and James, A. T. (1967a). In "Biochemistry of Chloroplasts" (T. W. Goodwin, ed.), Vol. II, pp. 677–690. Academic Press, London and New York.

Nichols, B. W., James, A. T. and Breuer, J. (1967b). *Biochem. J.* **104**, 486–496.

Niehaus, W. G. and Ryhage, R. (1967). *Tetrahedron Lett.* **49**, 5021–5026.

Niehaus, W. G. and Schroepfer, G. J. (1965). *Biochem. biophys. Res. Commun.* **21**, 271–275.

Niehaus, W. G., Kisic, A., Torkelson, A., Bednarczyk, D. J. and Schroepfer, G. J. (1970a). *J. biol. Chem.* **245**, 3790–3797.

Niehaus, W. G., Kisic, A., Torkelson, A., Bednarczyk, D. J. and Schroepfer, G. J. (1970b). *J. biol. Chem.* **245**, 3802–3809.

Noda, M. and Fujiwara, N. (1967). *Biochim. biophys. Acta* **137**, 199–201.

Norris, F. A. and Mattil, K. F. (1947). *J. Am. Oil Chem. Soc.* **24**, 274–275.

Nugteren, D. H. (1965). *Biochim. biophys. Acta* **106**, 280–290.

Numa, S. (1969). In "Methods in Enzymology" (J. M. Lowenstein, ed.), Vol. XIV, pp. 9–16. Academic Press, New York and London.

Nyman, B. (1965a). *Physiologia Pl.* **18**, 1085–1094.

Nyman, B. (1965b). *Physiologia Pl.* **18**, 1095–1104.

O'Connor, R. T. (1956). *J. Am. Oil Chem. Soc.* **33**, 1–15.

Oesterhelt, D., Bauer, H. and Lynen, F. (1969). *Proc. natn. Acad. Sci. U.S.A.* **63**, 1377–1382.

Ohnishi, T. and Ohnishi, T. (1962). *J. Biochem., Tokyo* **52**, 230–231.

Olney, C. E., Jensen, R. G., Sampugna, J. and Quinn, J. G. (1968). *Lipids* **3**, 498–502.

Ongun, A., and Mudd, J. B. (1968). *J. biol. Chem.* **243**, 1558–1566.

Ongun, A. and Mudd, J. B. (1970). *Pl. Physiol., Lancaster* **45**, 255–262.

Ongun, A., Thomson, W. W. and Mudd, J. B. (1968). *J. Lipid Res.* **9**, 409–415.

Ory, R. L., St. Angelo, A. J. and Altschul, A. M. (1960). *J. Lipid Res.* **1**, 208–213.

Ory, R. L., St. Angelo, A. J. and Altschul, A. M. (1962). *J. Lipid Res.* **3**, 99–105.

Ory, R. L., Baker, R. H. and Boudreaux, G. J. (1964). *Biochemistry, Easton* **3**, 2013–2016.

Ory, R. L., Kiser, J. and Pradel, P. A. (1969). *Lipids* **4**, 261–264.

Overath, P. and Stumpf, P. K. (1964). *J. biol. Chem.* **239**, 4103–4110.

Padley, F. B. (1969). *J. Chromat.* **39**, 37–46.

Panter, R. A. and Mudd, J. B. (1969). *FEBS Lett.* **5**, 169–170.

Parker, P. L., Van Baalen, C. and Maurer, L. (1967). *Science, N.Y.* **155**, 707–708.

Parkes, J. G. and Thompson, W. (1970). *Biochim. biophys. Acta* **196**, 162–169.

Paschke, R. F. and Wheeler, D. H. (1954). *J. Am. Oil Chem. Soc.* **31**, 81–85.

Patterson, G. W. (1970). *Lipids* **5**, 597–600.

Patton, S., Fuller, G., Loeblich, A. R. and Benson, A. A. (1966). *Biochim. biophys. Acta* **116**, 577–579.

Payne, S. N. (1964). *J. Chromat.* **15**, 173–179.

Peterson, J. A. and Coon, M. J. (1968). *J. biol. Chem.* **243**, 329–334.

Peterson, J. A., Kusunose, M., Kusunose, E. and Coon, M. J. (1967). *J. biol. Chem.* **242**, 4334–4340.
Phillips, B. E., Smith, C. R. and Tjarks, L. W. (1970). *Biochim. biophys. Acta* **210**, 353–359.
Pinsky, A. and Ordin, L. (1969). *Plant and Cell Physiol.* **10**, 771–785.
Pitt, G. A. J. and Morton, R. A. (1957). *In* "Progress in the Chemistry of Fats and Other Lipids" (R. T. Holman, W. O. Lundberg and T. Malkin, eds), Vol. 4, pp. 227–278. Pergamon Press, Oxford.
Placek, L. L. (1963). *J. Am. Oil Chem. Soc.* **40**, 319–329.
Plate, C. A., Joshi, V. C., Sedgwick, B. and Wakil, S. J. (1968). *J. biol. Chem.* **243**, 5439–5445.
Pohl, P., Glasl, H. and Wagner, H. (1970). *J. Chromat.* **49**, 488–492.
Pomeranz, Y. and Chung, O. (1965). *J. Chromat.* **19**, 540–550.
Popjak, G., French, T. H., Hunter, G. D. and Martin, A. J. P. (1951). *Biochem. J.* **48**, 612–618.
Porter, J. W. and Long, R. W. (1958). *J. biol. Chem.* **233**, 20–25.
Porter, J. W., Wakil, S. J., Tietz, A., Jacob, M. I. and Gibson, D. M. (1957). *Biochim. biophys. Acta* **25**, 35–41.
Powell, G. L., Elovson, J. and Vagelos, P. R. (1969). *J. biol. Chem.* **244**, 5616–5624.
Powell, R. G. and Smith, C. R. (1966). *Biochemistry, Easton* **5**, 625–631.
Powell, R. G., Smith, C. R., Glass, C. A. and Wolff, I. A. (1965). *J. org. Chem.* **30**, 610–615.
Powell, R. G., Smith, C. R. and Wolff, I. A. (1966). *J. org. Chem.* **31**, 528–533.
Powell, R. G., Smith, C. R. and Wolff, I. A. (1967a). *Lipids* **2**, 172–177.
Powell, R. G., Smith, C. R. and Wolff, I. A. (1967b). *J. org. Chem.* **32**, 1442–1446.
Pratt, H. K. and Goeschl, J. D. (1969). *A. Rev. Pl. Physiol.* **20**, 541–584.
Preiss, B. and Bloch, K. (1964). *J. biol. Chem.* **239**, 85–88.
Privett, O. S. and Blank, M. L. (1961). *J. Lipid Res.* **2**, 37–44.
Privett, O. S. and Blank, M. L. (1963). *J. Am. Oil Chem. Soc.* **40**, 70–75.
Privett, O. S., Nickell, C., Lundberg, W. O. and Boyer, P. D. (1955). *J. Am. Oil Chem. Soc.* **32**, 505–511.
Purdy, S. J. and Truter, E. V. (1963a). *Proc. Roy. Soc. Ser. B* **158**, 536–544.
Purdy, S. J. and Truter, E. V. (1963b). *Proc. Royl Soc. Ser. B* **158**, 544–553.
Purdy, S. J. and Truter, E. V. (1963c). *Proc. Roy. Soc. Ser B* **158**, 553–561.
Putt, E. D., Craig, B. M. and Carson, R. B. (1969). *J. Am. Oil Chem. Soc.* **46**, 126–129.
Quarles, R. H. and Dawson, R. M. C. (1969). *Biochem. J.* **112**, 787–794.
Radler, F. and Horn, D. H. S. (1965). *Aust. J. Chem.* **18**, 1059–1069.
Radunz, A. (1965). *Hoppe-Seyler's Z. physiol. Chem.* **341**, 192–203.
Radunz, A. (1968a). *Hoppe-Seyler's Z. physiol. Chem.* **349**, 303–309.
Radunz, A. (1968b). *Hoppe-Seyler's Z. physiol. Chem.* **349**, 1091–1094.
Radunz, A. (1969). *Hoppe-Seyler's Z. physiol. Chem.* **350**, 411–417.
Raju, P. K. and Reiser, R. (1966). *Lipids* **1**, 10–15.
Raju, P. K. and Reiser, R. (1967). *J. biol. Chem.* **242**, 379–384.
Raju, P. K. and Reiser, R. (1969). *Biochim. biophys. Acta* **176**, 48–53.
Rando, R. R. and Bloch, K. (1968). *J. biol. Chem.* **243**, 5627–5634.
Rasmussen, R. K. and Klein, H. P. (1967). *Biochem. biophys. Res. Commun.* **28**, 415–419.
Rebeiz, C. A. and Castelfranco, P. (1964). *Pl. Physiol., Lancaster* **39**, 932–938.

Rebeiz, C. A., Castelfranco, P. and Engelbrecht, A. H. (1965a). *Pl. Physiol.*, *Lancaster* **40**, 281–286.

Rebeiz, C. A., Castelfranco, P. and Breidenbach, R. W. (1965b). *Pl. Physiol.*, *Lancaster* **40**, 286–289.

Reid, E. H. and Jamieson, G. R. (1970). Unpublished data.

Reiesner, H., Finlayson, A. J., McConnell, W. B. and Ledingham, G. A. (1963). *Can. J. Biochem. Physiol.* **41**, 737–743.

Renkonen, O. and Bloch, K. (1969). *J. biol. Chem.* **244**, 4899–4903.

Richards, J. H. and Hendrickson, J. B. (1964). "The Biosynthesis of Steroids, Terpenes and Acetogenins." W. A. Benjamin Inc., New York.

Rilling, H. C. and Coon, M. J. (1960). *J. biol. Chem.* **235**, 3087–3092.

Rinne, R. W. (1969). *Pl. Physiol.*, *Lancaster* **44**, 89–94.

Rittenberg, D. and Bloch, K. (1945). *J. biol. Chem.* **160**, 417–424.

Roehm, J. N. and Privett, O. S. (1969). *J. Lipid Res.* **10**, 245–246.

Roehm, J. N. and Privett, O. S. (1970). *Lipids* **5**, 353–358.

Romani, R. J., Breidenbach, R. W. and van Kooy, J. G. (1965). *Pl. Physiol.*, *Lancaster* **40**, 561–566.

Rosenberg, A. (1963). *Biochemistry, Easton* **2**, 1148–1154.

Rosenberg, A. (1967a). *Science, N.Y.* **157**, 1187–1192.

Rosenberg, A. (1967b). *Science, N.Y.* **157**, 1192–1196.

Rosenberg, A. and Gouaux, J. (1967). *J. Lipid Res.* **8**, 80–83.

Rosenberg, A. and Pecker, M. (1964). *Biochemistry, Easton* **3**, 254–258.

Rosenberg, A., Gouaux, J. and Milch, P. (1966). *J. Lipid Res.* **7**, 733–738.

Ross, J., Gebhart, A. I. and Gerecht, J. F. (1949). *J. Am. chem. Soc.* **71**, 282–286.

Rossi, C. R., Galzigna, L. and Gibson, D. M. (1969). *In* "Methods in Enzymology" (J. M. Lowenstein, ed.), Vol. XIV, pp. 91–95. Academic Press, New York and London.

Rothstein, A. (1963). *In* "Functional Biochemistry of Cell Structures" (O. Lindberg, ed.), Vol. II, p. 40. Pergamon Press, London and Oxford.

Rothstein, M. and Götz, P. (1968). *Archs Biochem. Biophys.* **126**, 131–140.

Rotsch, E. and Debuch, H. (1965). *Hoppe-Seyler's Z. physiol. Chem.* **343**, 135–149.

Roubal, W. T. and Tappel, A. L. (1966). *Archs Biochem. Biophys.* **113**, 5–8.

Roughan, P. G. (1970). *Biochem. J.* **117**, 1–8.

Roughan, P. G. and Batt, R. D. (1968). *Analyt. Biochem.* **22**, 74–88.

Roughan, P. G. and Batt, R. D. (1969). *Phytochemistry* **8**, 363–369.

Rouser, G., Baumann, A. J., Nicolaides, N. and Heller, D. (1961). *J. Am. Oil Chem. Soc.* **38**, 565–581.

Rouser, G., Kritchevsky, G., Heller, D. and Lieber, E. (1963). *J. Am. Oil Chem. Soc.* **40**, 425–454.

Rouser, G., Galli, C., Lieber, E., Blank, M. L. and Privett, O. S. (1964). *J. Am. Oil Chem. Soc.* **41**, 836–840.

Rouser, G., Siakatos, A. N. and Fleischer, S. (1966). *Lipids* **1**, 85–86.

Ryder, E., Gregolin, C., Chang, H. C., Kleinschmidt, A. K., Warner, R. C. and Lane, M. D. (1967a). *Fedn Proc. Fedn Am. Socs exp. Biol.* **26**, 672–672.

Ryder, E., Gregolin, C., Chang, H. C. and Lane, M. D. (1967b). *Proc. Natn. Acad. Sci. U.S.A.* **57**, 1455–1462.

Safford, R. and Nichols, B. W. (1970). *Biochim. biophys. Acta* **210**, 57–64.

Saga, M., Tsutsumi, Y. and Nakano, M. (1969). *Biochim. biophys. Acta* **184**, 213–215.

St. Angelo, A. J. and Altschul, A. M. (1964). *Pl. Physiol.*, *Lancaster* **39**, 880–883.

Sambasivarao, K. and McCluer, R. H. (1963). *J. Lipid Res.* **4**, 106–108.
Samuel, D., Estroumza, J. and Ailhaud, G. (1970). *Eur. J. Biochem.* **12**, 576–582.
Sartorelli, L., Galzigna, L., Rossi, C. R. and Gibson, D. M. (1967). *Biochem. biophys. Res. Commun.* **26**, 90–94.
Sastry, P. S. and Kates, M. (1964a). *Biochim. biophys. Acta* **84**, 231–233.
Sastry, P. S. and Kates, M. (1964b). *Biochemistry, Easton* **3**, 1271–1280.
Sastry, P. S. and Kates, M. (1964c). *Biochemistry, Easton* **3**, 1280–1287.
Sastry, P. S. and Kates, M. (1965). *Can. J. Biochem.* **43**, 1445–1453.
Sastry, P. S. and Kates, M. (1966). *Can. J. Biochem.* **44**, 459–467.
Sastry, P. S. and Kates, M. (1969). *In* "Methods in Enzymology" (J. M. Lowenstein, ed.), Vol. XIV, pp. 204–208. Academic Press, New York and London.
Schauenstein, E. (1967). *J. Lipid Res.* **8**, 417–428.
Scheuerbrandt, G. and Bloch, K. (1962). *J. biol. Chem.* **237**, 2064–2068.
Scheuerbrandt, G., Goldfine, H., Baronowsky, P. E. and Bloch, K. (1961). *J. biol. Chem.* **236**, PC70–PC71.
Schlenk, H. and Gellerman, J. L. (1965). *J. Am. Oil Chem. Soc.* **42**, 504–511.
Schmalfuss, K. (1937). *Fette Seifen* **44**, 31–33.
Schneider, E. L., Sook, P. L. and Hopkins, D. T. (1968). *J. Am. Oil Chem. Soc.* **45**, 585–590.
Schroepfer, G. J. and Bloch, K. (1965). *J. biol. Chem.* **240**, 54–63.
Schroepfer, G. J., Niehaus, W. G. and McCloskey, J. A. (1970). *J. biol. Chem.* **245**, 3798–3801.
Schulz, H., Weeks, G., Toomey, R. E., Shapiro, M. and Wakil, S. J. (1969). *J. biol. Chem.* **244**, 6577–6583.
Schwartz, P. and Carter, H. E. (1954). *Proc. natn. Acad. Sci. U.S.A.* **40**, 499–508.
Schweizer, E., Lerch, I., Kroeplin-Reuff, L. and Lynen, F. (1970a). *Eur. J. Biochem.* **15**, 472–482.
Schweizer, E., Piccinini, F., Duba, C., Gunther, S., Ritter, E. and Lynen, F. (1970b). *Eur. J. Biochem.* **15**, 483–499.
Scott, P. G. W. (1967). *Process Biochem.* **2** (May), 16–19.
Scott, W. E. and Krewson, C. F. (1966). *J. Am. Oil Chem. Soc.* **43**, 466–468.
Scott, W. E., Krewson, C. F., Luddy, F. E. and Reimenschneider, R. W. (1963). *J. Am. Oil Chem. Soc.* **40**, 587–589.
Scott, R. P. W., Fowlis, I. A., Welti, D. and Wilkins, T. (1966). *In* "Gas Chromatography 1966" (A. B. Littlewood, ed.), pp. 318–336. Institute of Petroleum, Elsevier, Amsterdam.
Seigler, D. S., Mikolajczak, K. L., Smith, C. R., Wolff, I. A. and Bates, R. B. (1970). *Chem. Phys. Lip.* **4**, 147–161.
Senn, V. J. (1969). *J. Am. Oil Chem. Soc.* **46**, 476–478.
Sephton, H. H. and Sutton, D. A. (1956). *J. Am. Oil Chem. Soc.* **33**, 263–272.
Serck-Hanssen, K. (1958). *Chemy Ind.* 1554–1554.
Seubert, W., Lamberts, I., Cramer, R. and Ohly, B. (1968). *Biochim. biophys. Acta* **164**, 498–517.
Shannon, L. M., de Vellis, J. and Lew, J. Y. (1963). *Pl. Physiol., Lancaster* **38**, 691–697.
Shenstone, F. S. and Vickery, J. R. (1961). *Nature, Lond.* **190**, 168–169.
Shibuya, I. and Hase, E. (1965). *Pl. Cell Physiol., Tokyo* **6**, 267–283.
Shibuya, I. and Maruo, B. (1965). *Nature, Lond.* **207**, 1096–1097.
Shibuya, I., Maruo, B. and Benson, A. A. (1965). *Pl. Physiol., Lancaster* **40**, 1251–1256.

Shorland, F. B. (1963). *In* "Chemical Plant Taxonomy" (T. Swain, ed.), pp. 253–303. Academic Press, London and New York.

Siddiqi, A. M. and Tappel, A. L. (1957). *J. Am. Oil Chem. Soc.* **34**, 529–533.

Simoni, R. D. and Stumpf, P. K. (1969). *In* "Methods in Enzymology" (J. M. Lowenstein, ed.), Vol. XIV, pp. 84–88. Academic Press, New York and London.

Simoni, R. D., Criddle, R. S. and Stumpf, P. K. (1967). *J. biol. Chem.* **242**, 573–581.

Sims, R. P. A., McGregor, W. G., Plessers, A. G. and Mes, J. C. (1961). *J. Am. Oil Chem. Soc.* **38**, 273–276.

Singh, H. and Privett, O. S. (1970). *Biochim. biophys. Acta* **202**, 200–202.

Sjostrand, F. S. and Barajas, L. (1970). *J. Ultrastructure Res.* **32**, 293–306.

Skidmore, W. D. and Entenman, C. (1962). *J. Lipid Res.* **3**, 471–475.

Skipski, V. P., Smolowe, A. F., Sullivan, R. C. and Barclay, M. (1965). *Biochim. biophys. Acta* **106**, 386–396.

Slawson, V. and Stein, R. A. (1970). *Lipids* **5**, 713–717.

Smith, C. R. (1966). *Lipids* **1**, 268–273.

Smith, C. R. (1970). *In* "Progress in the Chemistry of Fats and Other Lipids" (R. T. Holman, ed.), Vol. 11, Pt. 1, pp. 139–177. Pergamon Press, Oxford.

Smith, C. R. and Wolff, I. A. (1966). *Lipids* **1**, 123–127.

Smith, C. R. and Wolff, I. A. (1969). *Lipids* **4**, 9–14.

Smith, C. R., Bagby, M. O., Lohmar, R. L., Glass, C. A. and Wolff, I. A. (1960a). *J. org. Chem.* **25**, 218–222.

Smith, C. R., Bagby, M. O., Miwa, T. K., Lohmar, R. L. and Wolff, I. A. (1960b). *J. org. Chem.* **25**, 1770–1774.

Smith, C. R., Wilson, T. L., Miwa, T. K., Zobel, H., Lohmar, R. L. and Wolff, I.A. (1961). *J. org. Chem.* **26**, 2903–2905.

Smith, C. R., Wilson, T. L., Bates, R. B. and Scholfield, C. R. (1962). *J. org. Chem.* **27**, 3112–3117.

Smith, C. R., Wilson, T. L., Melvin, E. H. and Wolff, I. A. (1960c). *J. Am. Chem. Soc.* **82**, 1417–1421.

Smith, C. R., Kleiman, R. and Wolff, I. A. (1968). *Lipids* **3**, 37–42.

Smith, G. N. and Bu'Lock, J. D. (1964). *Biochem. biophys. Res. Commun.* **17**, 433–436.

Smith, S. and Dils, R. (1966). *Biochim. biophys. Acta* **116**, 23–40.

Solberg, Y. J. (1960). *Acta chem. scand.* **14**, 2152–2160.

Sørenson, N. A. (1963). *In* "Chemical Plant Taxonomy" (T. Swain, ed.), pp. 219–252. Academic Press, London and New York.

Sowden, J. C. and Fischer, H. O. L. (1941). *J. Am. chem. Soc.* **63**, 3244–3251.

Spencer, A., Corman, L. and Lowenstein, J. M. (1964). *Biochem. J.* **93**, 378–388.

Spencer, G. F., Kleiman, R., Earle, F. R. and Wolff, I. A. (1969). *Lipids* **4**, 99–101.

Spencer, G. F., Kleiman, R., Earle, F. R. and Wolff, I. A. (1970). *Lipids* **5**, 285–287.

Sprecher, H. W., Maier, R., Barber, M. and Holman, R. T. (1965). *Biochemistry, Easton* **4**, 1856–1863.

Squires, C. L., Stumpf, P. K. and Schmid, C. (1958). *Pl. Physiol., Lancaster* **33**, 365–366.

Stanley, R. G. and Conn, E. E. (1957). *Pl. Physiol., Lancaster* **32**, 412–418.

Stansly, P. G. and Beinert, H. (1953). *Biochim. biophys. Acta* **11**, 600–601.

Stefansson, B. R. and Hougen, F. W. (1964). *Can. J. Pl. Sci.* **44**, 359–364.

Steinberg, D., Herndon, J. H., Uhlendorf, B. W., Mize, C. E., Avigan, J. and Milne, G. W. A. (1967). *Science, N.Y.* **156**, 1740–1742.

Stern, N., Shenberg, E. and Tietz, A. (1969). *Eur. J. Biochem.* **8**, 101–108.

Stodola, F. H., Vesonder, R. F. and Wickerham, L. J. (1965). *Biochemistry, Easton* **4**, 1390–1394.

Stodola, F. H., Deinema, M. H. and Spencer, J. F. T. (1967). *Bact. Rev.* **31**, 194–213.

Stoffel, W. and Caesar, H. (1965). *Hoppe-Seyler's Z. physiol. Chem.* **341**, 76–83.

Stoffel, W. and Schieffer, H. G. (1968). *Hoppe-Seyler's Z. physiol. Chem.* **349**, 1017–1026.

Stoffel, W., Ditzer, R. and Caesar, H. (1964). *Hoppe-Seyler's Z. physiol. Chem.* **339**, 167–181.

Stokke, O. (1969). *Biochim. biophys. Acta* **176**, 54–59.

Stokke, O., Try, K. and Eldjarn, L. (1967). *Biochim. biophys. Acta* **144**, 271–284.

Struik, C. B. and Beerthuis, R. K. (1966). *Biochim. biophys. Acta* **116**, 12–22.

Stumpf, P. K. (1956). *J. biol. Chem.* **223**, 643–649.

Stumpf, P. K. (1962). *Nature, Lond.* **194**, 1158–1160.

Stumpf, P. K. (1969). *A. Rev. Biochem.* **38**, 159–207.

Stumpf, P. K. and Barber, G. A. (1956). *Pl. Physiol., Lancaster* **31**, 304–308.

Stumpf, P. K. and Barber, G. A. (1957). *J. biol. Chem.* **227**, 407–417.

Stumpf, P. K. and Boardman, N. K. (1970). *J. biol. Chem.* **245**, 2579–2587.

Stumpf, P. K. and James, A. T. (1962). *Biochim. biophys. Acta* **57**, 400–402.

Stumpf, P. K. and James, A. T. (1963). *Biochim. biophys. Acta* **70**, 20–32.

Stumpf, P. K., Bove, J. M. and Goffeau, A. (1963). *Biochim. biophys. Acta* **70**, 260–270.

Stumpf, P. K., Brooks, J., Galliard, T., Hawke, J. C. and Simoni, R. (1967). *In* "Biochemistry of Chloroplasts" (T. Goodwin, ed.), Vol. II, pp. 213–239. Academic Press, London and New York.

Subbaram, M. R. and Youngs, C. G. (1964). *J. Am. Oil Chem. Soc.* **41**, 445–448.

Sumida, S. and Mudd, J. B. (1968). *Pl. Physiol., Lancaster* **43**, 1162–1164.

Sumida, S. and Mudd, J. B. (1970). *Pl. Physiol., Lancaster* **45**, 719–722.

Sumper, M., Oesterhelt, D., Riepertinger, C. and Lynen, F. (1969a). *Eur. J. Biochem.* **10**, 377–387.

Sumper, M., Riepertinger, C. and Lynen, F. (1969b). *FEBS Lett.* **5**, 45–49.

Synerholm, M. E. and Zimmerman, P. W. (1947). *Contr. Boyce Thompson Inst. Plant Res.* **14**, 369–374.

Tallent, W. H., Cope, D. G., Hagemann, J. W., Earle, F. R. and Wolff, I. A. (1966a). *Lipids* **1**, 335–340.

Tallent, W. H., Harris, J., Wolff, I. A. and Lundin, R. E. (1966b). *Tetrahedron Lett.* 4329–4334.

Tallent, W. H., Harris, J., Spencer, G. F. and Wolff, I. A. (1968). *Lipids* **3**, 425–430.

Tang, C. S. and Jennings, W. G. (1968). *J. Agric. Food Chem.* **16**, 252–254.

Tang, W. J. and Castelfranco, P. A. (1968). *Pl. Physiol., Lancaster* **43**, 1232–1238.

Tappel, A. L. (1963). *In* "The Enzymes" (P. D. Boyer, H. Lardy and K. Myrbäck, eds), Vol. 8, pp. 275–283. Academic Press, New York and London.

Tavener, R. J. A. and Laidman, D. L. (1970). *Biochem. J.* **113**, 32P–32P.

Thomas, D. R. and Stobart, A. K. (1970). *J. exp. Bot.* **21**, 274–285.

Thomas, P. J. and Law, J. H. (1966). *J. biol. Chem.* **241**, 5013–5018.

Thompson, A. C. and Hedin, P. A. (1965). *Crop Science* **5**, 133–135.

Thompson, A. C., Henson, R. D., Minyard, J. P. and Hedin, P. A. (1968). *Lipids* **3**, 373–374.

Tipton, C. L. and Swords, M. D. (1966). *J. Protozool.* **13**, 469–472.

Titchener, E. B. and Gibson, D. M. (1957). *Fedn. Proc. Fedn Am. Socs exp. Biol.* **16**, 262–262.

Tookey, H. L. and Balls, A. K. (1956). *J. biol. Chem.* **218**, 213–224.
Tookey, H. L., Wilson, R. G., Lohmar, R. L. and Dutton, H. J. (1958). *J. biol. Chem.* **230**, 65–72.
Toomey, R. E. and Wakil, S. J. (1966a). *Biochim. biophys. Acta* **116**, 189–197.
Toomey, R. E. and Wakil, S. J. (1966b). *J. biol. Chem.* **241**, 1159–1165.
Trosper, T. and Sauer, K. (1968). *Biochim. biophys. Acta* **162**, 97–107.
Tsujimoto, M. (1940). *J. Soc. Chem. Ind. (Japan)* **43**, 208–209.
Tulloch, A. P. (1960). *Can. J. Chem.* **38**, 204–207.
Tulloch, A. P. (1963). *Can. J. Biochem. Physiol.* **41**, 1115–1121.
Tulloch, A. P. (1965). *Can. J. Chem.* **43**, 415–420.
Tulloch, A. P. and Spencer, J. F. T. (1964). *Can. J. Chem.* **42**, 830–835.
Tulloch, A. P. and Weenink, R. O. (1969). *Can. J. Chem.* **47**, 3119–3126.
Tulloch, A. P., Spencer, J. F. T. and Gorin, P. A. J. (1962). *Can. J. Chem.* **40**, 1326–1338.
Tulloch, A. P., Hill, A. and Spencer, J. F. T. (1968). *Can. J. Chem.* **46**, 3337–3351.
Udelnova, T. M. and Boichenko, E. A. (1967). *Biokhimiya* **32**, 779–785.
Ukita, T. and Tanimura, A. (1961). *Chem. Pharm. Bull. (Tokyo)* **9**, 43–46.
Vagelos, P. R. (1960). *J. biol. Chem.* **235**, 346–350.
Vagelos, P. R. and Larrabee, A. R. (1967). *J. biol. Chem.* **242**, 1776–1781.
Vagelos, P. R. and Larrabee, A. R. (1969). *In* "Methods in Enzymology" (J. M. Lowenstein, ed.), Vol. XIV, pp. 81–83. Academic Press, New York and London.
Vagelos, P. R., Alberts, A. W. and Martin, D. B. (1963). *J. biol. Chem.* **238**, 533–540.
Vagelos, P. R., Alberts, A. W. and Majerus, P. W. (1969a). *In* "Methods in Enzymology" (J. M. Lowenstein, ed.), Vol. XIV, pp. 39–43. Academic Press, New York and London.
Vagelos, P. R., Alberts, A. W. and Majerus, P. W. (1969b). *In* "Methods of Enzymology" (J. M. Lowenstein, ed.), Vol. XIV, pp. 60–66. Academic Press, New York and London.
Vanaman, T. C., Wakil, S. J. and Hill, R. L. (1968a). *J. biol. Chem.* **243**, 6409–6419.
Vanaman, T. C., Wakil, S. J. and Hill, R. L. (1968b). *J. biol. Chem.* **243**, 6420–6431.
van Deenen, L. L. M. and De Gier, J. (1964). *In* "The Red Cell" (C. Bishop and D. M. Surgenor, eds), p. 243. Academic Press, New York and London.
van Deenen, L. L. M. and de Haas, G. H. (1966). *A. Rev. Biochem.* **35**, 157–194.
Vanden Driessche, T. (1961). *Annls Physiol. vég., Brux.*, **6**, 27.
Vanden Driessche, T. (1964). *Annls Physiol. vég., Brux.* **9**, 13.
van der Veen, J. and Olcott, H. (1967). *J. Agric. Food. Chem.* **15**, 682–684.
Vander Wal, R. J. (1960). *J. Am. Oil chem. Soc.* **37**, 18–20.
Vander Wal, R. J. (1964). *Adv. Lipid Res.* **2**, 1–16.
Vandor, S. L. and Richardson, K. E. (1968). *Can. J. Biochem.* **46**, 1309–1315.
Veldink, G. A., Vliegenthart, J. F. G. and Boldingh, J. (1968). *Biochem. J.* **110**, 58P–58P.
Veldink, G. A., Vliegenthart, J. F. G. and Boldingh, J. (1970a). *Biochim. biophys. Acta* **202**, 198–199.
Veldink, G. A., Vliegenthart, J. F. G. and Boldingh, J. (1970b). *FEBS Lett.* **7**, 188–190.
Veldink, G. A., Vliegenthart, J. F. G. and Boldingh, J. (1970c). *Biochem. J.* **120**, 55–60.
Vellis, J. de, Shannon, L. M. and Lew, J. Y. (1963). *Pl. Physiol., Lancaster* **38**, 686–690.

Verkade, P. E. (1938). *Chemy. Ind.* 704–711.

Vignais, P. M., Vignais, P. V. and Lehninger, A. L. (1963a). *Biochem. biophys. Res. Commun.* 11, 313–318.

Vignais, P. V., Vignais, P. M., Rossi, C. S. and Lehninger, A. (1963b). *Biochem. biophys. Res. Commun.* 11, 307–312.

Vioque, E. and Holman, R. T. (1962). *Archs Biochem. Biophys.* 99, 522–528.

Viswanathan, C. V. (1968). *Chromatog. Revs.* 10, 18.

Vogel, H. C., Doizaki, W. M. and Zieve, L. (1962). *J. Lipid Res.* 3, 138–140.

von Rudloff, E. (1956). *Can. J. Chem.* 34, 1413–1418.

von Rudloff, E. (1959). *Can. J. Chem.* 37, 1038–1042.

Wada, F., Shibita, H., Goto, M. and Sakamoto, Y. (1968). *Biochim. biophys. Acta* 162, 518–524.

Wagner, H. and Pohl, P. (1965). *Biochem. Z.* 341, 476–484.

Wagner, H., Zofcsik, W. and Heng, I. (1969a). *Z. Naturf.* 246, 922–927.

Wagner, H., Pohl, P. and Munzing, A. (1969b). *Z. Naturf.* 246, 360–360.

Wain, R. L. and Wightman, F. (1954). *Proc. Roy. Soc. Ser. B,* 142, 525–536.

Wakabayashi, K. and Shimazono, N. (1963). *Biochim. biophys. Acta* 70, 132–142.

Wakil, S. J. (1958). *J. Am. chem. Soc.* 80, 6465–6465.

Wakil, S. J. (1961). *J. Lipid Res.* 2, 1–24.

Wakil, S. J. and Ganguly, J. (1959). *J. Am. chem. Soc.* 81, 2597–2598.

Wakil, S. J. and Gibson, D. M. (1960). *Biochim. biophys. Acta* 41, 122–129.

Wakil, S. J., Porter, J. W. and Gibson, D. M. (1957). *Biochim. biophys. Acta* 24, 453–461.

Wallace, J. W. and Newman, D. W. (1965). *Phytochemistry* 4, 43–47.

Ward, P. F. V., Hall, R. J. and Peters, R. A. (1964). *Nature, Lond.* 201, 611–612.

Warshaw, J. B. and Kimura, R. E. (1970). *Biochem. biophys. Res. Commun.* 38, 58–64.

Weber, E. J. (1969). *J. Am. Oil Chem. Soc.* 46, 485–488.

Webster, D. E. and Chang, S. B. (1969). *Pl. Physiol., Lancaster* 44, 1523–1527.

Weeks, G. and Wakil, S. J. (1968). *J. biol. Chem.* 243, 1180–1189.

Weeks, G. and Wakil, S. J. (1969). *In* "Methods in Enzymology" (J. M. Lowenstein, ed.), Vol. XIV, pp. 66-73. Academic Press, London and New York.

Weenink, R. O. (1959). *New Zealand J. Sci.* 2, 273–274.

Weenink, R. O. and Shorland, F. B. (1964). *Biochim. biophys. Acta* 84, 613–614.

Weier, T. E. and Benson, A. A. (1967). *Am. J. Bot.* 54, 389–402.

Wheeldon, L. W. (1960). *J. Lipid Res.* 1, 439–445.

Whereat, A. F., Orishimo, M. W., Nelson, J. and Phillips, S. J. (1969). *J. biol. Chem.* 244, 6498–6506.

Wilde, P. F. and Stewart, P. S. (1968). *Biochem. J.* 108, 225–231.

Willebrands, A. F. and Van der Veen, K. J. (1966). *Biochim. biophys. Acta* 116, 583–585.

Willecke, K., Ritter, E. and Lynen, F. (1969). *Eur. J. Biochem.* 8, 503–509.

Willemot, C. and Boll, W. G. (1967). *Can. J. Bot.* 45, 1863–1876.

Willemot, C. and Stumpf, P. K. (1967a). *Pl. Physiol., Lancaster* 42, 391–397.

Willemot, C. and Stumpf, P. K. (1967b). *Can. J. Bot.* 45, 579–584.

Williamson, I. P. and Wakil, S. J. (1966). *J. biol. Chem.* 241, 2326–2332.

Wilson, T. L., Smith, C. R. and Mikolajczak, K. L. (1961). *J. Am. Oil Chem. Soc.* 38, 696–699.

Wintermans, J. F. G. M. (1960). *Biochim. biophys. Acta* 44, 49–54.

Wisnieski, B. J., Keith, A. D. and Resnick, M. R. (1970). *J. Bact.* 101, 160–164.

Wojtczak, L. and Lehninger, A. L. (1961). *Biochim. biophys. Acta* **51**, 442–456.
Wojtczak, L. and Zaluska, H. (1969). *Biochim. biophys. Acta* **189**, 455–456.
Wojtczak, L., Wlodawer, P. and Zborowski, J. (1963). *Biochim. biophys. Acta* **70**, 290–305.
Wolf, F. T., Coniglio, J. G. and Bridges, R. B. (1966). *In* "Biochemistry of Chloroplasts" (T. W. Goodwin, ed.), Vol. I, pp. 187–194. Academic Press, London and New York.
Wolff, I. A. (1966). *Science, N.Y.* **154**, 1140–1149.
Wolken, J. J. (1959). *A. Rev. Pl. Physiol.* **10**, 71–86.
Wood, B. J. B., Nichols, B. W. and James, A. T. (1965). *Biochim. biophys. Acta* **106**, 261–273.
Wren, J. J. and Merryfield, D. S. (1970). *J. Sci. Food Agric.* **21**, 254–257.
Wren, J. J. and Szczepanowska, A. D. (1964). *J. Chromat.* **14**, 405–410.
Yagi, T. and Benson, A. A. (1962). *Biochim. biophys. Acta* **57**, 601–603.
Yalpani, M., Willecke, K. and Lynen, F. (1969). *Eur. J. Biochem.* **8**, 495–509.
Yamada, M. (1957). *Sci. Pap. Coll. Gen. Educ., Tokyo* **7**, 97–99.
Yamada, M. and Stumpf, P. K. (1964). *Biochem. biophys. Res. Commun.* **14**, 165–171.
Yamada, M. and Stumpf, P. K. (1965a). *Pl. Physiol., Lancaster* **40**, 653–658.
Yamada, M. and Stumpf, P. K. (1965b). *Pl. Physiol., Lancaster* **40**, 659–664.
Yang, S. F. (1969). *In* "Methods in Enzymology" (J. M. Lowenstein, ed.), Vol. XIV, pp. 208–211. Academic Press, New York and London.
Yang, S. F. and Stumpf, P. K. (1965a). *Biochim. biophys. Acta* **98**, 19–26.
Yang, S. F. and Stumpf, P. K. (1965b). *Biochim. biophys. Acta* **98**, 27–35.
Yang, S. F., Freer, S. and Benson, A. A. (1967). *J. biol. Chem.* **242**, 477–484.
Yano, I., Furukawa, Y. and Kusunose, M. (1969). *Biochim. biophys. Acta* **187**, 166–168.
Yano, I., Furukawa, Y. and Kusunose, M. (1970a). *Biochim. biophys. Acta* **210**, 105–115.
Yano, I., Morris, L. J., Nichols, B. W. and James, A. T. (1970b). Unpublished data.
Yermanos, D. M. and Knowles, P. F. (1962). *Crop Sci.* **2**, 109–111.
Yermanos, D. M., Hall, B. J. and Burge, W. (1964). *Agron. J.* **56**, 582–585.
Youngs, C. G. (1961). *J. Am. Oil Chem. Soc.* **38**, 62–67.
Yuan, C. and Bloch, K. (1961). *J. biol. Chem.* **236**, 1277–1279.
Yung, K. H. and Mudd, J. B. (1966). *Pl. Physiol., Lancaster* **41**, 506–509.
Zill, L. P. and Cheniae, G. (1962). *A. Rev. Pl. Physiol.* **13**, 225–264.
Zimmerman, D. C. (1966). *Biochem. biophys. Res. Commun.* **23**, 398–402.
Zimmerman, D. C. and Klosterman, H. J. (1965). *J. Am. Oil Chem. Soc.* **42**, 58–62.
Zimmerman, D. C. and Vick, B. A. (1970). *Lipids* **5**, 392–397.

More Recent Publications Not Considered in the Text

CHAPTER 1. *Structure and Distribution of Plant Fatty Acids*

Albro, P. W. and Fishbein, L. (1970). Isolation and characterization of 5,8,12-trihydroxy-*trans*-9-octadecenoic acid from wheat bran. *Phytochemistry* 10, 631–636.

Litchfield, C. (1970). Taxonomic patterns in the fat content, fatty acid composition and triglyceride composition of Palmae seeds. *Chem. Phys. Lipids* 4, 96–103.

Vickery, J. R. (1971). The fatty acid composition of the seed oils of Proteaceae: a chemotaxonomic study. *Phytochemistry* 10, 123–130.

CHAPTER 2. *Plant Acyl Lipids*

Fujino, Y. and Ho, S. (1971). Existence of ceramide in alfalfa leaves. *Biochim. biophys. Acta* 231, 242–243.

Seigler, D., Seaman, F. and Mabry, T. J. (1971). New cyanogenetic lipids from *Ugnadia speciosa*. *Phytochemistry* 10, 485–487.

CHAPTER 3. *The Lipid and Fatty Acid Composition of Specific Tissues*

Adams, B. L., McMahon, V. and Seckbach, J. (1971). Fatty acids in the thermophilic alga, *Cyanidium caldarium*. *Biochem. biophys. Res. Commun.* 42, 359–365.

Allebone, J. E., Hamilton, R. J., Bryce, T. A. and Kelly, W. (1971). Anthraquinone in plant surface waxes. *Experientia* 27, 13–13.

Allen, C. F., Good, P. and Holton, R. W. (1970). Lipid composition of *Cyanidium*. *Pl. Physiol., Lancaster* 46, 748–751.

Allen, C. F., Good, P., Mollenhauer, H. H. and Totten, C. (1971). Studies on Seeds 4. Lipid composition of bean cotyledon vesicles. *J. Cell. Biol.* 48, 542–546.

Beach, D. H., Harrington, G. W. and Holz, G. G. (1970). The polyunsaturated fatty acids of marine and freshwater cryptomonads. *J. Protozool.* 17, 501–510.

Bishop, D. G., Andersen, K. S. and Smillie, R. M. (1971). The distribution of galactolipids in mesophyll and bundle sheath chloroplasts of maize and sorghum. *Biochim. biophys. Acta* 231, 412–414.

Dorrell, D. G. (1970). Distribution of fatty acids within the seed of flax. *Can. J. Plant Sci.* 50, 71–75.

Harrington, G. W., Beach, D. H., Dunham, J. E. and Holz, G. G. (1970). The polyunsaturated fatty acids of marine dinoflagellates. *J. Protozool.* 17, 213–219.

Jellum, M. D. (1970). Fatty acid composition of oil from four kernel fractions of corn (*Zea Mays* L) inbred lines. *Cereal Chem.* 47, 549–558.

Kleinschmidt, M. G. and McMahon, V. A. (1970). Effect of growth temperature on the lipid composition of *Cyanidium caldarium* 1. Class separation of lipids. *Pl. Physiol. Lancaster* 46, 286–289.

Kleinschmidt, M. G. and McMahon, V. A. (1970). Effect of growth temperature on the lipid composition of *Cyanidium caldarium* 2. Glycolipid and phospholipid components. *Pl. Physiol. Lancaster* **46**, 290–293.

Laur, M. H., and Quang, L. P. (1970). Sur les lipides neutres de trois Fucacees des cotes françaises : *Fucus serratus* L., *Fucus vesiculosus* L, et *Pelvetia canaliculata*. Analyse quantitative et qualitative des different composants. *C. r. Acad. Sci. (Paris) Ser. D.* **271**, 1752.

Litchfield, C. (1970). Taxonomic patterns in the fat content, fatty acid composition and triglyceride composition of Palmae seeds. *Chem. Phys. Lipids* **4**, 96–103.

Macmurray, T. A. and Morrison, W. R. (1970). Composition of wheat flour lipids. *J. Sci. Food Agric.* **21**, 520–528.

Mollenhauer, H. H. and Totten, C. (1971). Studies on seeds. 3. Isolation and structure of lipid-containing vesicles. *J. Cell. biol.* **48**, 533.

Pelloqui, A., Lai, R. and Busson, F. (1970). Comparative study of lipids of *Spirulina platensis* Geitler and *Spirulina-Geitleri* J. Detoni. *C. r. Acad. Sci. (Paris)* **271**, 932–935.

Schulman, Y. and Monselis, S. P. (1970). Some studies on the cuticular wax of citrus fruits. *J. Hort. Sci.* **45**, 471–478.

Vickery, J. R. (1971). The fatty acid composition of the seed oils of Proteaceae : a chemotaxonomic study. *Phytochemistry* **10**, 123–130.

CHAPTER 4. *Distribution of Individual Fatty Acids Between Lipid Classes*

Gray, I. K., Rumsby, M. G. and Hawkes, J. C. (1967). Fatty acid composition of the monogalactosyl diglyceride fraction from castor bean leaf tissues of different stages of maturity. *Phytochemistry* **6**, 107–113.

Nagai, J., Ohta, T. and Saito, E. (1971). Incorporation of propionate into wax esters by etiolated *Euglena*. *Biochem. biophys. Res. Commun.* **42**, 523–528.

Phillips, B. E. and Smith, C. R. (1970). Glycerides of *Monnina emarginata* seed oil. *Biochim. biophys. Acta* **218**, 71–82.

CHAPTER 5. *Biosynthesis of Plant Fatty Acids*

Adams, B. L., McMahon, V. and Seckbach, J. (1971). Fatty Acids in the thermophilic alga, *Cyanidium caldarium*. *Biochem. biophys. Res. Commun.* **42**, 359–365.

Brett, D., Howling, D., Morris, L. J. and James, A. T. (1971). Specificity of the fatty acid desaturases. The conversion of saturated to monoenoic acids. *Archs Biochem. Biophys.* **143**, 535–547.

Delo, J., Ernst-Fonberg, M. L. and Bloch, K. (1971). Fatty acid synthetases from *Euglena gracilis*. *Archs Biochem. Biophys.* **143**, 384–391.

Ernst-Fonberg, M. L. and Bloch, K. (1971). A chloroplast-associated fatty acid synthetase system in *Euglena*. *Archs Biochem. Biophys.* **143**, 392–400.

Givan, C. V. and Stumpf, P. K. (1971). Some factors regulating fatty acid synthesis by isolated spinach chloroplasts. *Pl. Physiol. Lancaster* **47**, 510–515.

Harwood, J. L. and Stumpf, P. K. (1970). Synthesis of fatty acids at the initial stage of seed germination. *Pl. Physiol. Lancaster* **46**, 500–508.

Harwood, J. L. and Stumpf, P. K. (1971). Control of fatty acid synthesis in germinating seeds. *Archs Biochem. Biophys.* **142**, 281–291.

Huang, K. P. and Stumpf, P. K. (1971). Fatty acid synthesis by a soluble fatty acid synthetase from *Solanum tuberosum*. *Archs Biochem. Biophys.* **143**, 412–427.

CHAPTER 6. *The Biosynthesis of Acyl Lipids*

Douce, R. and Guillot-Salomon, T. (1970). Sur l'incorporation de la radioactivite du sn-glycerol-3-phosphate-^{14}C dans le monogalactosyl diglyceride des plastes isoles *FEBS Lett.* **11**, 121–124.

Kolattukudy, P. E. (1970). Cutin biosynthesis in *Vicia faba* leaves. *Pl. Physiol. Lancaster* **46**, 759–760.

Kolattukudy, P. E. (1970). Biosynthesis of a lipid polymer, cutin: the structural component of plant cuticle. *Biochem. biophys. Res. Commun.* **41**, 299–305.

Kolattukudy, P. E., Walton, T. J. and Kushnaha, R. P. S. (1971). Epoxy acids in the lipid polymer cutin, and their role in the biosynthesis of cutin. *Biochem. biophys. Res. Commun.* **42**, 739–744.

Matson, R. S., Fei, M. and Chang, S. B. (1970). Comparative studies of the biosynthesis of galactolipids in *Euglena gracilis* strain Z. *Pl. Physiol. Lancaster* **45**, 531–532.

CHAPTER 7. *Lipolytic Enzymes*

Galliard, T. (1971). The enzymic deacylation of phospholipids and galacto-lipids in plants. Purification and properties of a lipolytic acylhydrolase from potato tubers. *Biochem. J.* **121**, 379–390.

Heller, M. and Arad, R. (1970). Properties of the phospholipase D from peanut seeds. *Biochim. biophys. Acta* **210**, 276–286.

Kartha, A. R. S. (1971). The identity and proportions of the lipases present in unripe oilseeds. *Chemy. Ind.* 506–507.

Noma, A., and Börgstrom, B. (1971). Acid lipase of castor beans-positional specificity and reaction mechanism. *Biochim. biophys. Acta* **227**, 106–115.

Ory, R. L. and St. Angelo, A. J. (1971). Lipolysis in castor seeds: a reinvesti-gation of the neutral lipase. *Lipids* **5**, 54–57.

CHAPTER 8. *Biological Degradation of Plant Fatty Acids*

Hutton, D. and Stumpf, P. K. (1971). The pathway of ricinoleic acid cata-bolism in the germinating castor bean (*Ricinus communis* L) and pea (*Pisum sativum* L). *Archs Biochem. Biophys.* **142**, 48–60.

Zimmerman, D. C. and Vick, B. A. (1970). Hydroperoxide isomerase. A new enzyme of lipid metabolism. *Pl. Physiol. Lancaster* **46**, 445–453.

CHAPTER 9. *Lipid and Fatty Acid Metabolism During Organogenesis and Senescence*

Adams, B. L., McMahon, V. and Seckbach, J. (1971). Fatty acids in the thermophilic alga, *Cyanidium caldarium*. *Biochem. biophys. Res. Commun.* **42**, 359–365.

Bishop, D. G., and Smillie, R. M. (1970). The effect of chloramphenicol and cycloheximide on lipid synthesis during chloroplast development in *Euglena gracilis*. *Archs Biochem. Biophys.* **137**, 179–189.

Costes, M. M. C. and Bazier, R. (1971). Les lipides des chloroplasts de la feuille de Tomate: effets de la lumiere sur leur composition en acides gras. *C. r. Acad. Sci. (Paris)* **272**, 492.

Davis, D. G. (1971). Scanning electron microscopic studies on wax formations on leaves of higher plants. *Can. J. Bot.* **49**, 543–546.

Draper, S. R. and Simon, E. W. (1970). Lipid biosynthesis in senescing cotyledons of cucumber. *Phytochemistry* **9**, 1997–2002.

Goldberg, I. and Ohad, I. (1970). Lipid and pigment changes during synthesis of chloroplast membranes in a mutant of *Chlamydomonas reinhardi* y-1. *J. Cell biol.* **44**, 563–571.

Gray, I. K., Rumsby, M. G. and Hawke, J. C. (1967). Fatty acid composition of the monogalactosyl diglyceride fraction from castor bean leaf tissues of different stages of maturity. *Phytochemistry* **6**, 107–113.

Hallam, N. D. (1970). Growth and regeneration of waxes on leaves of Eucalyptus. *Planta* **93**, 257–268.

Harwood, J. L. and Stumpf, P. K. (1970). Synthesis of fatty acids at the initial stage of seed germination. *Pl. Physiol. Lancaster* **46**, 500–508.

Harwood, J. L. and Stumpf, P. K. (1971). Control of fatty acid synthesis in germinating seeds. *Archs Biochem. Biophys.* **142**, 281–291.

Kolattukudy, P. E. (1971). Enzymatic synthesis of fatty alcohols in *Brassica oleracea*. *Archs Biochem. Biophys.* **142**, 701–709.

Kolattukudy, P. E. and Liu, T. Y. (1970). Direct evidence for biosynthetic interrelationships among hydrocarbons, secondary alcohols and ketones in *Brassica oleraceae*. *Biochem. biophys. Res. Commun.* **41**, 1369–1374.

Kuiper, P. J. C. (1970). Lipids in alfalfa leaves in relation to cold hardiness. *Pl. Physiol. Lancaster* **45**, 684–686.

Penner, D. and Meggitt, W. (1970). Herbicide effects on soybean seed lipids. *Crop Sci.* **10**, 553.

Skarsaune, S. K., Youngs, V. L. and Gilles, K. A. (1970). Changes in wheat lipids during seed maturation. Changes in lipid composition. *Cereal Chem.* **47**, 533–544.

Still, G. G., Davis, D. G. and Zander, G. L. (1970). Plant epicuticular lipids—alteration by herbicidal carbamates. *Pl. Physiol. Lancaster* **46**, 307–314.

Tremolières, A. and Lepage, M. (1971). Changes in lipid composition during greening of etiolated pea seedlings. *Pl. Physiol. Lancaster* **47**, 329–334.

Tremolières, A. and Mazliak, P. (1970). Formation des lipides au cours du developpement de la feuille de Trefle (*Trifolium repens* L). *Physiol. Veg.* **8**, 135–150.

Wilkinson, R. E. (1970). Sicklepod fatty acid response to photoperiod *Pl. Physiol. Lancaster* **46**, 463–465.

Chapter 10. *The Role of Lipids in Plant Metabolism*

Brand, J., Krogmann, D. W. and Crane, F. L. (1971). Lipid requirement for photosystem-1 activity in heptane-extracted spinach chloroplasts. *Pl. Physiol. Lancaster* **47**, 135–138.

Chen, L. F., Lund, D. B. and Richardson, T. (1971). Essential fatty acids and glucose permeability of lecithin membranes. *Biochim. biophys. Acta* **225**, 89–95.

Gurr, M. I. and Brawn, P. (1970). The composition of phosphatidyl choline species in *Chlorella vulgaris* during the formation of linoleic acid. *Eur. J. Biochem.* **17**, 19–22.

Hendler, R. W. (1971). Biological membrane ultrastructure. *Physiol. Rev.* **51**, 66–97.

Mitchell, J. W., Mandava, N., Worley, J. F. and Drowne, M. E. (1971). Fatty hormones in pollen and immature seeds of bean. *J. Agric. Food Chem.* **19**, 391–393.

Chapter 11. *The Analysis of Plant Lipids*

Macmurray, T. A. and Morrison, W. R. (1970). Composition of wheat flour lipids. *J. Sci. Fd. Agric.* **21**, 520–528.

Supplementary Reading

The following books deal wholly or partly with various general aspects of acyl lipid chemistry and/or plant biochemistry: they may be useful to those readers requiring introductory or supplementary reading. In the two books marked with an asterisk, Gunstone lists a more complete selection. Other continuing series of reviews include "Progress in the Chemistry of Fats and Other Lipids" (R. T. Holman, ed.) Pergamon Press, London and "Advances in Lipid Research" (R. Paoletti and D. Kritchevsky, ed.) Academic Press, London and New York.

(1971) "Lipid Biochemistry: an Introduction." M. I. Gurr and A. T. James. Chapman and Hall, London.

*(1970) "Topics in Lipid Chemistry", Vol. 1 (F. D. Gunstone, ed.). Logos Press, London.

(1970) "Lipid Metabolism" (S. J. Wakil, ed.). Academic Press, New York and London.

(1969) "Introduction to Lipids." D. Chapman. McGraw-Hill, London.

(1969) "Methods in Enzymology", Vol. XIV—Lipids (J. M. Lowenstein, ed.). Academic Press, London and New York.

(1969) Metabolism of Fatty Acids. P. K. Stumpf, In "Annual Review of Biochemistry", Vol. 38, pp. 159–212. Annual Reviews, Inc., Palo Alto.

(1968) Fatty Acid Oxidation and Biosynthesis. D. E. Green and D. W. Allmann, In "Metabolic Pathways" (3rd Edn.), Vol. 2 (D. M. Greenberg, ed.), pp. 1–67. Academic Press, London and New York.

(1968) "Fatty Acids" (2nd Edn.), Part V (K. S. Markley, ed.). Interscience, New York.

(1968) "Physiological Chemistry of Lipids in Mammals." E. J. Masoro. W. B. Saunders, Philadelphia.

(1968) "Progress in Phytochemistry", Vol. 1 (L. Reinhold and Y. Liwschitz, eds.). Interscience, New York.

(1967) "Biochemistry of Chloroplasts", Vols. I and II (T. W. Goodwin, ed.). Academic Press, London and New York.

*(1967) "An Introduction to the Chemistry and Biochemistry of Fatty Acids and their Glycerides" (2nd Edn.). F. D. Gunstone. Chapman and Hall, London.

(1967) "Lipid Chromatographic Analysis", Vol. 1 (G. V. Marinetti, ed.). Edward Arnold, London.

(1967) Fat Metabolism in Plants. J. B. Mudd, In "Annual Review of Plant Physiology", Vol. 18, pp. 229–252. Annual Reviews, Inc., Palo Alto.

(1966) "Biological Chemistry". H. R. Mahler and E. H. Cordes. Harper and Row (New York, Evanston and London)/John Weatherill, Inc. (Tokyo).

(1966) "A Dictionary of Flowering Plants and Ferns" (7th Edn.). J. C. Willis. Cambridge University Press, Cambridge.

(1965) "Plant Biochemistry" (J. Bonner and J. E. Varner, eds). Academic Press, New York and London.

(1965) "The Structure of Lipids by Spectroscopic and X-ray Techniques." D. Chapman. Methuen, London.

(1964) "Phospholipids: Chemistry, Metabolism and Function." G. B. Ansell and J. N. Hawthorne. Elsevier, Amsterdam. (2nd Edn. in preparation).

(1964) "The Chemical Constitution of the Natural Fats" (4th Edn.). T. P. Hilditch and P. N. Williams. Chapman and Hall, London.

Author Index

The numbers in italics refer to the page in the Bibliography in which the reference appears.

A

Abdelkader, A. B., 65, 66, 257, 258, *306*
Abraham, S., 116, 118, *307*
Abrahamsson, S., 28, *306*
Acker, L., 196, *306*
Ackman, R. G., 8, 297, 298, *306*
Agranoff, B. W., 177, 178, *315*
Ahlers, N. H. E., 15, *306*
Ailhaud, G., 101, *327*
Akamatsu, Y., 174, *306*
Aladjeni, E., 197, *315*
Alam, A. V., 53, *306*
Alberts, A. W., 103, 104, 105, 106, 107, 117, *306, 314, 321, 330*
Allen, C. F., 49, 70, 76, 90, 92, 285, 294, *306*
Allman, D. W., 98, 101, 102, 103, 105, 201, *314*
Altschul, A. M., 193, 195, 207, *306, 317, 324, 326*
Anderson, R. L., 204, *306*
Aneja, R., 44, *306*
Anh, N'G. X., 242, *312*
Antia, N. J., *306*
Antony, G. J., 217, *306*
Appelman, D., 271, *307*
Appelqvist, L.-A., 63, 127, 246, 261, 262, 267, *307*
Appleby, R. S., 92, 157, 186, 280, *323*
Applewhite, T. H., *308*
Argoudelis, C. J., 299, *307*
Arnaud, A., 20, *307*
Arvidson, G. A. E., 180, 188, *307*
Asai, J., 272, *314*
Asselineau, C., 28, 140, *307*
Asselineau, J., *307*
Avigan, J., 219, *308, 322, 328*
Avins, L. R., 219, *307*
Awasthi, Y. C., 269, *307*
Axelrod, B., 228, 229, *310, 318*
Ayling, J., 116, *307*
Aylward, F., 299, *307*
Azerad, R., 174, *317*

B

Badami, R. C., 16, *307*
Baddiley, J., 98, *307*
Bagby, M. O., 12, 15, 19, 21, 27, 64, *303, 322, 328*
Bailey, J. L., 68, *307*
Baker, E. A., 26, *307*
Baker, R. H., 194, *324*
Bakerman, H. A., 112, *319*
Balls, A. K., 197, *330*
Barajas, L., 268, *328*
Baranska, J., 155, *307*
Barber, G. A., 125, 205, 211, *329*
Barber, H. N., 73, 254, *320*
Barber, M., 21, 43, 290, *328*
Barclay, A. S., 19, 27, 166, *312*
Barclay, M., 283, 284, *328*
Barenholz, Y., 218, *313*
Barnes, E. M., 113, 114, *307*
Baronowsky, P. E., 131, 132, *327*
Barrett, C. B., 290, 294, *307*
Barron, E. J., 101, 122, 123, 126, 176, 211, *307, 322*
Bartels, C. T., 156, 158, 189, 190, 196, *307*
Bartley, J. C., 116, 118, *307*
Bates, R. B., 50, 51, *322*
Bates, R. W., 25, *328*
Batt, R. D., 66, 67, 73, 295, *326*
Bauer, H., 113, *324*
Baumann, A. J., 293, *326*
Bednarczyk, D. J., 172, 173, *324*
Beerthuis, R. K., 204, *329*
Beevers, H., 207, 208, 209, 210, 211, 275, *307, 308, 310*
Beinert, H., 102, *328*
Benson, A. A., 44, 46, 47, 49, 71, 76, 180, 181, 182, 183, 185, 196, 198, 200, 249, 265, 266, 269, 270, 271, 272, 288, *307, 308, 311, 312, 317, 320, 322, 324, 331, 332*
Bereyney, R., 269, *307*
Bergelson, L. D., 50, *308*

Berner, D. L., *308*
Bersten, A. M., 139, 159, *318*
Betts, B. E., 56, *310*
Bezard, J., 138, 143, 155, *317*
Bhatty, M. K., 12, 298, *308*
Bilinski, E., *306*
Binder, R. G., 16, 25, *308*
Birge, C. H., 133, *308*
Bishop, N. I., 267, *308*
Blank, M. L., 294, *308, 325, 326*
Blass, J. P., 219, *308, 322*
Blecker, H. H., 78, 271, *316*
Bloch, K., 12, 13, 79, 80, 92, 102, 108, 109, 113, 122, 129, 130, 131, 132, 133, 134, 135, 136, 137, 139, 142, 143, 144, 147, 148, 149, 154, 157, 159, 160, 174, 177, 185, 221, 251, 260, 271, *308, 309, 310, 312, 313, 314, 315, 317, 318, 320, 321, 322, 323, 325, 326, 327, 332*
Bloomfield, D. K., 134, 149, *308*
Boardman, N. K., 127, *329*
Boatman, S. G., 243, *308*
Boch, R., 11, *316*
Bohlmann, F., 17, 167, *308*
Boichenko, E. A., 49, *330*
Bokman, A. H., 213, *317*
Boldingh, J., 229, 231, 234, *330*
Boll, W. G., 180, *331*
Bomstein, R. A., 44, *308*
Bonner, J., 276, *312*
Bonner, W. D., 268, *308*
Bonsen, P. P. M., 197, *308*
Booth, V. H., 53, *308*
Borgström, B., 195, *308*
Borkenhagen, L. F., 180, *308*
Bourdreaux, G. J., 194, *324*
Bove, J. M., 127, *329*
Bowen, D. M., 218, *308*
Boyer, P. D., 227, 230, *325*
Boynton, J. E., 246, *307*
Brady, R. O., 103, *308*
Brand, I. v. d., 58, 191, 192, 276, *315*
Branton, D., 265, *308*
Breidenbach, R. W., 101, 207, 211, *308, 326*
Bressler, R., 113, *309*
Brett, D., *309*
Breuer, J., 69, 144, *324*
Bridges, R. B., 71, 73, 74, *332*
Brierley, G., *313*

Brindley, D. N., 108, 109, 113, 122, *309, 321*
Bringi, N. V., 51, *318*
Broadbeer, C., 177, *308*
Brock, D. J. H., 132, 133, *309, 315, 318*
Brockerhoff, H., 83, 84, 85, 86, *309*
Brodie, J. D., 104, *309*
Brodnity, M. H., 224, *309*
Brody, M., 270, *309*
Brody, S. S., 270, *309*
Brooks, J., 127, *329*
Brooks, J. L., 127, *309*
Brooks, S., 56, *310*
Broughton, M. E., 63, *312*
Brown, C. M., 154, *309*
Brown, J. L., 283, *309*
Bücking, H., 196, *306*
Bu' Lock, J. D., 17, 19, 164, 165, 166, 167, 175, *309, 328*
Burge, W., 261, *332*
Burgher, R. D., 297, *306*
Burkhardt, H. J., 89, *309*
Burton, D., 117, 125, 127, *309*
Butler, R. D., 258, *309*
Butterworth, P. H. W. 104, *317*
Byrne, W. L., 274, *312,*

C

Caesar, H., 204, *329*
Callely, A. G., 212, *309*
Canvin, D. T., 24, 64, 138, 153, 169, 170, 207, 239, *309, 312*
Cardoso, H. T., 30, 64, *310*
Carr, A. H., 303, *316*
Carroll, K. K., 283, *309*
Carroll, W. R., 111, *319*
Carson, R. B., 261, 262, *325*
Carter, H. E., 23, 24, 48, 54, 55, 56, 57, 83, 168, 282, 283, 287, 300, *309, 327*
Castelfranco, P., 101, 206, 211, 213, *310, 325, 326*
Castelfranco, P. A., 257, *329*
Catesson, A. M., 65, 66, 257, *306*
Cerletti, P., 274, *310*
Chadha, J. S., 44, *306*
Chaikoff, I. L., 116, 118, *307*
Chang, H. C., 104, 117, *326*
Chang, S. B., 49, 184, 249, 260, 270, 271, *308, 310, 331*
Chang, Y. Y., 182, 185, *310*

Channon, H. J., 252, *310*
Chapman, D., 301, *310*
Chasson, R. M., 255, *319*
Cheniae, G., 185, *332*
Cheniae, G. M., 122, 129, 176, 178, *310*
Cherry, J. H., 207, *310*
Chibnall, A. C., 252, *310*
Ching, T. M., 193, 194, 242, *310*
Chisholm, M. J., 5, 7, 11, 12, 13, 14, 15, 16, 18, 19, 27, 51, 64, 129, 165, *310, 316*
Chisum, P., 49, 70, 76, *306*
Christiansen, K., 122, *310*
Christie, W. W., 229, 230, 231, 232, 290, *310, 316*
Christopher, J., 229, *310*
Chuecas, L., 8, *310*
Chung, A. E., 174, 188, 274, *310*
Chung, O., 62, *325*
Clark, A. H., 16, *320*
Clarke, N., 44, 244, *311*
Clayton, T. A., 287, *310*
Clermont, H., *310*
Click, R. E., 255, *310*
Cobern, D., 225, 226, *310*
Cole, H. I., 30, 64, *310*
Coleman, M. H., 82, *310*
Colli, W., 122, *310*
Collins, F. D., 47, *310*
Collins, F. I., 261, *317*
Comber, R., 243, *311*
Conacher, H. B. S., 20, 27, *310*
Coniglio, J. G., 71, 73, 74, *332*
Conn, E. E., 206, *328*
Constantopoulos, G., 92, 130, 159, 160, 251, 260, *308, 310*
Contopoulou, R., 213, *310*
Coon, M. J., 204, 222, *319, 320, 324, 325, 326*
Cooper, T. G., 207, 208, 209, 210, 211, *310*
Coots, R. H., 204, *306*
Cope, D. G., 27, *329*
Coppens, N., 169, *310*
Cordes, E. H., 98, 201, *321*
Corman, L., 118, *328*
Couch, J. R., 53, *306*
Coulon-Morelec, M. J., 47, *311*
Craig, B. M., 12, 63, 129, 261, 262, 298, *308, 311, 312, 325*
Cramer, R., 122, *327*

Crane, F. L., 269, *307*
Creger, C. R., 53, *306*
Creveling, R. K., 17, *317*
Criddle, R. S., 128, *328*
Crombie, L., 15, *311*
Crombie, W. M., 68, 69, 71, 239, 240, 243, *308, 311, 315*
Cronan, J. E., 133, *311*
Crossley, A., 17, *311*
Crouchman, M. L., 173, *322*
Csupor, L., 259, *314*
Czygan, F., 53, *319*

D

Dahlen, J. V., 121, 122, *311*
Dallas, M. S. J., 290, *307*
Daniel, H., 49, 288, *308, 311*
Das, M. L., 222, *311*
Davenport, J. B., 48, *311*
Davey, K. W., 135, *314*
Davidoff, F., 12, 169, 204, *311*
Davidson, F. M., 197, 199, 280, *311*
Davies, W. E., 218, *311*
Davies, W. H., 185, *311*
Davis, E. N., 172, *311*
Davis, H. F., 49, 70, 76, 90, 92, 285, 294, *306*
Davison, V. L., 8, 13, *319*
Dawson, R. M. C., 44, 180, 182, 198, 199, 244, *311, 325*
Debuch, H., 46, 47, 69, 71, 90, 157, *311, 326*
De Gier, J., 90, 155, 270, *311, 314, 315, 330*
de Haas, G. H., 84, 197, *311, 330*
Deinema, M. H., 25, *329*
Delaumeny, J.-M., 124, 174, *313*
Devine, J., 17, *311*
Devor, K. A., 128, *311*
de Vries, B., 290, *311*
Diamond, M. J., 16, *308*
Dickson, L. G., 119, *311*
Dils, R., 116, 117, 118, *312, 328*
Dimick, P. S., 221, *311*
Dittmer, J. C., 293, *311*
Ditzer, R., 204, *329*
Doizaki, W. M., 283, *331*
Dolev, A., 229, 231, 233, *311*
Donaldon, W. E., 159, *311*
Donley, J., 274, *321*

Döring, G., 270, *309*
Dorsey, J. A., 118, *311*
Douce, R., 47, 182, *310*, *311*
Downey, R. K., 129, 261, 262, *312*, *319*
Downing, D. T., 299, *312*
Draper, S. R., 258, *312*
Drapron, R., 242, *312*
Drennan, C. H., 138, *312*
Drysdale, G. R., 201, *312*
Duba, D., 111, 115, *327*
Dunphy, P. J., 293, *312*
Dupont, J., 182, 205, *311*, *312*
Duttera, S. M., 274, *312*
Dutton, H. J., 8, 13, 225, 226, 229, 231,
233, 297, *311*, *313*, *319*, *323*, *330*

E

Earle, F. R., 4, 7, 8, 12, 13, 19, 21, 27,
29, 166, 290, 299, *312*, *318*, *319*, *322*,
328, *329*
Easter, D. J., 117, *312*
Eberhagen, D., 7, 75, 76, 77, *319*
Eberhardt, F. M., 127, *312*
Echlin, P., 76, *312*
Eglinton, G., 26, *312*
Egwin, P. O., 229, 230, 231, 232, *316*
Eichenberger, W., 191, *312*
Eldjarn, L., 219, *329*
Elix, J. A., 21, 173, *312*
Elke, J. H. von, 228, *315*
Elovson, J., 105, 147, *312*, *325*
Endo, K., 133, *312*
Engelbrecht, A. H., 206, 211, *326*
English, J., 276, *312*
Entenman, C., 293, *328*
Eriksson, C. E., 229, *312*
Ernster, L., 222, *311*
Erwin, J., 79, 80, 129, 130, 135, 136, 141,
142, 143, 157, 271, *312*, *317*
Estroumza, J., 101, *327*
Ettre, L. S., 303, *312*
Evans, C. D., 82, 225, 226, 290, *312*, *313*
Evans, R. S., 305, *314*
Ewing, D. F., 51, *316*

F

Fagerson, I. S., 224, *309*
Fales, H. M., 29, 219, *314*, *322*
Fallis, A. G., 21, *312*
Farquhar, M., *320*

Faulkner, R. N., 226, *312*
Faure, M., 182, *311*
Fawcett, C. H., 21, 205, *312*
Ferrans, V. J., 197, *314*
Ferrari, R. A., 49, 180, 181, 183, 249,
270, 288, *308*, *312*
Fielding, L., 180, *308*
Finleyson, A. J., 212, *320*, *326*
Fischer, H. O. L., 88, *328*
Fisher, N., 62, 63, *312*
Flavin, M., 212, *312*
Fleischer, B., 272, *313*
Fleischer, S., 272, 295, *313*, *314*, *326*
Fogerty, A. C., 30, 139, 155, 159, *317*,
318
Follansbee, C., 219, *322*
Fontell, K., 16, *323*
Fowler, S. D., 49, 70, 76, 90, 92, 285,
294, *306*
Fowlis, I. A., 303, 304, *313*, *327*
Frankel, E. N., 225, 226, *313*
Frear, D. S., 173, *315*, *317*
Freer, S., 182, 198, *332*
Freidinger, R. M., 27, *322*
French, R. B., 63, *313*
French, T. H., 102, *325*
Fries, R., 116, *307*
Frost, D. J., 303, *313*
Fujü, K., 179, *316*
Fujiwara, N., 93, *324*
Fulco, A. J., 12, 124, 134, 135, 140, 149,
155, 159, 217, 271, *307*, *313*
Fuller, G., 76, 271, *324*
Funahashi, S., 191, 244, *316*, *318*
Furukawa, Y., 169, 174, 217, *332*

G

Galanos, D. S., 46, 47, 57, 283, 295, *309*,
313
Galli, C., *326*
Galliard, T., 49, 60, 61, 64, 65, 127, 170,
196, 200, 257, 278, *313*, *329*
Galloway, R. A., 119, *311*
Galzinga, L., 101, 274, *313*, *326*, *327*
Gan, M. V., 122, *310*
Ganguly, J., 102, *331*
Ganoya, M. C., 274, *312*
Gardner, H. W., 228, 229, 231, 234, *313*
Garton, G. A., 69, *313*
Gastambide-Odier, M., 124, 174, *313*

Gatt, S., 218, *313*
Gebhart, A. I., 224, 225, *326*
Gellerman, J. L., 7, 73, 74, 75, *313*
Gerecht, J. F., 224, 225, *326*
Gerloff, E. D., 154, *313*
Gibble, W. P., 125, *313*
Gibson, D. M., 101, 102, 103, 274, *313*,
  325, 326, 327, 329, 331
Gigg, R. H., 55, 56, *309, 310*
Gini, B., 234, *313*
Giordano, M. G., 274, *310*
Giovanelli, J., 30, 211, *313, 318*
Giovenco, M. A., 274, *310*
Giovenco, S., 274, *310*
Glasl, H., 286, *325*
Glass, C. A., 20, 27, *325, 328*
Goeschl, J. D., 278, *325*
Goffeau, A., 127, *329*
Goldblatt, L. A., 16, *308*
Goldfine, H., 131, 132, *313, 327*
Goldman, D. S., 124, *318*
Good, P., 49, 70, 76, 90, 92, 285, 294, *306*
Goodwin, J. C., 172, *311*
Goodwin, T. W., 185, *311*
Gorin, P. A. J., 25, *330*
Goto, M., 222, *331*
Götz, P., 142, *326*
Gouaux, J., 92, 245, 246, *326*
Grace, N. H., 205, *314*
Graham, A. B., 274, *314*
Green, D. E., 98, 101, 102, 103, 105, 201,
  272, 274, *314, 318*
Green, M. J., 303, *313*
Green, T. G., 62, 275, *314*
Greene, R. S., 299, *312*
Greenspan, M. D., 106, *314*
Greenwood, F. L., 54, *309*
Gregolin, C., 104, 117, *326*
Greville, G. D., 201, *320*
Grob, E. C., 259, *314*
Guchait, R. B., 121, *314*
Guilbot, A., 242, *312*
Guinand, M., 24, *314*
Gunstone, F. D., 15, 18, 19, 20, 25, 27,
  64, 82, 161, 290, 291, 292, *366, 310,*
  314
Gunther, S., 111, 115, *327*
Gurr, M. I., 71, 135, 144, 147, 188, 250,
  273, *314*
Guss, P. L., 228, 235, *314*

H

Haagen-Smit, A. J., 276, *312*
Haahti, E. O. A., 29, 305, *314*
de Haas, G. H., 197, *308*
Haberlandt, G., *314*
Hack, M. H., 197, *314*
Hacker, M., 194, *314*
Hackett, D. P., 255, *310*
Hadaway, H. C., 169, *317*
Haeffner, E. W., 13, *314*
Haest, C. W. M., 155, *314*
Hagedorn, D. J., 228, *315*
Hagemann, J. W., 27, *329*
Haigh, W. G., 7, 156, 167, *315*
Haining, J. L., 228, *318*
Hajra, A. K., 177, 178, 218, *311, 315*
Hale, S. A., 228, *315*
Hall, B. J., 261, *332*
Hall, C. W., 184, *323*
Hall, D. M., 277, *315*
Hall, G. E., 303, *313*
Hall, R. J., 23, *331*
Hall, S. W., 15, 24, 25, 30, 168, 170, 172,
  175, 276, 290, *315, 322, 323*
Halpin, R. A., 274, *321*
Hamberg, M., 225, 227, 229, 230, 231,
  232, 234, *315*
Hamilton, R. J., 26, 52, 291, *312, 314,*
  315
Hammonf, E. G., *309*
Hammond, E. W., 12, 27, 298, *323*
Hamza, Y., 50, *318*
Hanks, D. P., 17, *316*
Hardman, E. E., 239, 240, 243, *311,*
  315
Harlan, W. R., 121, 122, *315*
Harris, J., 15, *329*
Harris, P., 136, 138, 143, 144, 153, 155,
  158, 248, *315, 317, 324*
Harris, R. A., 272, 314
Harris, R. V., 76, 136, 139, 143, 144, 146,
  188, 286, 287, *315, 323, 324*
Hart, P., 197, *319*
Hartmann, G. R., 173, *315*
Harvey, B. L., 262, *312*
Hase, E., 246, *327*
Hatch, M. D., 125, 126, 211, 212, *315*
Hatt, H. H., 19, *315*
Haverkate, F., 12, 47, 79, 90, 93, 157,
  182, 292, *315*

Hawke, J. C., 116, 127, 128, 135, 248, 315, 319, 329
Hawthorne, J. N., 56, 310
Hedin, P. A., 65, 329
Heinen, W., 58, 191, 192, 276, 315
Heinstein, P. F., 103, 117, 125, 315
Heinz, D. E., 17, 164, 317
Heinz, E., 49, 223, 315
Heller, D., 285, 293, 326
Heller, M., 197, 315
Helmkamp, G. M., 133, 315
Helmkamp, G. M., Jr., 312
Hemington, N., 180, 182, 199, 311
Hendrickson, H. S., 54, 57, 283, 309
Hendrickson, J. B., 149, 150, 326
Hendry, R. A., 48, 309
Heng, I., 57, 331
Herndon, J. H., 219, 328
Herndon, J. H. Jr., 219, 322
Henson, R. D., 65, 329
Heyes, J. K., 7, 316
Heywang, B. W., 318
Hilditch, T. P., 1, 2, 11, 15, 17, 28, 33, 35, 62, 64, 68, 97, 152, 275, 311, 314, 316
Hill, A., 25, 330
Hill, R. L., 105, 330
Hinkle, P. C., 122, 310
Hirayama, O., 179, 316
Hirschmann, H., 83, 316
Hitchcock, C., 23, 168, 214, 215, 216, 217, 220, 255, 300, 303, 304, 316, 322
Hobbs, J. S., 225, 226, 310
Hoelle, C., 214, 319
Hoffman, R. L., 290, 312
Holden, M., 227, 259, 316
Holdt, M. M. von, 7, 18, 320
Holloway, P. W., 13, 128, 145, 273, 316, 318
Holm, G. A. L., 54, 318, 328
Holman, R. T., 16, 17, 21, 43, 88, 122, 229, 230, 231, 232, 234, 290, 310, 316, 321, 323, 331
Holton, R. W., 78, 271, 316
Hooper, N. K., 175, 274, 316
Hopfer, U., 269, 316
Hopkins, C. Y., 5, 7, 11, 12, 13, 14, 15, 16, 18, 19, 27, 30, 51, 64, 129, 165, 310, 316
Hopkins, D. T., 300, 327

Hörhammer, L., 288, 316
Horn, D. H. S., 7, 18, 52, 325
Horowitz, B., 31, 261, 316
Hon, C. T., 191, 316
Hougen, F. W., 261, 328
Herndon, J. H., 328
Howard, C. F., 122, 317
Howe, R., 25, 222, 318
Howell, R. W., 261, 317
Howlett, M. D. D., 303, 317
Howling, D., 138, 145, 309, 317
Howton, D. R., 157, 322
Hoyle, R. J., 85, 309
Hsu, R. Y., 104, 317
Huang, K. P., 101, 317
Huber, R. E., 239, 317
Hulanicka, D., 79, 80, 142, 157, 271, 312, 317
Hulme, A. C., 257, 278, 313
Humphreys, T. E., 213, 317
Hunter, G. D., 102, 325
Hutt, H. H., 44, 317
Hutton, D., 209, 210, 317

I

Ichihara, K., 222, 317
Ingram, J. M. A., 205, 312
Inkpen, J. A., 134, 138, 281, 317

J

Jaakonmäki, I., 305, 314
Jacklin, A. G., 15, 311
Jackman, M. E., 255, 317
Jacks, T. J., 207, 317
Jackson, L. L., 173, 317
Jacob, M. I., 102, 325
James, A. T., 4, 7, 15, 23, 32, 64, 65, 68, 69, 71, 76, 90, 127, 130, 131, 135, 136, 138, 139, 142, 143, 144, 145, 146, 147, 153, 155, 156, 158, 167, 168, 169, 170, 188, 189, 214, 215, 220, 248, 255, 285, 286, 287, 295, 303, 307, 309, 315, 316, 317, 321, 323, 324, 329, 332
Jamieson, G. R., 68, 69, 76, 300, 317, 326
Jann, B., 57, 283, 309
Jauréguiberry, G., 174, 317
Jennings, W. G., 17, 164, 221, 315, 317, 329
Jensen, R. G., 195, 324

Jevans, A. W., 11, 19, 30, 165, *316, 317*
Ji, T. H., 265, *317*
Johnson, A. R., 30, 139, 155, 159, 175, 300, *318*
Johnson, A. W., 17, 20, 21, *318*
Johnson, G. T., 116, *320*
Johnston, J. M., 283, *309*
Jones, D. F., 25, 222, 223, *318*
Jones, E. R. H., 17, 21, 173, *312, 318*
Jones, P. D., 145, 273, 274, *318*
Jones, Q., 7, 290, *318*
Jones, R. L., 277, *315*
Joshi, V. C., 112, 117, *318, 325*
Junk, J. W., 222, *320*
Jurriens, G., 82, 292, *318*
Jurtshuk, P., 274, *318*

K

Kahn, A., 207, *308*
Kaneda, T., 174, *318*
Kanemasa, Y., 124, *318*
Kaneshiro, T., 180, *318*
Kapoulas, V. M., 46, 47, 295, *313*
Karlsson, K.-A., 54, 300, *318*
Kartha, A. R. S., 13, 62, 275, *318*
Kasbekar, M. G., 51, *318*
Kass, L. R., 132, *309, 318*
Katayama, M., 244, *318*
Kates, M., 23, 24, 68, 75, 76, 77, 90, 91, 93, 127, 168, 176, 177, 178, 180, 181, 183, 190, 197, 198, 199, 200, 249, 271, 280, 287, 289, 300, 312, 318
Kaufmann, H. P., 50, 292, *318*
Kauss, H., 191, *318*
Kawanami, J., 169, 174, *318*
Keith, A. D., 146, *331*
Kelly, W., 139, 146, *323*
Kennedy, E. P., 180, 182, *308, 310*
Kennet, B. H., *318*
Kenyon, C., 130, 159, 160, *308*
Kenyon, C. N., 33, 80, 271, *318*
Kerr, P. C., 129, 176, *310*
Khan, R. A., 13, *318*
Kies, M. W., 228, *318*
Kilgast, D., 19, *314*
Kimura, A., 169, 174, *318*
Kimura, R. E., 122, *331*
King, G., 26, *318*
Kircher, H. W., *318*
Kiser, J., 195, *324*

Kishimoto, Y., 296, *318*
Kisic, A., 57, 172, 173, 187, *309, 324*
Kleiman, R., 4, 7, 8, 12, 13, 21, 29, 299, *318, 319, 328*
Klein, H. P., 117, *325*
Kleinig, H., 53, *319*
Kleinschmidt, A. K., 117, *319, 326*
Klenk, E., 7, 75, 76, 77, *319*
Klopfenstein, W. E., 248, *319*
Klouwen, H., *313*
Knaggs, J. A., 44, *306*
Knipprath, W. G., 7, 75, 76, 77, 155, *319*
Knoche, H. W., 169, 172, *319*
Knoop, F., 201, *319*
Knowles, P. F., 261, 262, *319, 332*
Koch, R. B., 228, 231, 234, *313, 319*
Koike, M., 111, *319*
Kolattukudy, P. E., 72, 73, 190, 222, 252, 253, 254, 255, 256, *319*
Koob, J. L., 23, 24, 54, 57, 168, 300, *309*
Koof, H. P., 7, 75, 76, 77, *319*
Korn, E. D., 8, 9, 12, 78, 129, 157, 169, 204, 263, *311, 319*
Kosterman, H. J., 242, *332*
Kreger, D. R., 253, *319*
Krewson, C. F., *327*
Kritchevsky, G., 285, *326*
Kroeplin-Reuff, L., 111, 115, *327*
Kroesen, A. C. J., 82, 292, *318*
Krzymanski, J., 262
Kuhn, R., 53, *319*
Kulkarni, N. D., 184, 185, 249, 270, *310*
Kumar, S., 116, 122, *323*
Kurtz, E. B., 125, *313*
Kusunose, E., 222, *317, 319, 325*
Kusunose, M., 169, 174, 217, 222, *317, 319, 325, 332*

L

Laidman, D. L., 194, 242, *329*
Lamberts, I., 122, *327*
Land, M. D., 117, *326*
Landau, B. R., 217, *306*
Lands, W. E. M., 84, 197, *319*
Lane, M. D., 104, 117, *319, 326*
Lardy, H. A., 201, *312*
Larrabee, A. R., 105, 112, *319, 330*
Laties, G. G., 214, *319*
Lau, Y. C., *306*
Laudat, P., *322*

Launay, B., 242, *312*
Launay, B., *312*
Law, J. H., 55, 174, 175, 180, 188, 274, *309, 310, 316, 317, 318, 329*
Law, J. L., 174, *306*
Lawrence, R. C., 116, *319*
Leane, J. B., 303, *313*
Lederer, E., 53, 124, 174, *313, 317, 319*
Ledingham, G. A., 212, *326*
Lee, A., 25, *308*
Lee, B., 277, *320*
Lee, S. G., 255, *319*
Legler, G., 25, *320*
Lehninger, A. L., 201, 269, 272, 273, *316, 320, 330, 332*
Lemieux, R. U., 26, 168, *320*
Lenfant, M., 174, *317*
Lennarz, W. J., 28, 147, 174, 269, *316, 320*
Lepage, M., 47, 49, 51, 64, 65, 287, *311, 320*
LeQuan, M., 21, *312*
Lerch, I., 111, 115, *327*
Lester, R. L., 293, *311*
Levis, G. M., 217, 218, *320*
Levy, H. R., 117, *322*
Levy, R., 12, 159, *313*
Lew, J. Y., 126, *330*
Lewis, H. L., 116, *320*
Lieber, E., 285, *326*
Light, R. J., 147, *320*
Ligthelm, S. P., 7, 18, *320*
Lippel, K., 218, *320*
List, G. R., 82, *312*
Litchfield, C., 7, 292, *320*
Lloyd, D., 212, *309, 320*
Loeblich, A. R., 76, 271, *324*
Lohmar, R. L., 12, 25, 27, *303, 328, 330*
Long, C., 197, 199, 280, *311, 320*
Long, R. W., 118, *325*
Lowenstein, J. M., 118, *328*
Lu, A. Y. H., 222, *320*
Lucas, R. A., 225, 226, *310*
Luddy, F. E., 173, *327*
Lundberg, W. O., 227, 230, *325*
Lundin, K., 271, *310*
Lundin, R. E., 15, *329*
Lust, G., 114, 118, *320*
Lynen, F., 102, 103, 104, 108, 109, 110, 111, 112, 113, 114, 115, 116, 117, 118, 201, *307, 320, 321, 324, 327, 329, 331, 332*

M

McClelland, M. J., 299, *320*
McCloskey, J. A., 172, 174, 299, 302, *317, 320*
McCluer, R. H., 300, *327*
McConnell, D. G., 82, 225, 226, 290, *312, 313*
McConnell, W. B., 212, *320, 326*
McDaniel, E. G., 112, *319*
MacDonald, R. C., 218, *320*
McElroy, F. A., 116, *320*
Macey, M. J. K., 73, 254, *320*
McFadden, W. H., 303, *312*
Macfarlane, M. G., 47, *320*
McGregor, W. G., 239, *328*
McKenna, E. J., 222, *320*
Mackenzie, D. J., 225, 226, *310*
McKillican, M. E., 93, 238, *320*
McLachlan, J., 8, *306*
McLean, J., 16, *320*
McMahon, V., 126, 143, 154, 240, 244, *321*
McManus, T. T., 127, 128, 143, *323*
MacMurray, T. A., 287, *310*
Mahler, H. R., 98, 201, *321*
Maier, R., 21, 43, 88, 290, *321, 328*
Majerus, P. W., 104, 105, 106, 107, *306, 321, 330*
Malins, D. C., 293, *321*
Malkin, T., 44, 57, *317, 321*
Mandersloot, J. G., 270, *311*
Mangold, H. K., 50, 293, *318, 321*
Mapson, L. W., 235, 278, *321*
Marcel, Y., 122, *310*
March, J. F., 278, *321*
Marcus, A., 206, *321*
Marechal, J., 182, *311*
Markley, K. S., 1, 5, 11, *321*
Marsh, J. B., 147, *321*
Marshall, M. O., 12, 13, 18, 27, 161, 162, 165, 298, *323*
Martin, A. J. P., 102, 295, *317, 325*
Martin, D. B., 117, *330*
Martin, J. T., 26, 71, *307, 321*
Martin, R. O., 214, *321*
Martonosi, A., 274, *321*
Maruo, B., 46, 68, 71, *308*
Mathias, M. M., 205, *312*
Matic, M., 25, 26, 58, *321*
Matsuhashi, M., 103, 117, *321*

Matsuhashi, S., 117, *321*
Matsumura, S., 108, 109, 113, 122, 128, *309*, *321*
Mattil, K. F., 81, *324*
Mattson, F. H., 81, 82, 193, *321*
Maurer, L., 78, *324*
Mazliak, P., 33, 65, 66, 177, 178, 257, 258, 277, *306*, *321*, *322*
Mead, J. F., 155, 157, 218, *311*, *319*, *320*, *322*
Meara, M. L., 68, *316*
Mecham, D. K., 281, *322*
Mendelowitz, A., 15, 64, *316*
Mercer, E. I., 185, *311*
Merryfield, D. S., 50, 281, *332*
Mes, J. C., 239, *328*
Meyer, F., 142, 154, *322*
Michel, G., 24, *314*
Mikolajczak, K. L., 19, 21, 26, 27, 29, 50, 51, 64, 173, 175, *322*, *331*
Milch, P., 92, 245, *326*
Miller, A. L., 117, *322*
Miller, J. A., 49, 288, *308*
Miller, R. W., 7, 290, *322*
Milne, G. W. A., 219, *322*, *328*
Minyard, J. P., 65, *329*
Mitsuda, H., 231, *322*
Miwa, G. C., *322*
Miwa, T. K., 12, 25, *303*, *322*, *328*
Miyachi, S., 196, 200, 275, *322*
Miyano, M., 46, 49, 182, *308*, *311*
Mize, C. E., 219, *322*, *328*
Mizugaki, M., 107, 133, *322*
Mohammad, A., 281, *322*
Mohrhauer, H., 122, *310*
Montrozier, H., 140, *307*
Mooney, L. A., 122, 123, *307*, *322*
Moorhouse, R., 93, 190, 250, 292, *324*
Morré, D. J., 181, *322*
Morris, I., 76, *312*
Morris, L. J., 12, 13, 15, 16, 18, 19, 23, 24, 25, 26, 27, 30, 88, 89, 138, 139, 145, 146, 148, 156, 161, 162, 165, 167, 168, 170, 171, 172, 173, 175, 215, 216, 276, 290, 291, 292, 293, 298, *307*, *315*, *316*, *317*, *322*, *323*, *332*
Morrison, W. R., 287, 296, *310*, *323*
Morton, R. A., 301, *325*
Moss, J., 117, *319*
Mounts, T. L., 297, *311*, *323*

Moustafa, E. M., 235, *321*
Mudd, J. B., 49, 65, 71, 127, 128, 135, 143, 182, 183, 184, 185, 188, 190, 191, 211, 249, *311*, *323*, *324*, *329*, *332*
Mumma, R. O., 47, 49, *311*, *320*
Munzing, A., 57, *331*
Murray, K. E., *318*
Murty, N. L., 63, *311*
Mutwakil, A., 261, 262, *319*
Myhre, D. V., 49, *323*

N

Nagai, J., 130, 133, 134, 136, 137, 144, 147, 148, 149, 159, 160, *308*, *323*
Nakagawa, Y., 169, 174, *318*
Nakamura, M., 191, *316*
Nakano, M., 218, *326*
Nakayama, T., 55, 57, 283, *309*
Nakazawa, Y., 57, 283, *309*
Nandedkar, A. K. N., 116, 122, *323*
Nawar, W. W., 224, *309*
Nelson, J., 122, *331*
Nervi, A. M., 103, 104, *306*
Neufeld, E. F., 184, *323*
Newcomb, E. H., 124, 213, *317*, *323*
Newman, D. W., 191, 247, 248, 259, 260, *312*, *323*, *331*
Nichamen, M. Z., 293, *323*
Nichols, B., 57, 283, *309*
Nichols, B. W., 7, 12, 32, 33, 64, 65, 66, 68, 69, 71, 73, 74, 75, 76, 78, 80, 90, 92, 93, 94, 144, 157, 158, 181, 186, 188, 189, 190, 246, 247, 248, 249, 250, 251, 271, 274, 280, 282, 285, 286, 287, 288, 292, 293, 298, *307*, *317*, *323*, *324*, *326*, *332*
Nickell, C., 227, 230, *325*
Nicolaides, N., 293, *326*
Niehaus, W. G., 172, 173, 299, *324*
Noda, M., 93, *324*
Nojima, S., 48, *309*
Norris, F. A., 81, *324*
Nugteren, D. H., 121, *324*
Numa, S., 103, *324*
Nyman, B., 194, *324*
Nyquist, S., 181, *322*

O

O'Connor, R. T., 302, *324*
Ochoa, S., 201, 212, *312*, *320*

Odavic, R., 197, *320*
Oesterhelt, D., 113, 116, 117, *324, 329*
Ogrodnik, J. A., 15, *316*
Ohly, B., 122, *327*
Ohnishi, T., 272, *324*
Ohnishi, T., 272, *324*
Ohno, K., 48, *309*
Olcott, H., 279, *330*
Oldham, N. M., 293, *323*
Olney, C. E., 195, *324*
Olson, R. E., 293, *323*
Ongun, A., 49, 65, 70, 71, 184, 185, 191, 249, *324*
Onore, M., 78, 271, *316*
Ordin, L., 274, *325*
Orishimo, M. W., 122, *331*
Orrenius, S., 222, *311*
Ory, R. L., 193, 194, 195, *308, 324*
Otsuka, H., 169, 174, *318*
Overath, P., 126, 128, *324*

P

Padley, F. B., 290, 305, *307, 314, 324*
Page, C. B., 21, *312*
Panter, R. A., 211, *324*
Park, R. B., 265, *308*
Parker, P. L., 78, *324*
Parkes, J. G., 269, *324*
Parmar, S. S., 218, *311*
Paschke, R. F., 7, *324*
Patterson, G. W., 119, 154, *311, 324*
Patton, S., 76, 221, 271, *311, 324*
Payne, S. N., 294, *324*
Pearson, J. A., 30, 139, 155, 159, *317, 318*
Pecker, M., 245, 246, 247, 267, *326*
Peluffo, R. O., 145, 273, *318*
Penniston, J. T., 272, *314*
Pennock, J. F., 293, *312*
Perkins, E. J., 299, *307*
Peters, H. M., 290, *312*
Peters, R. A., 23, *331*
Peterson, J. A., 222, *324, 325*
Phillids, S. J., 122, *331*
Phillips, B. E., 15, 16, *325*
Piccinini, F., 111, 115, *327*
Pieringer, R. A., 84, *319*
Pinsky, A., 274, *325*
Pistorius, E., 228, 229, *310, 318*
Pitt, G. A. J., 301, *325*

Pittman, R. C., 219, *322*
Placek, L. L., 12, 64, *325*
Plate, C. A., 112, 117, *318, 325*
Plessers, A. G., 239, *328*
Pohl, P., 57, 76, 286, *325, 331*
Pomeranz, Y., 62, *325*
Poole, A. G., 44, 57, *317, 321*
Popjak, G., 102, *325*
Porter, J. W., 102, 104, 118, 121, 122, *309, 311, 314, 317, 325, 331*
Powell, G. L., 105, *325*
Powell, R. G., 15, 16, 18, 19, 20, 27, *314, 325*
Power, D. M., 52, *315*
Pradel, P. A., 195, *324*
Pratt, H. K., 278, *325*
Preiss, B., 221, *325*
Priestley, J. H., 277, *320*
Prince, L., 12, *316*
Privett, O. S., 179, 227, 230, 239, 240, 241, 287, 294, 299, *308, 325, 326, 328*
Prokazova, N. V., 50, *308*
Promé, J. C., 140, *307*
Pullman, M. E., 122, *310*
Purdy, S. J., 52, 252, *325*
Putt, E. D., 261, 262, *325*
Putz, G. R., 121, *314*
Pynadath, T. I., 116, *323*

Q

Quackenbush, F. W., 134, 138, 281, *317*
Quarles, R. H., 44, 198, 244, *311, 325*
Quinn, J. G., 195, *324*
Qureshi, M. I., 291, 292, *314*

R

Radin, N. S., 218, 296, *308, 311, 318*
Radler, F., 52, *325*
Radunz, A., 29, 73, 92, *325*
Raju, P. K., 30, 155, 159, *325*
Rando, R. R., 133, *315, 325*
Rao, C. V. N., 299, *307*
Rasmussen, R. K., 117, *325*
Rebeiz, C. A., 101, 206, 211, *325, 326*
Reed, L. J., 111, *319*
Reid, E. H., 68, 69, 76, *317, 326*
Reiesner, H., 212, *326*
Reimenschneider, R. W., 173, *327*
Reiser, R., 30, 155, 159, 292, *320, 325*
Renkonen, O., 177, 185, *326*
Resnick, M. R., 146, *331*

Rhodes, M. J. C., 257, 278, *313, 321*
Rhodes, R. A., 172, *311*
Richards, J. H., 149, 150, *326*
Richardson, K. E., 180, *330*
Richardson, T., 154, 228, 235, *313, 314, 315*
Richter, G., 288, *316*
Riepertinger, C., 113, 116, 117, *329*
Riley, J. P., 8, *310*
Rilling, H. C., 204, *326*
Rinne, R. W., 154, *326*
Rittenberg, D., 102, *326*
Ritter, E., 111, 115, *327, 331*
Rivera, E., 181, *322*
Robinson, M. P., 71, 144, 188, *314*
Roehm, J. N., 179, 239, 299, *326*
Rogers, M. F., 21, *322*
Rohwedder, W. K., 172, 229, 231, 233, *311*
Rose, A. H., 154, *309*
Rosenberg, A., 79, 80, 92, 245, 246, 247, 267, 275, *326*
Ross, J., 224, 225, *326*
Rossi, C. R., 101, 274, *313, 326, 327*
Rossi, C. S., 273, *331*
Rothstein, A., *326*
Rothstein, M., 142, *326*
Rotsch, E., 46, 47, 90, *311, 326*
Roubal, W. T., 234, *326*
Roughan, P. G., 66, 67, 73, 145, 188, 250, 252, 270, 295, *326*
Rouser, G., 285, 293, 295, *336*
Ruzicka, F. J., 269, *307*
Ryder, E., 104, 117, *326*
Ryhage, R., 299, *324*

S

Safford, R., 7, 93, 94, 188, 190, 250, *315, 326*
Saga, M., 218, *326*
St. Angelo, A. J., 193, 195, *306, 324, 326*
Sakamoto, Y., 222, *331*
Sambasivarao, K., 300, *327*
Sampugna, J., 195, *324*
Samuel, D., 101, *327*
Samuelsson, B., 225, 227, 229, 230, 231, 232, 234, *315*
Sargent, E. J., 197, *320*
Sargent, M. V., 21, 173, *312*
Sartorelli, L., 101, 274, *313, 327*

Sastry, P. S., 23, 24, 68, 90, 91, 93, 168, 176, 177, 178, 180, 183, 199, 200, 249, 287, *318*
Sauer, K., 267, *330*
Schauenstein, E., 226, *327*
Scheuerbrandt, G., 13, 131, 132, 133, 174, *327*
Schieffer, H. G., 268, *329*
Schirmer, E. W., 116, *323*
Schlenk, H., 7, 73, 74, 75, *313*
Schmalfuss, K., 260, *327*
Schmid, C., 125, *328*
Schmidt, J. A., 294, *308*
Schneider, E. L., 300, *327*
Scholfield, C. R., 25, 82, *312, 328*
Schroeder, D. H., 228, *318*
Schroepfer, G. J., 139, 172, 173, *324*
Schubrook, D. C., 21, *312*
Schuly, H., 118, 167, *308*
Schwartz, H. M., 7, 18, *320*
Schwartz, P., 83, *327*
Schweizer, E., 111, 115, *327*
Scott, P. G. W., 303, *327*
Scott, R. P. W., 304, *327*
Scott, W. E., 173, *327*
Sealy, A. J., 18, 19, *314*
Sedgwick, B., 117, *325*
Seigler, D. S., 50, 51, *322*
Sekuzu, I., 274, *318*
Selke, E., 225, 226, *313*
Selvaraj, Y., 13, *318*
Senn, V. J., 89, *327*
Sephton, H. H., 224, 225, *327*
Seubert, W., 122, *327*
Serck-Hanssen, K., 25, *327*
Shannon, L. M., 126, *330*
Shapiro, B., 197, *315*
Shapiro, M., 118, *327*
Shenberg, E., 159, *328*
Shenstone, F. S., 30, 68, 139, 155, 159, *317, 318*
Shibita, H., 222, *331*
Shibuya, I., 49, 68, 71, 246, *311*
Shigley, J. W., 248, *319*
Shimazono, N., 221, *331*
Shively, J. M., 169, *319*
Shorland, F. B., 2, 7, 33, 35, 90, *316, 328, 331*
Siakatos, A. N., 295, *326*
Siddiqi, A. M., 229, 232, *328*

Siegl, W. O., 12, *307*
Silbert, D. F., 133, *308*
Simoni, R., 127, *329*
Simoni, R. D., 128, *328*
Sims, R. P. A., 239, *328*
Singh, H., 240, 241, 275, 287, *328*
Singh, S. P., 62, *318*
Sjostrand, F. S., 268, *328*
Skidmore, W. D., 293, *328*
Skopski, V. P., 283, 284, *328*
Slakey, P. M., 84, *319*
Slautterback, D. B., *313*
Slawson, V., 226, *328*
Smalley, H. M., 164, *309*
Smith, C. R., 8, 11, 12, 13, 15, 16, 18,19,
    20, 21, 24, 25, 26, 27, 29, 50, 51, 62,
    64, 168, 173, 175, 275, *307, 322, 325,*
    328, 331, 332
Smith, G. N., 17, 19, 165, 166, 167, 175,
    309, 328
Smith, L. M., 296, *323*
Smith, S., 116, 118, *328*
Smolowe, A. F., 283, 284, *328*
Solberg, Y. J., 26, *328*
Sook, P. L., 300, *327*
Sørenson, N. A., 17, 20, *328*
Sowden, J. C., 88, *328*
Spencer, A., 118, *328*
Spencer, D. M., 21, *312*
Spencer, G. F., 7, 12, 13, 15, 21, 299,
    319, 328, 329
Spencer, J. F. T., 24, 25, 223, *315, 329,*
    330
Sprecher, H. W., 21, 43, 290, *328*
Squires, C., 126, *307*
Squires, C. L., 125, *328*
Stahmann, M. A., 154, 228, 235, *313, 314*
Stainsby, W. J., 62, 275, *314*
Ställberg-Slenhagen, S., 28, *306*
Stanacer, N. Z., 48, *309*
Stanier, R. Y., 33, 80, 271, *318*
Stanley, R. G., 206, *318*
Stansly, P. G., 102, *328*
Stefansson, B. R., 261, *328*
Stein, R. A., 226, *328*
Steinberg, D., 219, *308, 322, 328*
Stenhagen, E., 28, *306*
Stern, N., 159, *328*
Stewart, H. B., 116, *320*
Stewart, P. S., 25, *331*

Stobart, A. K., 246, 247, *329*
Stockenius, W., 272, *313*
Stodola, F. H., 25, 26, *329*
Stoffel, W., 204, 268, *329*
Stohr, H., 194, *314*
Stokke, O., 219, *329*
Strickland, E. H., 46, 47, *308*
Strobach, D. R., 56, *310*
Strom, R., 274, *310*
Struik, C. B., 204, *329*
Stubbs, J. M., 68, *324*
Stumpf, P. K., 4, 101, 103, 116, 117,
    119, 124, 125, 126, 127, 128, 129, 135,
    138, 143, 148, 154, 157, 170, 171, 176,
    177, 178, 205, 206, 207, 209, 210, 211,
    212, 213, 214, 220, 221, 240, 244, 246,
    255, 257, 258, *307, 308, 309, 310, 313,*
    315, 317, 321, 323, 324, 328, 329, 331,
    332
Subbaram, M. R., 82, 291, *329*
Subbarao, R., 15, *314*
Sugarman, P. M., 271, *307*
Sullivan, R. C., 283, 284, *328*
Sumida, S., 182, 183, *329*
Sumper, M., 113, 116, 117, *329*
Sutton, D. A., 224, 225, *327*
Svensson, S. G., 229, *312*
Sweeley, C. C., 293, *323*
Swindell, A. C., 113, 133, *307, 322*
Swords, M. D., 181, 186, *329*
Synerholm, M. E., 205, *329*
Szezepanowska, A. D., 281, *332*

T

Talamo, B., 106, 107, *306*
Tallent, W. H., 15, 27, *329*
Tang, C. S., 221, *329*
Tang, W. J., 257, *329*
Tanimura, A., 50, *330*
Tappel, A. L., 227, 229, 231, 232, 234,
    326, 328, 329
Tavener, R. J. A., 194, 242, *329*
Taylor, G. M., 19, *314*
Thaller, V., 21, *312*
Thomas, D. R., 246, 247, *329*
Thomas, P. J., 174, *329*
Thompson, A. C., 65, *329*
Thompson, W., 269, *324*
Thomson, W. W., 65, 71, *324*
Tietz, A., 102, 159, *325, 328*

Tipton, C. L., 48, 181, 186, *309*, *329*
Titchener, E. B., 102, *313*, *329*
Tjarks, L. W., 15, 16, 51, *322*, *325*
Tocher, C. S., 8, *306*
Tookey, H. L., 197, *330*
Toomey, R. E., 106, 107, 118, 133, *322*, *327*, *330*
Torkelson, A., 172, 173, *324*
Toubiana, R., 174, *317*
Triffet, A. C. K., 19, *315*
Trosper, T., 267, *330*
Truter, E. V., 52, 252, *325*
Try, K., 219, *329*
Tsujimoto, M., 68, *330*
Tsutsumi, Y., 218, *326*
Tulloch, A. P., 24, 25, 27, 49, 173, 223, *315*, *330*

U
Udelnova, T. M., 49, *330*
Uhlendorf, B. W., 219, *328*
Ukita, T., 50, *330*
Umemura, Y., 191, *316*

V
Vagelos, P. R., 103, 104, 105, 106, 107, 112, 117, 133, 213, *306*, *308*, *312*, *314*, *319*, *321*, *325*, *330*
Vanaman, T. C., 105, *330*
Van Baalen, C., 78, *324*
van Deenen, L. L. M., 12, 47, 84, 90, 93, 155, 157, 182, 184, 188, 190, 197, 270, 292, *307*, *308*, *311*, *314*, *323*, *330*
Vanden Driessche, T., 261, *330*
Van der Veen, K. J., 204, *331*
van der Veen, J., 279, *330*
van Vliet, H. H. D. M., 184, 188, 190, *323*
Van der Wal, R. J., 82, 88, *330*
Vandor, S. L., 180, *330*
Van Steveninck, R. F. M., 255, *317*
Vaver, V. A., 50, *308*
Velasco, J., 206, *321*
Veldink, G. A., 229, 231, 234, *330*
Vellis, J. de, 126, *330*
Venables, S. E., 212, *320*
Verkade, P. E., 221, *330*
Vesonder, R. F., 26, *329*
Vick, B. A., 229, *332*
Vickery, J. R., 68, *327*

Vignais, P. M., 272, 273, *330*, *331*
Vignais, P. V., 272, 273, *330*, *331*
Vihko, R., 305, *314*
Vioque, E., 234, *331*
Viswanathan, C. V., 292, *331*
Vliegenhart, J. F. G., 229, 231, 234, *330*
Vogel, H. C., 283, *331*
Volcani, B. E., 75, 76, 77, 181, 271, 289, *318*
Volpenheim, R. A., 81, 82, 193, *321*
von Rudloff, E., 25, 299, *331*
von Wettstein, D., 127, 246, *307*

W
Wada, F., 222, *331*
Wagner, H., 57, 76, 286, 288, *316*, *325*, *331*
Wailes, P. C., 19, *315*
Wain, R. L., 21, 205, *312*, *331*
Wakabayashi, K., 221, *331*
Wakil, S. J., 13, 102, 103, 105, 106, 107, 112, 113, 114, 117, 118, 119, 121, 122, 128, 133, 145, 273, 274, *307*, *309*, *313*, *315*, *316*, *318*, *322*, *325*, *327*, *330*, *331*
Walker, N. J., 221, *311*
Wallace, J. W., 247, 260, *331*
Wallen, L. L., 172, *311*
Ward, P. F. V., 23, *331*
Wardale, D. A., 278, *321*
Warner, R. C., 117, *326*
Warshaw, J. B., 122, *331*
Wasson, G., 104, *309*
Watt, P. R., 44, *317*
Webb, J. P. W., 169, *317*
Weber, E., 55, *309*
Weber, E. J., 236, 237, *331*
Webster, D. E., 49, *331*
Weeks, G., 107, 118, 133, *322*, *327*, *331*
Weenink, R. O., 90, *330*, *331*
Weier, T. E., 265, 266, *331*
Weisleder, D., 229, 231, *313*
Welti, D., 303, 304, *313*, *317*, *327*
Wessels, H., 292, *318*
Wharry, D. M., 27, *323*
Wheeldon, L. W., 284, *331*
Wheeler, D. H., 7, *324*
Whereat, A. F., 122, *331*
Whittle, K. J., 293, *312*
Whyborn, A. G., 68, *307*

Wickerham, L. J., 26, *329*
Wightman, F., 205, *331*
Wilde, P. F., 25, *331*
Wilkins, T., 304, *327*
Willebrands, A. F., 204, *331*
Willecke, K., 115, *331, 332*
Willemot, C., 119, 180, 255, 257, 258, *331*
Williams, P. N., 1, 2, 11, 28, 33, 35, 97, 152, *316*
Williamson, I. P., 106, *331*
Wilson, R. G., *330*
Wilson, T. L., 25, 29, 64, 175, *331, 328*
Winter, G., 31, 261, *316*
Wintermans, J. F. G. M., 66, 70, 71, 288, 295, *308, 331*
Winterstein, A., 53, *319*
Wiser, R., 49, 288, *308*
Wisnieski, B. J., 146, *331*
Witham, P. M., 21, *312*
Wlodawer, P., 155, 273, *307, 332*
Wojtczak, L., 273, *332*
Wolf, F. T., 71, 73, 74, *332*
Wolff, I. A., 4, 7, 8, 11, 12, 13, 15, 16, 18, 19, 20, 21, 24, 25, 26, 27, 29, 50, 51, 62, 64, 90, 166, 168, 173, 275, 290, 299, *307, 312, 318, 319, 322, 325, 328, 329, 332*
Wolken, J. J., 245, *332*
Wolmark, N., 85, *309*
Wood, B. J. B., 7, 78, 271, *324*
Wood, G. C., 274, *314*

Wooltorton, L. S. C., 257, 278, *313, 326*
Wren, J. J., 50, 281, *332*

Y

Yagi, T., 49, 196, 200, *311, 332*
Yalpani, M., 115, *332*
Yamada, M., 101, 170, 171, 193, 207, 211, 221, 240, *332*
Yamamoto, A., 231, *322*
Yang, S. F., 116, 126, 129, 138, 148, 171, 182, 198, 199, *332*
Yano, I., 169, 174, 217, *332*
Yasumoto, K., 231, *322*
Yatsu, L. Y., 207, *313*
Yermanos, D. M., 261, 262, *332*
Youngs, C. G., 82, 87, 88, 291, *329, 332*
Yuan, C., 134, 142, 148, *332*
Yung, K. H., 249, *332*
Yurkowski, M., 85, 86, *309*

Z

Zalik, S., 239, *317*
Zaluska, H., 273, *332*
Zborowski. J., 273, *332*
Zieve, L., 283, *331*
Zill, L. P., 185, *332*
Zimmerman, D. C., 229, 234, 242, 276, *332*
Zimmerman, P. W., 205, *329*
Zobel, H., 25, *328*
Zofcsik, W., 57, *331*
Zschocke, A., 84, *319*

Subject Index

Pages are given in italics when reference is made to a table or figure. Fatty acids (as ions) and plants are listed under their trivial names (e.g. stearate, nasturtium) where these are commonly used; their systematic names (e.g. octadecanoate, *Tropaeolum majus*) are also included. Systematic names are used in all other cases. This index does not contain the names of those higher plants which are mentioned only in Chapter 1 as sources of unusual fatty acids; such plant species are listed by families on pp. 36–39 and by genera on pp. 40–42.

A

Acer plantanoides, ester formation, 259
Acetate, *see also* Acetyl-CoA
 conversion to acetyl-CoA, 97–98, 101
 conversion to fatty acids, 102, 104–119
 elongation of fatty acids, 119–124
 glyoxylate cycle, 201, *203*, *208*
 production by β-oxidation, 201, 202, *202*, *208*
 rate of metabolism, 209
 TCA cycle, 201, *203*, *208*
Acetoacetate, in fatty acid biosynthesis, 98, 104, 106, *106*
Acetoxy acids, occurrence, 26
Acetyl-CoA, *see also* Acetate
 as "starter" for fatty acid biosynthesis, 104–119
 :carbon dioxide ligase, *see* Acetyl-CoA carboxylase
 carboxylation, 102–104, *103*
 conversion to sugars (glyoxylate cycle), 201, *203*, *208*
 function, 98
 oxidation to CO_2 (TCA cycle), 201, *203*, *208*
 production by β-oxidation, 201, *202*, *208*
 reactions, 98, *100*
 structure, 98, *99*
 synthetase, *see* Thiokinase
Acetyl-CoA carboxylase
 decarboxylation of malonyl-CoA, 123
 fatty acid synthetase, *106*
 higher plants, 117, 125–126, 210
 inhibitor of, 125, 126, 127
 isolation, 103
 mechanism of carboxylation, *103*

12

mechanism of transcarboxylation, *104*
 properties, *100*, 102–104, *103*, 116, 117
 rate of fatty acid biosynthesis, 116
Acetylenes, biosynthesis, 164, *164*
Acetylenic acids
 biosynthesis, 164–168
 hydroxyderivatives, 17, 23, *165*, 167, *167*, 168
 occurrence, 17–21
 structures, *10*, *18*, *20*
 "unusual" fatty acid class, 9
ω6-Acetylenic bonds, *see* Crepenynic family
ω9-Acetylenic bonds, *see* Stearolic family
Acidic lipids, isolation, 285, *285*, 287
Aconitase, *203*, 207, 208, *208*
cis-Aconitate, *203*
ACP, *see* Acyl carrier protein
Activation
 of carboxylic acids, 97–101
 of long-chain fatty acids, 98–101, 143, 201, 206, 211
 of malonic acid, 126
 of medium-chain fatty acids, 98, 201
 of short-chain fatty acids, 98–101, 102, 126, 201
Acyl adenylate, as intermediate during acyl-CoA biosynthesis, 98, *101*
Acyl carrier protein (ACP)
 as acyl-transfer cofactor, 98, 105
 biosynthesis, 105
 disulphide form in chloroplast grana, 127
 E. coli, 105, 128
 fatty acid synthetase complex, 111–113

Acyl carrier protein (ACP)—*cont.*
 hydrolysis, 105
 plants, 126, 127, 128
 properties, 105, 128
 structure, *99*, 128
Acyl-CoA : acetyl-CoA   C-acyltransferase, *see* 3-Ketoacyl thiolase
Acyl-CoA synthetase, *see* Thiokinase
Acyl dehydrogenases, 102, 201, *202*,
  209, *210*
Acyl kinase, 101
Acyl transfer between lipids, 250
Acyl transferases
 acyl-CoA : acetyl-CoA,   *see*   3-Keto-
  acyl-thiolase
 acyl-CoA : carnitine, 211
 CoA : ACP : desaturase, *138*, 149
 diglyceride, 179
 fatty acid synthetase, 105, 106
 glyceryl : phosphoryl, 179
 inhibition by sterculate, 156, *156*
 lipid : desaturase, 144, 151
Aerobic pathway, *see also* Desaturation
 unsaturated fatty acid biosynthesis,
  130, *130*, 134–156
"Ageing", 255
 and lipid metabolism, 255, 257, 258
Aleurites montana, seed fatty acids, *64*
Alfalfa, sulpholipid hydrolases, 200
Algae, *see also systematic names*
 blue-green, fatty acids, 7, 32, 33, 78, *78*
 blue-green, lipids, 76, 78
 brown, fatty acids, 32, 76, *77*
 Chrysophyceae, fatty acids, *8*, 9
 effect of growth conditions on composition, 32, *79*, 153
 evolution, 33
 fatty acid biosynthesis, 129, *130*, 136–
  140, 142–146, 153–158, 188–190
 fatty acids, 7, 8, *8*, 9, 31–33, 73–80, *75*,
  77, *78*, *79*, *94*
 fatty acids in individual lipids, 90–93,
  94
 freshwater, fatty acids, *75*, 76
 green, fatty acids, 7–9, 73–76, *75*, *77*
 green, lipids, 73–76
 lipid metabolism, 248–252
 lipids, 73–80
 marine, fatty acids, 8, *8*, 76, *77*
 oxidation of propionate, 212, *212*

red, fatty acids, 76, *77*
 taxonomy, 30–33, *32*, 80
Alkanoates, *see* Fatty acids
Alkenoates, *see* Enoates *and* Fatty
  acids
Alkenyl phospholipids, *see* Plasmalogens
Allenic acids
 as precursors of acetylenic acids, 166
 occurrence, 21, *43*, 44
Allium spp., *see* Onion
Alpha-oxidation, *see* α-Oxidation
Anabaena cylindrica
 fatty acids, 33, *78*
 lipid metabolism, 250, *251*
 monogalactosyl diglyceride, *94*
Anabaena flos-aquae
 fatty acids, *78*
 monogalactosyl diglyceride, *94*
Anabaena variabilis, separation of lipids,
  286
Anacystis nidulans
 fatty acids, 33, *78*
 Hill reaction, 271
 lipid metabolism, 250, 251
 separation of lipids, *286*
Anaerobic Pathway
 and the "plant" pathway, 135, *136*
 unsaturated fatty acid biosynthesis,
  129, 130–134, *131*, *132*, 140
Anchusa, leaf monogalactosyl diglyceride, *94*
Angelate (2-Methyl-*cis*-2-butenoate),
  occurrence, 29
Anomodum rostratus, chloroplast fatty
  acids, *74*
Anteiso-acids
 biosynthesis, 108, 174
 occurrence, 28–29
 structure, 28
Antioxidants, 223, 229, 281
Antirrhinum
 chloroplast fatty acids, *71*
 leaf fatty acids, *69*
Apium petroselinum (*Petroselinum sativum*), *see* Parsley
Apple parenchyma
 lipid biosynthesis in "ageing", 257
 lipids, 60, *61*
 mitochondrial fatty acids, *66*
 phosphatidic acid biosynthesis, 177

Arachidate (Eicosanoate)
  biosynthesis, 102–129, 124, 128, 243
  desaturation, 139
  occurrence, *3,4,4,8,24,63,69,72,74,77*
  structure, *4*
Arachidonate (*cis*-5,*cis*-8,*cis*-11,*cis*-14-
  Eicosatetraenoate)
  autoxidation, 226
  biosynthesis, 141, *141*, 157
  elongation, *123*
  occurrence, 7, *8*, 73, *74*, *75*, *77*, 78, *79*
  oxidation by lipoxygenase, *227*, *230*
  relationship to linoleate, *6*
  structure, *4*
Arachis hypogaea, see Peanut
α-Artemesate, *see* Coriolate
Arthrobacter viscosus, ACP in, 128
Arthrobacterium simplex, α-hydroxylation in, 169
Aster alpinus, seed oil fatty acids, 90
Autoxidation
  fatty acids, 223–226
  linoleate, 224, *224*, *225*
  linolenate, 224, *225*, 226
  mechanism, *223*
  oleate, 224, *224*, *225*
  on storage, 27, 281
  saturated acids, 224
Avocado mesocarp
  absence of glyoxysomes, 209
  ACP, 128
  elongation of fatty acids, 129
  fatty acid biosynthesis, *109*, 125, 126,
    130, 135, 138
  galactosyl diglyceride biosynthesis,
    184
  glyceride biosynthesis, 176
  lipids, 61
  β-oxidation, 125, 206
  oxidative phosphorylation, 125
  TCA cycle activity, 125
  thiokinase, 101, 211

B
Bacillus licheniformis, biosynthesis of
  dienoic acids, 140
Bacteria
  fatty acid biosynthesis, 101, 103–108,
    106, *109*, 117, *129*, 130–134, *130*,
    131, *132*, 140, 159, 174
  12*

fatty acids, 28–29, 124, 140, 153
  α-oxidation, 174, 217
  ω-oxidation, 222
  oxidation of propionate, *212*, 213
Barley (*Hordeum vulgare*)
  fatty acid biosynthesis, 128
  lipid synthesis in leaves, 246
  phospholipase B, 196
Basidiomycetes, acetylenic compounds,
  17
Behenate (Docosanoate)
  biosynthesis, 102–129, 124, 128, 243
  desaturation, 139
  occurrence, *4*, *8*, *24*, *63*, *69*, *71*, *72*, *74*, *77*
  structure, *4*
Beta vulgaris, see Sugar beet
Beta-oxidation, *see* β-Oxidation
Biotin
  function 103, *103*, 104, *104*
  structure, *99*
Blue-green algae, *see* Algae, blue-green
Bolekate (*cis*-13,17-Octadecadien-9,11-
  diynoate)
  biosynthesis, *165*, 166
  occurrence, 19
  structure, *18*
Boleko oil, *see* Isano oil
Boraginaceae, leaf fatty acids, 68
Brachythecium, fatty acids, 7, *74*
Branched-chain acids
  biosynthesis, 173–175
  occurrence, 28–30
  structures, *10*
  "unusual" fatty acid class, 11
Brassica spp. *see* Broccoli, Cabbage,
  Cauliflower, Rape *and* Turnip
Briza spp. seed lipids, 62
Broad bean (*Vicia faba*)
  chloroplast fatty acids, *71*
  dark-grown leaf fatty acids, *69*
  leaf fatty acids, *69*
Broccoli, wax ester synthesis, 190–191
Brown algae (Phaeophyceae), *see* Algae,
  brown
Bryophyta
  leaf fatty acids, 7, 73, *74*
  leaf lipids, 73
Bulbs
  effect of temperature on fatty acids,
    153

Bulbs—*cont.*
  fatty acids, 65
  fatty acids of individual lipids, 90
  lipids, 64
Bush bean, malonate in roots, 126
erythro-Butane-2,3-diol, occurrence, 50
Butyrate
  as intermediate in fatty acid bio-
    synthesis, 104, *106*, 107
  as "starter" for fatty acid biosynthe-
    sis, 108
  biosynthesis by elongation, 122

C

Cabbage (*Brassica oleracea*)
  leaf lipids, 68
  leaf wax biosynthesis, 252–255, *253,
    256*
  leaf waxes, 71, *72*, 252
  phospholipase D, 198
Calcium
  effect on fatty acid biosynthesis, 261
  effect on lipoxygenase, 229, 231
Calendate (*trans*-8,*trans*-10,*cis*-12-
    Octadecatrienoate)
  biosynthesis, 160–164, *161, 162, 163*
  occurrence, 16
  structure, *14*
Candida utilis
  elongation of fatty acids, 124
  α-oxidation, 217
Caprate (Decanoate)
  biosynthesis, 102–129, *131*
  elongation, 121, 132, *132*
  occurrence, 3, *8, 62*
  ω-oxidation, 222
  structure, *4*
Caproate (Hexanoate)
  biosynthesis, 102–129
  elongation, 121, *132*
  occurrence, 3
  structure, *4*
Caprylate (Octanoate)
  biosynthesis, 102–129, *131*
  elongation, 121, 124, 132, *132*
  occurrence, 3, *62*
  ω-oxidation, 222
  structure, *4*
Cardiolipin, *see* Diphosphatidyl glycerol
Carnitine, 122, 211

Carotenoid esters, 53
Carrot, fatty acids, 12, *66*
Carthamus tinctorius, *see* Safflower
Castor (*Ricinus communis*)
  leaf, fatty acids, in individual lipids, 90
  leaf, light-induced changes, 247–248,
    247
  lipases, *see* Castor bean lipases
  seed, fatty acid biosynthesis, *130,*
    147, 169–170, *171*, 239
  seed, fatty acids, 24, 26, *64*, 153
  seed, glyoxysomes, 207, 209
  seed, β-oxidation, 207, 209, 210
  seed structure, 60
  seed, TCA cycle activity, 207
  seed, thiokinase, 101, 211
  seed, triglyceride biosynthesis, 239
Castor bean lipases
  cofactors, 194
  mechanism, 195
  pH optima, 193
  substrate specificity, 195
Catalase
  glyoxysomal, 207, 209, 210
  inhibition by cyanide, 210
Catalpate (*trans*-9,*trans*-11,*cis*-13-
    Octadecatrienoate)
  biosynthesis, 160–164, *162, 163*
  occurrence, 16
  structure, *14*
Cation transport, role of lipids, 269–
    270
Cauliflower
  CDP-diglyceride biosynthesis, 182
  galactosyl diglyceride biosynthesis,
    184
  phosphatidyl glycerol biosynthesis,
    182
Cellulose synthetase, lipid requirement,
    274
Cephalin, *see* Phosphatidyl ethanol-
    amine
Ceramide monoglycoside, *see* Cerebro-
    side
Ceramide phosphate polysaccharide
  isolation, 287
  structure and occurrence, 57
Ceramium rubrum, fatty acids, *77*
Cereals, solvents for lipid extraction,
    281

Cerebronate (2-hydroxytetracosanoate)
biosynthesis, 168, 213–221, 218
occurrence, 23, 24, 91
α-oxidation, 218
Cerebrosides
fatty acids, 23, 24, 90, 91, 215, 220
isolation, 285
occurrence, 61, 66, 76
structure, 54
Chain elongation, see Fatty acids,
elongation
Chaulmoogra oil, fatty acids, 30
Chaulmoograte (13-[2-Cyclopentenyl]-
tridecanoate), occurrence, 30
Chlorella ellipsoidea
glycolipid hydrolases, 200
phospholipase B, 196
Chlorella pyrenoidosa
acetate metabolism, 249
fatty acids, 7, 75
galactosyl diglyceride synthesis, 183
lipid metabolism, 249
Chlorella vulgaris
desaturation, 136, 139, 143, 153
fatty acid biosynthesis, 130, 136–140,
143–146, 153, 157–158, 188–190,
273–274
fatty acid metabolism, 249–250
fatty acids, 75, 76
fatty acids in individual lipids, 90, 92
fatty acids of monogalactosyl diglyceride fractions, 93–95, 94
glycosyl glyceride metabolism, 249,
250, 251
lipid and fatty acid biosynthesis
during greening, 246–247
phospholipid metabolism, 181, 188–
190, 249, 250, 251
separation of lipids, 286
Chlorogloea fritschii, fatty acids, 78
Chlorophenoxyalkanoates, β-oxidation
in plants, 205
Chlorophyceae, see Algae, green
Chlorophyll
in chloroplast development, 246
in chloroplast membranes, 265, 266,
267
interaction with sulpholipid, 246, 265,
267
occurrence, 61

precursor of branched acids, 29, 219,
220
Chloroplasts, 68–71, 245–248
acetyl-CoA carboxylase, 125–126
ACP-disulphide, 127
fatty acid biosynthesis, 109, 127, 135,
143, 158
fatty acids, 12, 71
grana, 264, 265
lamellae, 264–267, 266
light-induced synthesis in, 127
lipids, 70
structure, 264–267
Chrysanthemum floculosum, polyacetylene biosynthesis, 167
Chrysophyceae, see Algae
Citrate
biosynthesis, 98, 203, 208
effect on carboxylase, 116
glyoxylate cycle, 203, 208
regulation of fatty acid biosynthesis,
118
TCA cycle, 203, 208
Citrate synthetase, 100, 203, 207, 208
Citric acid cycle, see Tricarboxylic acid
cycle
Citrullus vulgaris, see Water melon
Claviceps purpurea (Ergot), ricinoleate
biosynthesis, 170–172, 171
Clostridium butyricum, fatty acid biosynthesis, 131, 132, 174, 274
Clover leaf
fatty acids in individual lipids, 90
lipids, 67
Clubmoss, fatty acids, 25, 27
CoA, see Coenzyme A
Cocoa butter
fatty acids, 86
optically active glycerides, 89
triglyceride species, 87
Codium fragile, fatty acids, 77
Coenzyme A (CoA)
as acyl-transfer cofactor, 97
function, 98
structure, 98, 99
thiolesters, biosynthesis, 97–101, 201
Coix lachrima, lipids, 50
Coixenolide, 50, 51
Condensing enzyme, see 3-Ketoacyl
synthetase

Conjugated double bonds
  definition, 14
  formation by lipoxygenase, 227
Conjugated ethylenic acids
  biosynthesis, 160–164, 234
  hydroxyderivatives, 14–17, 23
  occurrence, 14–17
  structures, *10*, *14*
  taxonomy and, 31
  "unusual" fatty acid class, 9
Cordia verbenacea, lipids, 50
Coreopsis lanceolate, polyacetylene biosynthesis, 167
Coriolate   (13-Hydroxy-*cis*-9,*trans*-11-
  octadecadienoate)
  biosynthesis, 160, 161, *161*, 162, *162*,
    163
  enantiomers, 15
  occurrence, 15, 16
  peroxy-analogue, 234
  structure, *14*
Corms
  fatty acids, 65
  fatty acids in individual lipids, 90
  lipids, 64
Corn, *see* Maize
Coronarate (*cis*-9,10-Epoxy-*cis*-12-
  octadecenoate)
  as precursor of acetylenic acids, *167*
  as precursor of conjugated unsatu-
    rated acids, 161, *161*
  configuration, 27
  occurrence, 20, *22*, 27
Corynebacterium diphtheriae, biosynthesis of unsaturated fatty acids, *130*, 139
Corynebacterium simplex, α-hydroxylation, 169, 217
Cotton (*Gossypium hirsutum*)
  flower lipids, 65
  seed, cyclic acids, 30
  seed, effects of nutrients on composi-
    tion, 261
  seed, lipase, 193
  seed triglycerides, isolation of species,
    291
Crambe abyssinica
  seed, fatty acid biosynthesis, 238–239
  seed, lipid biosynthesis, 238–239
Crepenynate (*cis*-9-Octadecen-12-
  ynoate)

biosynthesis, 166, *167*
occurrence, 13, 19, 20, *64*
oxidation by lipoxygenase, 230, *230*
precursor of acetylenic acids, 164–168
  167
precursor of helenynolate, 164, 167,
  167
relationship to linoleate and vernolate, 166
structure, *10*
Crepenynic family of acetylenic acids
  biosynthesis, 164, 167, *167*, 168
  occurrence, 19–20, *20*
  structures, 19
Crepis foetida, seed fatty acids, 19, 27,
  64
Crepis rubra, biosynthesis of acetylenic
  acids, 166
Crithidia fasculata, α-oxidation, 219
Crotonase, *see* Enoyl hydratase
Cruciferae, 37
  positional distribution of acids in
    oils, 82
Cucumber leaf
  fatty acid metabolism, in senescing,
    258
  lipid metabolism, in senescing, 258
Cucumis sativus, *see* Cucumber
Cucurbita pepo, *see* Vegetable marrow
Cuticle
  formation, 277
  function, 277
  lipid metabolism, 252–255
  lipids, 57, 71–73, *72*
Cutin
  biosynthesis,   173,   191–192,   235,
    277–278
  fatty acids, 25, 26
  structure, 57–58, *58*
Cyanolipids, structures, 50, 51, *51*
Cyanophyceae, *see* Algae, blue-green
Cyclic acids, biosynthesis, 173–175
Cyclopentene fatty acids, occurrence,
  30, 31, *64*
13-(2-Cyclopentenyl) tridecanoate, *see*
  Chaulmoograte
13-(2-Cyclopentenyl)-6-tridecenoate,
  see Gorlate
11-(2-Cyclopentenyl)-undecenoate, *see*
  Hydnocarpate

Cyclopropane fatty acids
 biosynthesis, 174–175, 274
 occurrence, 28–30
Cyclopropene fatty acids
 biosynthesis, 175, 220, 274
 effect on unsaturated fatty acid accumulation, 137, 143, 149, 155–156
 occurrence, 28–30, *64*, 68
Cyclotella cryptica, fatty acids, *77*
Cyrtomium falcatum, chloroplast fatty
  acids, *74*
Cytochrome P-450, ω-oxidation, 222

D

Damaged tissue, traumatic acid in,
  28, 173, 235, 276–277
Daucus carota, see Carrot
Deacylated lipids
 analysis, 288–289, 295
 preparation, 288
2,4-Decadienoate
 biosynthesis, 164
 occurrence, 17, 21, *43*, 44
Decanoate, *see* Caprate
1-Decene-1,10-dicarboxylate, *see* Traumatic acid
cis-3-Decenoate, as intermediate in
  unsaturated fatty acid biosynthesis, 128, *131*, 132, *132*
cis-4-Decenoate, occurrence, *11*, 12
cis-2-Decen-4,6,8-triynoate, *see* Dehydromatricaria ester
3-Decynoyl-*N*-acetylcysteamine, as
  inhibitor of 3-hydroxydecanoyl
  thiolester dehydratase, 132
Dehydromatricaria ester (methyl *cis*-2-
  decen-4,6,8-triynoate), occurrence,
  20, *20*
Dehydrophytosphingosine, structure,
  54, *55*
Delta-notation, definition, 5
Densipolate (12-Hydroxy-*cis*-9,*cis*-15-
  octadecadienoate), occurrence, *22*,
  25
Δ3-Desaturase
 biosynthesis of *trans*-3-hexadecen-
  oate, 155, *156*, 157–158
 Δ9-desaturase and, 151, 160
 mechanism of action, *156*, 158
 of *Euglena*, 159, *160*

Δ5-Desaturase
 bacteria, 140
 Δ9-desaturase and, 151, 160
 mechanism of action, 159
 temperature control, 155
Δ7-Desaturase
 Δ9-desaturase and, 151, 159
 transfer of substrate to, 156
Δ9-Desaturase, 134–140
 biosynthesis of palmitoleate, 128, 160
 lack of, in yeast mutant, 146
 mechanism, 146–152, *156*, 158, 159
 of *Chlorella*, 139, 140
 of *Euglena*, 134, 136
 of rat liver, 135
 of soybean, 134
 specificity, 138–140, 151
 substrates, 137, 149
 transfer of substrate to, 149, 151, 156,
  158, 188
Δ10-Desaturase, bacterial, 140
Δ12-Desaturase
 biosynthesis of polyunsaturated
  acids, 141–146, *141*, *144*
 lack of, in animals, 141
 lipid requirement, 144
 mechanism, 151
 of *Chlorella*, 143–146
 specificity, 145
 substrates, 144, 149
 transfer of substrate to, 151, 156, 188
ω6-Desaturase
 biosynthesis of unsaturated acids,
  141–146, *141*, *144*
 lack of, in animals, 141
 lipid requirement, 144
 mechanism, 151
 of *Chlorella*, 143–146
 specificity, 145
 substrates, 144, 149
 transfer of substrate to, 151, 156
Desaturases, *see* Desaturation
Desaturation, *see also* Aerobic pathway
  and the individual Desaturases
 biosynthesis of unsaturated fatty
  acids, 134–156
 control, 152–156
 effect of Triton X, 127
 electron transport during, 148, *148*
 hydrogenation and, 97, 158

Desaturation—*cont.*
  induction of activity, 154–155
  inhibition by anaerobiosis, 130, 134, 135, 142, 146, 158
  lipid requirement, 138, 144, *144*, 151, *156*, 158, 188–189, 273–274
  mechanism, 146–152
  of saturated acids, 134–140
  of unsaturated acids, 137, 140–146, *141*, *144*
  "plant" pathway, 135, *136*
  plants, 135–146
  specificity, 138–140, 145–146
  speculative schemes for mechanism, 149–152, *150*
  stereospecificity of, 139, 146
Detection of lipids during chromatography, 292–293
Diatoms (Bacillariophyceae)
  fatty acids, 9, *77*
  taxonomy, 32, *32*
Dicarboxylic acids
  biosynthesis, 173
  occurrence, 25, 28
  yeasts, 25
Dichapetalum toxicarum, see Ratsbane
Dichlorophenoxyalkanoates, β-oxidation in plants, 205
Dictyostelium discoideum, biosynthesis of 3-hydroxyacids, 169
Digalactosyl diglyceride
  as energy reserve, 275
  biosynthesis, 177, 183, 184, 185, 186, *187*, 241, 246, 257
  chloroplast membranes, 265, 266, 267, 271, 272
  co-oxidation by lipoxygenase, 228, 235
  enzymic hydrolysis, 199–200, 259
  fatty acids, *91*, 92, *92*, *247*
  identification, 289
  in "ageing", 257
  in chloroplast development, 245–246, 247, 265
  in electron transport, 271
  in Hill reaction, 271–272
  in photosynthesis, 249, 250, 251, 265
  in repair mechanisms, 277
  in senescence, 258–259
  in sugar transport, 249, 270, 271
  isolation, 285, 286, 287, 288

  metabolism, 241, 250, 251, 258, 259
  occurrence, 48, *61*, 65, 66, *67*, 68, *70*, 71, 76
  quantitative determination, 295
  structure, 48, *48*
Diglyceride acyl transferase, *see* Acyl transferases
Diglyceride phosphokinase, 177
Diglycerides
  biosynthesis, 178–179
  glycosyl glyceride biosynthesis, 184–185, 186–188, *187*
  isolation, 282, 283, 284, 287
  occurrence, 66
  phosphatidic acid biosynthesis, 177
  phosphoglyceride biosynthesis, 180–183, 186–188, *187*
  structure, *43*, 44
Dihydromalvalate (8,9-Methylene-margarate)
  biosynthesis, 175
  occurrence, 30
Dihydrosphingosine, structure, 54, *55*
Dihydrosterculate (9,10-Methylene-stearate)
  biosynthesis, 175
  occurrence, 29, 30
Dihydroxyacetone phosphate, glyceride biosynthesis from, 177
vic-Dihydroxyacids
  biosynthesis, 173
  occurrence, *22*, 26
11,12-Dihydroxyarachidate, occurrence, *22*, 26
13,14-Dihydroxybehenate, occurrence, *22*, 26
15,16-Dihydroxylignocerate, occurrence, *22*, 26
12,13-Dihydroxyoleate, biosynthesis, 173
10,16-Dihydroxypalmitate, occurrence, *22*, 26
15,16-Dihydroxypalmitate (Ustilate A)
  occurrence, *22*, 26
  α-oxidation, 168
9,10-Dihydroxystearate
  biosynthesis, 173
  occurrence, *22*, 26, 27
10,18-Dihydroxystearate, occurrence, *22*, 26

2,4-Dimethylbehenate
  biosynthesis, 174
  occurrence, 28
α-Dimorphecolate   (9-Hydroxy-*trans*-
  10,*cis*-12-octadecadienoate)
  biosynthesis, 160, 161, *161*, 162, *162*,
  163, 164
  occurrence, 15, 16
  peroxy-analogue, 234
  structure, *14*
β-Dimorphecolate   (9-Hydroxy-*trans*-
  10,*trans*-12-octadecadienoate)
  biosynthesis, 160, *161*, *162*, *163*
  occurrence, 13, 16
  structure, *14*
Diol lipids, structure and occurrence,
  50–51
Diphosphatidyl glycerol (Cardiolipin)
  biosynthesis, 182, *187*
  isolation, 285, 288
  metabolism, 244
  occurrence, *61*, *67*
  structure, *45*, 47
Divinylmethane structure, *see* Methylene-interrupted double bonds
cis-5,*cis*-13-Docosadienoate, occurrence,
  11, 12
cis-13,*cis*-16-Docosadienoate, occurrence, 8, *63*
cis-4,*cis*-7,*cis*-10,*cis*-13,*cis*-16,*cis*-19-
  Docosahexaenoate
  occurrence, 8, 9, *77*
  oxidation by lipoxygenase, *227*, *230*
  relationship to α-linolenate, *6*
  structure, *4*
Docosanoate, *see* Behenate
cis-4,*cis*-7,*cis*-10,*cis*-13,*cis*-16-Docosa-
  pentaenoate, occurrence,*8*, *77*, *79*
cis-7,*cis*-10,*cis*-13,*cis*-16,*cis*-19-Docosa-
  pentaenoate, occurrence, 8, 9, *77*,
  79
cis-7,*cis*-10,*cis*-13,*cis*-16-Docosatetra-
  enoate
  biosynthesis, *123*
  occurrence, 8
cis-10,*cis*-13,*cis*-16,*cis*-19-Docosatetra-
  enoate, occurrence, 8
cis-10,*cis*-13,*cis*-16-Docosatrienoate
  occurrence, 8
  oxidation by lipoxygenase, *227*

cis-5-Docosenoate, occurrence, *11*, 12
cis-9-Docosenoate, biosynthesis, 139
cis-13-Docosenoate, *see* Erucate
2,4-Dodecadienoate
  biosynthesis, 164
  occurrence, 17
Dodecanoate, *see* Laurate
cis-4-Dodecenoate, occurrence, *11*, 12
Dotriacontanoate
  biosynthesis, 102–129, 108, 124
  occurrence, *72*
Douglas fir, seed lipases, 194

E

E. coli, *see* Escherichia coli
cis-8,*cis*-11-Eicosadienoate
  biosynthesis, *141*
  occurrence, 8, *63*, *74*, *75*, *77*, *79*
cis-11,*cis*-14-Eicosadienoate
  biosynthesis, *141*, 157
  occurrence, 7, 8, *63*, *74*, *75*, *77*, *79*
cis-2,*cis*-5,*cis*-8,*cis*-11,*cis*-14,*cis*-17-
  Eicosahexaenoate, occurrence, *79*
Eicosanoate, *see* Arachidate
cis-5,*cis*-8,*cis*-11,*cis*-14,*cis*-17-Eicosa-
  pentaenoate
  occurrence, 8, 9, *74*, *75*, *77*, *79*
  oxidation by lipoxygenase, *227*, *230*
cis-5,*cis*-8,*cis*-11,*cis*-14-Eicosatetraen-
  oate, *see* Arachidonate
cis-5,*cis*-11,*cis*-14,*cis*-17-Eicosatetraen-
  oate, occurrence, *11*, 12
cis-8,*cis*-11,*cis*-14,*cis*-17-Eicosatetraen-
  oate, occurrence, 8
cis-5,*cis*-8,*cis*-11-Eicosatrienoate
  biosynthesis, 141, *141*
  oxidation by lipoxygenase, *227*, *230*
cis-5,*cis*-11,*cis*-14-Eicosatrienoate,   occurrence, *11*, 12
cis-8,*cis*-11,*cis*-14-Eicosatrienoate
  biosynthesis, 122, 141, *141*, 157
  occurrence, 8, *74*, *75*, *77*, *79*
  oxidation by lipoxygenase, *227*, 231,
  232
cis-11,*cis*-14,*cis*-17-Eicosatrienoate
  biosynthesis, 122, *123*
  occurrence, 7, 8, *74*, *75*, *77*, *79*
  oxidation by lipoxygenase, *230*
cis-5-Eicosenoate, occurrence, *11*, 12
trans-6-Eicosenoate, occurrence, *11*, 13

cis-9-Eicosenoate
  biosynthesis, 139
  occurrence, *63, 74*
cis-11-Eicosenoate
  biosynthesis, 129
  occurrence, 5, *8, 63, 74*
cis-13-Eicosenoate
  occurrence, *11*, 13
Elaeis guineensis, *see* Oil palm
Elaidate (*trans*-9-Octadecenoate)
  elongation, 121
  β-oxidation, 204
Electron transport
  in desaturase reaction, 148, *148*
  lipid involvement, 271–272
α-Eleostearate   (*cis*-9,*cis*-11,*trans*-13-
    Octadecatrienoate)
  autoxidation, 226
  biosynthesis, 160, *161, 162*, 163, *163*
  distinction from β-eleostearate, 16
  occurrence, *3*, 15, *64*
  structure, *10, 14*
β-Eleostearate   (*trans*-9,*trans*-11,*trans*-
    13-Octadecatrienoate)
  biosynthesis, 160, *163*
  occurrence, 16
  structure, *14*
Elongation of fatty acids, *see* Fatty
    acids, elongation
Endoplasmic   reticulum,   *see* Micro-
    somes
Energy storage, lipids and, 201, 241–
    242, 274–275
Energy transduction, role of lipids,
    272–273
cis-2-Enoates, β-oxidation, 204
trans-2-Enoates
  as intermediates in chain elongation,
    121, *121*
  as intermediates in fatty acid bio-
    synthesis, *106*, 107, 132, 133
  as intermediates in β-oxidation, 201,
    202, *202*, 204
cis-3-Enoates
  as intermediates in unsaturated fatty
    acid biosynthesis, 132, *132*, 133,
    159
  isomerization to *trans*-2-enoates, 204
  β-oxidation, 204
  γ-oxidation, 205

trans-3-Enoates, *see also trans*-3-Hexa-
    decenoate
  isomerization to *trans*-2-enoates, 204
  β-oxidation, 204
cis-3-Enoyl:*trans*-2-enoyl   isomerase,
    β-oxidation, 204
trans-3-Enoyl:*trans*-2-enoyl isomerase,
    β-oxidation, 204
Enoyl hydratase
  bacterial, 128, 133
  β-oxidation, 202, *202*, 204
  reactions, *100, 106*, 107, 133, 202
Enoyl reductase
  and fatty acid synthetase, 102, *106*,
    107
  NADPH-dependent, 122
Enteromorpha compressa, fatty acids, *77*
Environment, effect on fatty acid
    synthesis, 31, 152, 259–261
Enzymic reactions, role of lipids, 271–
    274
Epoxidation
  of linoleate, *167*, 173
  of oleate, 172
Epoxyacids
  as precursors of unsaturated acids,
    147, 161, *161*
  biosynthesis, 172–173
  occurrence, 27
9,10-Epoxy-18-hydroxystearate, occur-
    rence, 26
cis-15,16-Epoxylinoleate
  biosynthesis, 173
  occurrence, *22*, 27
cis-9,10-Epoxy-*cis*-12-octadecenoate,
    see Coronarate
cis-12,13-Epoxy-*cis*-9-octadecenoate,
    see Vernolate
cis-9,10-Epoxy-12-octadecynoate
  biosynthesis, *167*
  occurrence, 20, *22*, 27
cis-12,13-Epoxyoleate, *see* Vernolate
cis-9,10-Epoxystearate
  biosynthesis, 172
  configuration, 27
  hydration, 173
  metabolism, 172–173
  occurrence, *22*, 27
Equisetum arvense, fatty acids, *74*
Ergot, *see Claviceps purpurea*

Erucate (*cis*-13-Docosenoate)
  accumulation in seeds, 7, 12, *63*, 238–239, 261, 262
  biosynthesis, 129, 261, 262
  occurrence, *3*, 7, *8*, 12, *63*, *77*, *238*
  relationship to oleate, 5, *6*
  structure, *4*
Escherichia coli
  acetyl-CoA carboxylase, 103, 117
  ACP, 105, 128
  fatty acid synthetase, 104–108, *106*, *109*, 117, 133
  glyceride biosynthesis, 176
  palmitoyl thiolesterase, 113–114
  thiokinase, 101
  unsaturated fatty acid biosynthesis, *129*, *131*, 133
Essential fatty acids, of animal nutrition, 142
Estolide triglyceride, 43
  biosynthesis, 172
  ricinoleic acid in, 25
Ethylene generation, lipid requirement, 278
Ethylmalonate, biosynthesis, 125
Euglena gracilis
  *Δ*3-desaturase, hypothetical, 159, *160*
  *Δ*9-desaturase, 134, 137, 144
  fatty acids, 9, 78–80, *79*, 129
  fatty acid biosynthesis, *109*, 129, *130*, 134, 137, *141*, 149, 247, 251
  fatty acids in developing chloroplasts, 246–247
  galactosyl diglyceride biosynthesis, 177, 185
  galactosyl diglycerides as energy reserves, 275
  lecithin biosynthesis, 181, 186
  lipids in developing chloroplasts, 245–247
  wax esters, 246, 275
Euglenaceae, taxonomy, *32*
Euphorbia lagascae, 173
Evolution, of plants, 30, 33
Exocarpate (*trans*-13-Octadecen-9,11-diynoate)
  cis-analogue, see *cis*-13-Octadecen-9,11-diynoate
  biosynthesis, *165*, 166
  occurrence, 19

  structure, *18*
Extraction and storage of plant lipids, 279–281
  prevention of autoxidation, 223, 229, 281
  solvents, 280–281

F

Families of fatty acids, *see*
  Crepenynic family of acetylenic acids
  cis-9-Family of conjugated unsaturated acids
  trans-9-Family of conjugated unsaturated acids
  cis-12-Family of conjugated unsaturated acids
  trans-12-Family of conjugated unsaturated acids
  ω7-Family of unsaturated acids
  *Δ*9-Family of unsaturated acids
  ω9-Family of unsaturated acids
  Linoleic (ω6) family of unsaturated acids
  Linolenic (ω3) family of unsaturated acids
  Stearolic family of acetylenic acids
cis-9-Family of conjugated unsaturated acids
  biosynthesis, 160–164
  occurrence, 15
  structures, *14*
trans-9-Family of conjugated unsaturated acids
  biosynthesis, 160–164
  occurrence, 16
  structures, *14*
cis-12-Family of conjugated unsaturated acids
  biosynthesis, 160–164
  occurrence, 16
  structures, *14*
trans-12-Family of conjugated unsaturated acids
  biosynthesis, 160–164
  occurrence, 16
  structures, *14*
ω3-Family of unsaturated acids, *see* Linolenic family
ω6-Family of unsaturated acids, *see* Linoleic family

ω7-Family of unsaturated acids, occurrence, 13

Δ9-Family of unsaturated acids
definition, 5
occurrence, 5, *8*
relationship to "unusual" acids, 9
structures, *6*

ω9-Family of unsaturated acids
biosynthesis, 123
definition, 5
occurrence, 5, *8*
relationship to "unusual" acids, 9
structures, *6*

Fatty acid composition, determination, 295–300
by argentation-TLC, 298
by GLC, 295–298
location of double bonds, 299
specific methods, 299–303

Fatty acid desaturases, *see* Desaturases

Fatty acid synthetase
complex, 102, 108–113
complex, mechanism, *110, 112*
complex, structure, *109*
E. coli, 104
higher plants, 126–128, 210
individual reactions, 105–108, *106*
pigeon liver, 102, 104
products, *109,* 113
properties, 104–119, *109*
rate 116–119
termination reactions, 113–116
yeast, 104, 108, 114

Fatty acids, *see under individual trivial or systematic name of anion and also*
Families of fatty acids
Free fatty acids
"Major" fatty acids
"Minor" fatty acids
"Unusual" fatty acids
activation, 97–101, 134, 201
autoxidation, 223–226
biosynthesis, 96–175
comparison between anabolic and catabolic pathways, 119, *120*
degradation, 201–221
elongation, 119–124, 128–129, 160
"essential", 142
industrial production, 2, *3*

light-induced biosynthesis, 127, 245–248
lipoxygenase-catalysed oxidation, 226–235
monounsaturated, biosynthesis, 130–140
occurrence, 1–42
oxidation, 201–235
α-oxidation, 213–221
β-oxidation, 201–213
ω-oxidation, 221–223
polyunsaturated, biosynthesis, 140–146
saturated, biosynthesis, 102–129
structures, 1–42
unsaturated, biosynthesis, 129–156

Fatty alcohols, 4
acetylenic, 20, 21
cuticular waxes, 26, *72*

Fatty methyl esters
identification, 300–303
isolation, 284
preparation, 295–296
separation, 295–300

Ferns
fatty acids, 7, 73, *74*
lipids, 73

Flax seed
fatty acid biosynthesis, 125
fatty acids, effect of temperature, 153
fatty acids, phytoglycolipid, 23, *24*
genetic control of composition, 261–262
linoleate hydroperoxide isomerase, *228, 234*
linseed oil, fatty acids, *86*
linseed oil, triglyceride species, *87*
lipid biosynthesis, 242, 243
lipid metabolism, *242,* 243
lipoxygenase, 229, 234
β-oxidation, 205

Flowers
fatty acids, 29
lipids, 65

18-Fluoro-oleate, occurrence, 23
16-Fluoropalmitate, occurrence, 23
Forget-me-not, leaf fatty acids, *69*
Free fatty acids
formation during extraction, 279
isolation, 282, 283, 287

Free radicals
  autoxidation, 223, *224*
  chain reaction, *223*
  lipoxygenase action, 231, 232, *232*,
    233, *233*
French bean (*Phaseolus vulgaris*), leaf
  fatty acids, 23, *24*
Fruit coat
  cuticle, 71–73
  lipids, 1, 60
Fruits *see also* Fruit coat
  fat accumulation, 236–241
  structure, 59
Fucus platycarpus, fatty acids, *77*
Fucus serratus, fatty acids, *77*
Fucus vesiculosus, fatty acids, *77*
Fumarase, *203, 208,* 209
Fumarate, *203, 208*
Function of lipids in membranes, 263–
  274
Fungi
  acetylenes, biosynthesis, 164
  acetylenic compounds, 17, 19
  allenic compounds, 21
  ricinoleic acid, 25
  unsaturated fatty acid biosynthesis,
    142
Furans, substituted
  biosynthesis, 173
  occurrence, 20, 21

G

Galactolipid, *see* Mono- *and* Di-
  galactosyl diglycerides
Galactosyl diglycerides, *see* Mono- *and*
  Di-galactosyl diglycerides
Genetic control of fat composition, 31,
  261–262
Geranyl stearate, 53
Germination, *see* Seed, germination
Ginkgo biloba
  leaf fatty acids, 73, *74*
  leaf lipids, *67*
Glycerides, *see* Diglycerides Glycosyl
  glycerides, Phosphoglycerides *and*
  Triglycerides
Glycerol ester hydrolase, *see* Lipase
Glycerol kinase, in glyceride biosyn-
  thesis, 176–177
L-α-Glycerophosphate

glyceride biosynthesis, 176–177, *187*
  reaction with thiolesters, 177
Glyceryl-phosphoryl acyl transferase,
  see Acyl transferases
Glycolate oxidase, glyoxysomal, 207
Glycolipid hydrolases, 199–200, 259
Glycosyl diglycerides (*see also* Mono-
  and Di-galactosyl diglycerides, *and*
  Sulphoquinovosyl diglyceride)
  as energy reserves, 275
  biosynthesis, 183–190
  enzymic hydrolysis of, 199–200, 259
  fatty acid conversions in, 188–190,
    272
  fatty acids in individual lipids, 89–
    93
  in "ageing", 257
  in chloroplast development, 245–
    247
  in maturing seeds, 237–241
  in membrane permeability, 270
  in membrane structure, 265
  in photosynthesis, 250–252, 265
  in repair mechanisms, 277
  in senescence, 258, 259
  isolation, 282, 284
  molecular species, isolation, 292
  positional distribution of fatty acids,
    93–95, *94*
  quantitative determination, 295
  separation from triglycerides, 282
  structure, 48–49, *48*
Glyoxylate, *203, 208,* 209
Glyoxylate cycle
  conversion of acetate to sugar, 201,
    203, 206, 207, *208*
  maximum activity in seedlings, 207,
    209
Glyoxysomes
  castor bean endosperm, 207
  enzymes, 207, *208*
  β-oxidation, 207–211
Gorlate   (13-[2-Cyclopentenyl]-6-tri-
    decenoate), occurrence, 30
Grape berry, cuticular waxes, *72*
Grasses
  leaf fatty acids, *69*
  leaf lipids, *67*
Green algae, *see* Algae, green
Greening cells, 245–248

H

Helenium bigelowii, seed fatty acids, 12, 64

Helenynolate (9-Hydroxy-*trans*-10-octadecen-12-ynoate)
biosynthesis, 164, 167, *167*
occurrence, 12

Helianthus spp., *see* Sunflower

cis-4,*cis*-7,*cis*-10,*cis*-13,*cis*-16-Heneicosapentaenoate, occurrence, 9

Hentriacontanoate
biosynthesis, 108, 124, 220
occurrence, *72*

Heptacosanoate
biosynthesis, 108, 124, 220
occurrence, 4, *72*

cis-5,*cis*-9-Heptadecadienoate, occurrence, *11*, 12

8,11-Heptadecadienoate, biosynthesis, 145

cis-9,*cis*-12-Heptadecadienoate
biosynthesis, 145
occurrence, *8*
oxidation by lipoxygenase, *230*

trans-10,16-Heptadecadien-8-ynoate
biosynthesis, *165*, 166–168
occurrence, 19
structure, *18*

Heptadecanoate, *see* Margarate

cis-8,*cis*-11,*cis*-14-Heptadecatrienoate (Norlinolenate)
biosynthesis, 160, 220, 243
occurrence, 8, 13, 243

cis-9,*cis*-12,*cis*-15-Heptadecatrienoate
biosynthesis, 243
occurrence, *8*, 243

8-Heptadecenal, intermediate in α-oxidation of oleate, 214

8-Heptadecenoate, intermediate in α-oxidation of oleate, 214

cis-8-Heptadecenoate, desaturation, 145

cis-9-Heptadecenoate
biosynthesis, 139, 243
occurrence, 8, *65*, *79*

trans-10-Heptadecen-8-ynoate, *see* Pyrulate

Heredopathia atactica polyneuritiformis, *see* Refsum's disease

Hexacosanoate
biosynthesis, 102–129, 124, 128, 243
occurrence, 4, *72*
α-oxidation, 217, 243

cis-17-Hexacosenoate, occurrence, 7

cis-5,*cis*-9-Hexadecadienoate, occurrence, *11*, 12

cis-5,*cis*-10-Hexadecadienoate, biosynthesis, 140

cis-7,*cis*-10-Hexadecadienoate
biosynthesis, 145
occurrence, *8*, 13, *74*, *75*, 77, *78*, *79*, *94*

cis-9,*cis*-12-Hexadecadienoate
biosynthesis, 145
occurrence, 5

1,16-Hexadecandioate
biosynthesis, 222
occurrence, 25, 28

Hexadecane, as carbon source, 25

Hexadecanoate, *see* Palmitate

cis-4,*cis*-7,*cis*-10,*cis*-13-Hexadecatetraenoate
biosynthesis, 251
occurrence, 7, *8*, *75*, 77, *79*

cis-4,*cis*-7,*cis*-10-Hexadecatrienoate, occurrence, *8*

cis-7,*cis*-10,*cis*-13-Hexadecatrienoate
occurrence, 7, *8*, 13, 68, *69*, 71, *74*, *75*, 77, *79*, *92*, *94*
relationship to α-linolenate, *6*
structure, *4*

trans-3-Hexadecenoate
biosynthesis, *156*, 157–159, 188, 189
effect of sterculate on biosynthesis, 155
mechanism of biosynthesis, 158
occurrence, 9, *11*, 12, *64*, 68, *69*, *75*, 77, *78*, *79*, *92*, *247*, 248
reduction to palmitate, 158, 189
specific location on phosphatidyl glycerol, 12, 90
structure, *10*, 157

cis-5-Hexadecenoate
biosynthesis, 140, 159
occurrence, *11*, 12

trans-5-Hexadecenoate, occurrence, *11*, 12

cis-6-Hexadecenoate, occurrence, *11*, 13

trans-6-Hexadecenoate, occurrence, *11*, 13

cis-7-Hexadecenoate
  biosynthesis, *131*, 139
  intermediate in α-oxidation of oleate,
    214
  occurrence, 8, 9, 13, *63*, *65*, *66*, 68, *69*,
    71, 73, *74*, *75*, 77, *79*, *91*, *94*, *247*
8-Hexadecenoate, biosynthesis, 159
cis-9-Hexadecenoate, *see* Palmitoleate
10-Hexadecenoate, biosynthesis, 140,
  159
cis-11-Hexadecenoate
  biosynthesis, 159
  occurrence, 73, *74*
Hexanoate, *see* Caproate
10-(2-Hexylcyclopropyl)decanoate, *see*
  Lactobacillate
Higher plants, *see also individual
  systematic or trivial names*
  index (Chapter 1) of species by
    genera, 40–42
  index (Chapter 1) of species by
    families, 36–39
Hill reaction
  algal composition and, *32*
  involvement of fatty acids, 251, 252,
    271, 272
  involvement of lipids, 271, 272
Holly
  leaf, fatty acids, *69*
  leaf, fatty acids in individual lipids,
    90
Hordeum vulgare, *see* Barley
Hydnocarpate (11-[2-Cyclopentenyl]-
  undecenoate)
  occurrence, 30, *64*
  structure, *10*
Hydnocarpus wightiana, seed fatty
  acids, 30, *64*
Hydration, of epoxy acids, 173
Hydrocarbons
  acetylenic, 20, 21
  biosynthesis, 253-256
  energy reserves, 62, 275
  isolation, 283, 284
  leaf wax, 26, *72*
Hydrogenation of fatty acids, 97
Hydroperoxide isomerase, *see* Isomerase
Hydroperoxides, fatty acid
  decomposition, 226, 234
  enzymic formation, *225*, 226–235

in cutin biosynthesis, 173, 191–192,
  277–278
isomerization, *228*, 234, 276
non-enzymic formation, 223–226,
  224, *225*
15-Hydroperoxy-*cis*-8,*cis*-11,*trans*-13-
  eicosatrienoate, biosynthesis by
  lipoxygenase, 231
9-Hydroperoxy-*trans*-10,*cis*-12-octade-
  cadienoate
  biosynthesis by lipoxygenase, *225,*
    227, *228*, 229
  configuration of biosynthetic, 231
  degradation, 234
  isomerization, 234
13-Hydroperoxy-*cis*-9,*trans*-11-octade-
  cadienoate
  biosynthesis by lipoxygenase, *225*,
    227, *228*, 229
  configuration of biosynthetic, 230
  degradation, 234
  isomerization, 234
13-Hydroperoxy-*cis*-9,*trans*-11,*cis*-15-
  octadecatrienoate, biosynthesis
  from linolenate, *225*, *227*
Hydrophobic interactions
  in fatty acid biosynthesis, 114–116
  in fatty acid desaturation, 139, 145,
    151
  in membrane structure, 263–269
  lipoxygenase, 231
3-Hydroxyacyl dehydrase *see* Enoyl
  hydratase *and* 3-Hydroxydecanoyl
  thiolester dehydrase
3-Hydroxyacyl dehydrogenase, β-oxidation, 202, *202*, 204
3-Hydroxyacyl epimerase, β-oxidation,
  204
2-Hydroxyalkanoates
  biosynthesis, 168–169, 220
  intermediates in α-oxidation, *215*,
    215–219
  occurrence, 22, 23–24, *24*
3-Hydroxyalkanoates
  as intermediates in anabolic and
    catabolic reactions, 119, *120*
  as intermediates in chain-elongation,
    121, *121*, 123
  as intermediates in fatty acid bio-
    synthesis, *106*, 107, 119, 132

3-Hydroxyalkanoates—*cont.*
as intermediates in β-oxidation, 201, 202, *202*
biosynthesis, 106, 107, 113–114, 133, 169
epimerization, 204
2-Hydroxyarachidate
biosynthesis, 168
occurrence, 23, *24, 91*
3-Hydroxyarachidate
biosynthesis, 169
occurrence, *22,* 24
2-Hydroxybehenate
biosynthesis, 168
occurrence, 23, *24, 91*
8-Hydroxybolekate (8-Hydroxy-*cis*-13,17-octadecadien-9,11-ynoate)
biosynthesis, *165,* 166–168
occurrence, 19
structure, *18*
4-Hydroxycaprate, formation in plants, 221
7-Hydroxycaprate, occurrence, *22,* 25
3-Hydroxydecanoyl thiolester dehydrase, anaerobic biosynthesis of unsaturated fatty acids, *131,* 132–134
14-Hydroxy-*cis*-11-eicosenoate, *see* Lesquerolate
8-Hydroxyexocarpate (8-Hydroxy-*trans*-13-octadecen-9,11-diynoate)
biosynthesis, *165,* 166–168
occurrence, *18*
structure, *18*
2-Hydroxyheneicosanoate
biosynthesis, 168
occurrence, 23, *24, 91*
7-Hydroxy-*trans*-10,16-heptadecadien-8-ynoate
biosynthesis, *165,* 166–168
occurrence, 19
structure, *18*
7-Hydroxy-*trans*-10-heptadecen-8-ynoate, *see* 7-Hydroxypyrulate
2-Hydroxyhexacosanoate
biosynthesis, 168, 217
occurrence, 23, *24, 91*
8-Hydroxyisanate, *see* Isanolate
9-Hydroxy-10-keto-*cis*-12-octadecen-

oate, biosynthesis, *228,* 234
9-Hydroxy-12-keto-*trans*-10-octadecenoate, biosynthesis, *228,* 234
13-Hydroxy-10-keto-*trans*-11-octadecenoate, biosynthesis, *228,* 234
13-Hydroxy-12-keto-*cis*-9-octadecenoate, biosynthesis, *228,* 234
2-Hydroxylactobacillate, biosynthesis, 169
α-Hydroxylation, biosynthesis of 2-hydroxyacids, 168, 213–220
12-Hydroxylaurate
biosynthesis, 222
occurrence, *22,* 25
2-Hydroxylignocerate, *see* Cerebronate
2-Hydroxylinolenate
biosynthesis, 168, 220
occurrence, 24
2-Hydroxy-13-methylmyristate, biosynthesis, 169, 174
3-Hydroxymyristate, 133, 169
11-Hydroxymyristate, occurrence, *22,* 25
14-Hydroxymyristate
biosynthesis, 222
occurrence, *22,* 25
2-Hydroxynonadecanoate
biosynthesis, 168
occurrence, 23, *24, 91*
9-Hydroxy-*trans*-10,*cis*-12-octadecadienoate, *see* α-Dimorphecolate
9-Hydroxy-*trans*-10,*trans*-12-octadecadienoate, *see* β-Dimorphecolate
11-Hydroxy-*cis*-9,*cis*-12-octadecadienoate, as hypothetical intermediate, 161
12-Hydroxy-*cis*-9,*cis*-15-octadecadienoate, *see* Densipolate
13-Hydroxy-*cis*-9,*trans*-11-octadecadienoate, *see* Coriolate
13-Hydroxy-*trans*-9,*trans*-11-octadecadienoate
biosynthesis, 160, *163*
occurrence, 16
structure, *14*
8-Hydroxy-*trans*-11,17-octadecadien-9-ynoate
biosynthesis, *165,* 166–168
occurrence, 19
structure, *18*

8-Hydroxy-*cis*-13,17-octadecadien-9,
11-ynoate, *see* 8-Hydroxybolekate
8-Hydroxy-9,11-octadecadiynoate
biosynthesis, *165*, 166–168
occurrence, 19
structure, *18*
18-Hydroxy-*cis*-9,*trans*-11,*trans*-13-
octadecatrienoate, *see* α-Kamlolenate
8-Hydroxy-*cis*-13-octadecen-9,11-
diynoate
biosynthesis, *165*, 166–168
occurrence, 19
structure, *18*
8-Hydroxy-*trans*-13-octadecen-9,11-
diynoate, *see* 8-Hydroxyexocarpate
8-Hydroxy-17-octadecen-9,11-diyn-
oate, *see* Isanolate
9-Hydroxy-*cis*-12-octadecenoate, occurrence, *22*, 25
10-Hydroxy-*cis*-12-octadecenoate, biosynthesis, 172
12-Hydroxy-*cis*-9-octadecenoate, *see*
Ricinoleate
8-Hydroxy-*trans*-11-octadecen-9-
ynoate, *see* 8-Hydroxyximenynate
9-Hydroxy-*trans*-10-octadecen-12-
ynoate, *see* Helenynolate
8-Hydroxy-5,6-octadienoate, occurrence, 21, *43*, 44
12-Hydroxyoleate, *see* Ricinoleate
17-Hydroxyoleate, occurrence, *22*, 25
18-Hydroxyoleate, occurrence, *22*, 25,
26
2-Hydroxypalmitate
biosynthesis, 168, 214–218
formation from phytosphingosine,
218
in wax biosynthesis, 255
intermediate in α-oxidation of palmitate, 214–218, *215*, *216*
occurrence, 23, *24*, *91*
3-Hydroxypalmitate
biosynthesis, 169
occurrence, *22*, 24
8-Hydroxypalmitate, occurrence, *22*, 25
9-Hydroxypalmitate, occurrence, *22*, 25
10-Hydroxypalmitate
biosynthesis, 172
degradation, 221

11-Hydroxypalmitate, occurrence, *22*,
25
15-Hydroxypalmitate, occurrence, *22*,
25
16-Hydroxypalmitate
biosynthesis, 222
occurrence, *22*, 25, 26
12-Hydroxypalmitoleate, occurrence,
22, 25
2-Hydroxypentacosanoate
biosynthesis, 168
occurrence, 23, *24*, *91*
2-Hydroxyphytanate, intermediate in
α-oxidation of phytanate, 219
3-Hydroxypropionate, propionate oxidation, 211
7-Hydroxypyrulate (7-Hydroxy-*trans*-
10-heptadecen-8-ynoate)
biosynthesis, *165*, 167–168
occurrence, 19
structure, *18*
2-Hydroxystearate
biosynthesis, 168, 214, 217
intermediate in α-oxidation of stearate, 213–218
occurrence, 23, *24*, *91*
3-Hydroxystearate
biosynthesis, 169
occurrence, *22*, 24
9-Hydroxystearate
conversion to oleate, 147
occurrence, *22*, 25
10-Hydroxystearate
biosynthesis, 172
conversion to oleate, 147
17-Hydroxystearate
biosynthesis, 221–223
occurrence, *22*, 25
18-Hydroxystearate
biosynthesis, 221–223
occurrence, *22*, 25, 26
2-Hydroxysterculate
biosynthesis, 168, 175, 220
occurrence, 24, 30
2-Hydroxytetracosanoate, *see* Cerebronate
2-Hydroxytricosanoate
biosynthesis, 168
occurrence, 23, *24*, *91*
Hypnum cupressiforme, fatty acids, *74*

I

Ilex spp., *see* Holly
Index of plants, *see* Higher plants
Inhibition
  of arachidonate biosynthesis, 157
  of chain elongation, 254
  of crepenynate biosynthesis by ster-
    culate, 166
  of desaturation by anaerobiosis, 130,
    134, 135, 146, 158
  of desaturation by sterculate, 137,
    138, 143, 149, 155–156, 158,
    159
  of desaturation in leaves, 260
  of glyoxysomal catalase by cyanide,
    210
  of 3-hydroxydecanoyl thiolester de-
    hydratase, 132–133
  of lipolysis, 280
  of lipoxygenase by alcohols, 231
  of lipoxygenase by thiol reagents,
    229
  of oxidation, *see* Antioxidants
  of α-oxidation by imidazole, *215*
  of TCA cycle, 206
Iodine value, measure of unsaturation
  of fatty acids, 97, 152
Iron
  cofactor in desaturation, 134, 148,
    148, 159
  cofactor in α-oxidation, 218, *218*
  effect on fatty acid biosynthesis, 260,
    261
  ferredoxin, 134, 136, *148*
  rubredoxin, 222
Isanate (17-Octadecen-9,11-diynoate),
  18, 19, *165*, 166
Isanolate (8-Hydroxy-17-octadecen-9,
  11-diynoate), *18*, 19, *165*, 166
Isano oil, fatty acids, 19, 26, 27
Iso-acids
  biosynthesis, 108, 174
  occurrence, 28–29
  structure, 28
Isocitrate, *203*, *208*, 209
Isocitrate dehydrogenase
  fatty acid synthetase, 102
  mitochondrial, *208*, 209
  TCA cycle, *203*, *208*
Isocitrate lyase, *203*, 207, 208, *208*, 209

Isoleucine, as precursor of branched-
  chain acids, 174
Isomerase, enoyl, β-oxidation, 204
Isomerase, linoleate hydroperoxide,
  228, 234, 276

J

Jacarate (*cis*-8,*trans*-10,*trans*-12-Octa-
  decatrienoate)
  biosynthesis, 160, *161*, *162*, *163*
  occurrence, 16
  structure, *14*

K

Kalenchoe crenata, lipid metabolism,
  246–247
Kamala oil, fatty acids, 15
α-Kamlolenate (18-Hydroxy-*cis*-9,
  trans-11,*trans*-13-octadecatrieno-
  ate)
  biosynthesis, 160, 221–223
  occurrence, 15
  structure, *14*
Ketoacids, occurrence, *22*, 26
3-Ketoacyl reductase, fatty acid synthetase, *106*, 107
3-Ketoacyl synthetase (Condensing
  enzyme)
  fatty acid synthetase, 106, *106*
  rate-determining step in fatty acid
    biosynthesis, 117
3-Ketoacyl thiolase
  β-oxidation, 202, *202*, 204
  reactions, *100*, 202
3-Ketoalkanoates
  accumulation during fatty acid bio-
    synthesis, 115, 116
  as intermediates in anabolic and
    catabolic reactions, 119, *120*
  as intermediates in chain elongation,
    121, *121*, 123
  as intermediates in β-oxidation, 201,
    202, *202*
4-Keto-α-eleostearate (4-Keto-*cis*-9,
  trans-11,*trans*-13-octadecatrieno-
  ate), *see* α-Licanate
α-Ketoglutarate, *203*, *208*
α-Ketoglutarate dehydrogenase, *203*,
  208

17-Keto-*cis*-20-hexacosenoate, occurrence, *22*, 26

Ketols
in waxes, *72*
isomerization of hydroperoxides, *228*, 234

Ketones
acetylenic, 20, 21
in waxes, 26, *72*

19-Keto-*cis*-22-octacosenoate, occurrence, *22*, 26

9-Keto-*trans*-10,*trans*-12-octadecadienoate
biosynthesis, 160
occurrence, 16
structure, *14*

9-Keto-11,13-octadecadienoate, biosynthesis, 234

13-Keto-9,11-octadecadienoate, biosynthesis, 160, 234

13-Keto-*trans*-9,*trans*-11-octadecadienoate
biosynthesis, 160
occurrence, 16
structure, *14*

4-Keto-*cis*-9,*trans*-11,*trans*-13,*cis*-15-octadecatetraenoate, *see* 4-Keto-α-parinarate

4-Keto-*cis*-9,*trans*-11,*trans*-13-octadecatrienoate, *see* 4-Keto-α-eleostearate

13-Keto-*trans*-9-octadecenoate, occurrence, 16

2-Ketopalmitate, intermediate in α-oxidation of palmitate, 214–217, *215*, *216*

8-Ketopalmitate, occurrence, 25

4-Keto-α-parinarate (4-Keto-*cis*-9, *trans*-11,*trans*-13,*cis*-15-octadecatetraenoate)
biosynthesis, 160
occurrence, 15
structure, *14*

15-Keto-*cis*-18-tetracosenoate, occurrence, *22*, 26

3-Ketothiolase, *see* 3-Ketoacyl thiolase

β-Ketothiolase, *see* 3-Ketoacyl thiolase

L

Lactuca sativa, *see* Lettuce

Laurate (Dodecanoate)
biosynthesis, 102–129
elongation, 121, 132, *132*
occurrence, 2, *3*, *8*, *62*, *72*
ω-oxidation, 222
structure, 2, *3*

Laurel (*Laurus nobilis*), exocarp lipids, 61

Leaves
cuticular lipids, 71–73, *72*
fatty acid biosynthesis, 127, *130*, 135, 142, 143
fatty acid function, 264–268
fatty acids, 7, 23, *24*, 25, 29, 68–69, *69*, 247–248, *248*
fatty acids in individual lipids, 90–93
lipid biosynthesis, 176–192
lipid function, 264–278
lipid metabolism, 248–252
lipids, 23, *24*, 66–68
lipoxygenase, 227
α-oxidation in, 214–217, *215*

Lecithin, *see* Phosphatidylcholine

Leptospira canicola, desaturation in, 159

Lesquerolate (14-Hydroxy-*cis*-11-eicosenoate)
biosynthesis, 122, 129, 148, *171*
occurrence, *22*, 25

Lettuce (*Lactuca sativa*)
chloroplasts, fatty acid biosynthesis, *109*, 127
leaf lipids, *67*

Leucine, as precursor of branched-chain acids, 174

α-Licanate (4-Keto-*cis*-9,*trans*-11, *trans*-13-octadecatrienoate)
biosynthesis, 160
occurrence, 15
structure, *14*

Lichens, fatty acids, 26

Light, effect on fatty acid biosynthesis, 127, 247–248, *247*, 251
effect on lipid biosynthesis, 184, 245–247, 251

Lignocerate (Tetracosanoate)
biosynthesis, 102–129, 124, 128, 243
occurrence, 4, *8*, *24*, *63*, *69*, *72*, *74*
structure, *4*

Linoleate (*cis*-9,*cis*-12-Octadecadienoate)

Linoleate—*cont.*
  as precursor of acetylenic acids, 166, 167, *167*
  as precursor of conjugated unsatu-rated acids, 161, *161*, *162*, *163*
  autoxidation, 224, *224*, *225*, 226
  biosynthesis, 141–146, *141*, 188, 262
  desaturation, 141, *141*, 143, 145, 188
  effect of sterculate on biosynthesis, 156
  effect of temperature on accumula-tion, 152–154
  elongation, 121, *141*, 157
  in cutin biosynthesis, 192
  in repair mechanisms, 276, 277
  occurrence, 2, *3*, 8, 17, *62*, *63*, *64*, *65*, *66*, 68, *69*, *71*, *74*, 75, *77*, *78*, *79*, *91*, *92*, *94*, *247*
  β-oxidation, 204
  oxidation by lipoxygenase, 192, 226–231, *227*, *228*, *230*, 276
  relationship to "minor" acids, 5, *6*
  relationship to vernolate and cre-penynate, 166, *167*
  structure, 2, *3*
Linoleic (ω6) family of unsaturated acids
  biosynthesis, 123, 141, *141*
  definition, 5
  occurrence, 7, *8*
  relationship to "unusual" acids, 9
  structures, *6*
α-Linolenate (*cis*-9,*cis*-12,*cis*-15-Octa-decatrienoate)
  autoxidation, 224, *225*, 226
  biosynthesis, 141–143, *141*, 145, 188, 251, 260, 261
  effect of temperature on accumula-tion, 152
  elongation, 121, 122, *123*, 141
  in chloroplast development, 247
  in cutin biosynthesis, 192
  in ethylene generation, 278
  in Hill reaction, 251, 267, 271–272
  in photosynthesis, 251–252, 267, 271–272
  in repair mechanisms, 276, 277
  occurrence, 2, *3*, 8, 17, *62*, *63*, *65*, *66*, 68, *69*, *71*, *74*, *75*, *77*, *78*, *79*, *91*, *92*, *94*, *247*
  α-oxidation, 160, 168, 220

oxidation by lipoxygenase, 192, *227*, *230*, 276
relationship to "minor" acids, 5, *6*
structure, 2, *3*
γ-Linolenate (*cis*-6,*cis*-9,*cis*-12-Octa-decatrienoate)
  biosynthesis, 141, *141*, 157
  elongation, 121, 122, 141, *141*, 157
  occurrence, 7, *8*, 68, *69*, *74*, *78*, *94*, 271
  oxidation by lipoxygenase, *227*, *230*
  relationship to linoleate, *6*
  structure, 2, *3*, *4*, *6*
α-Linolenic (ω3) family of unsaturated acids
  biosynthesis, 123, 141, *141*
  definition, 5
  occurrence, 7, *8*
  relationship to "unusual" acids, 9
  structures, *6*
α-Linolenic pathway of polyenoic acid biosynthesis, 141, *141*, 142
  taxonomy and, 32–33
γ-Linolenic pathway of polyenoic acid biosynthesis, 141, *141*, 142, 157
  arachidonate biosynthesis, 157
  taxonomy and, 32–33
Linseed, *see* Flax
Linum usitatissimum, *see* Flax
Lipase (pancreatic), in structural studies, 81–89
Lipases (plant)
  cellular location, 194
  in germination, 241–245
  in senescence, 258–259
  in spherosomes, 207
  sources, 193–195
Lipid-bound substrates
  biosynthesis of cyclic acids in plants, 175
  biosynthesis of lactobacillate, 174
  biosynthesis of tuberculostearate, 174
  biosynthesis of unsaturated fatty acids, 138, 144, 151, *156*, 158, 188–189
Lipophilic interactions, *see* Hydro-phobic interactions
Lipoproteins, 264, 265, 269
Liposomes, permeability, 270
Lipoxidase, *see* Lipoxygenases

Lipoxygenases
  amino-acids of, 232
  biosynthesis of conjugated unsaturated acids, 161
  co-oxidation, 228, 234
  effect of calcium, 229, 231
  ethylene generation, 278
  function, 234–235
  in senescence, 259
  isoenzymes, 228
  mechanism, 231–233, *232, 233*
  multiplicity of, 228, 229
  oxidation of fatty acids, 223, *225*, 226–235
  repair mechanisms, 276
  sources, 227, 229
  soybean, 227–231
  stereospecificity, 230
  substrate specificity, 227–230, *227, 230*
Liver
  desaturase, 135, 147, 155
  fatty acid synthetase, 102, 104, 109, 117
Lolium perenne, see Grasses
Lycopodium complanatum, fatty acids, 25, 27
Lysophosphatidic acid, formation, 177
Lyso-phospholipids
  acylation, 190
  degradation, 196
  formation, 196, 257
  in "ageing", 257
  in mitochondrial swelling, 273
  involvement in fatty acid biosynthesis, 138, 144, 151, *156*, 158, 174, 175, 188–189
  occurrence, 50
  structure, *50*

M

Maidenhair tree, *see Ginkgo biloba*
Maize (*Zea mays*)
  germ, linoleate hydroperoxide isomerase, *228*, 234
  germ, lipoxygenase, 229, 234
  leaf, lipids, *67*
  seed, fatty acid biosynthesis, 236–238
  seed, fatty acids, *63, 86*
  seed, lipid biosynthesis, 236–238, *237*

"Major" fatty acids
  biosynthesis, 96–156, 157
  definition, 1, 2
  distribution in commercial fats, *3*
  elongation, 123, 128, 141, 157
  "minor" acids, relationship to, 3, 5, *6*
  occurrence, 2
  structures, 2, *3*
  "unusual" acids, relationship to, 9
Malabar tallow, optical asymmetry, *89*
Malate, *203, 208*
Malate dehydrogenase, *203*, 207, *208*
Malate synthetase, *203*, 207, 208, *208*, 209
Malonate, *see also* Malonyl-CoA
  activation by thiokinase, 126
  elongation of fatty acids dependent on, 121–122, *121*
  elongation of fatty acids independent of 122–123
  fatty acid biosynthesis, 97, 102, 104–119
  inhibition of TCA cycle, 206
  propionate oxidation, 211
Malonyl-CoA, *see also* Malonate
  as intermediate in fatty acid biosynthesis, 102, 104–119
  biosynthesis, 102–104, 125, 126
  transcarboxylation, 125
Malonyl-CoA decarboxylase, *see* Acetyl-CoA carboxylase
Malvalate (7-[2-Octylcycloprop-1-enyl] heptanoate)
  biosynthesis, 175, 220
  occurrence, 29, 30
Margarate (Heptadecanoate)
  biosynthesis, 108, 213–219, 220, 243
  desaturation, 139
  occurrence, 4, *8, 66, 77*
  structure, *4*
Mastigocladus laminosus, fatty acids, *78*
Matteucia struthiopteris, fatty acids, *74*
Membranes
  cation transport, 269–270
  chloroplast, 264–267
  composition, 263, 264
  lipid function in, 263–274
  permeability, 269–271
Methionine, as precursor of branched-chain acids, 174, 175

Methyl esters, *see* Fatty methyl esters

Methylation (reductive), during biosynthesis of branched-chain acids, 174

2-Methyl-*cis*-2-butenoate, *see* Angelate

2-Methyl-*trans*-2-butenoate, *see* Tiglate

2-Methylbutyrate, as "starter" for fatty acid synthetase, 174

3-Methylbutyrate, as "starter" for fatty acid synthetase, 174

Methylene-interrupted double bonds
  action of lipoxygenase, 227, 229
  biogenesis, 141, 145
  definition, 2, 14
  distinction from conjugated, 14
  symbols for, 5
  ubiquity, 2, 97

8,9-Methylene-8-octadecen-17-ynoate, *see* Sterculynate

cis-9,10-Methylenepalmitate, occurrence, 29

10-Methylenestearate, metabolism, 174

cis-9,10-Methylenestearate, *see* Dihydrosterculate

cis-11,12-Methelenestearate, *see* Lactobacillate

16-Methylheptadecanoate, occurrence, 29

Methylmalonate, biosynthesis, 125, 174, 205, 211

10-Methylmargarate, occurrence, 28

12-Methylmyristate, occurrence, 28

13-Methylmyristate
  as precursor of 2-hydroxy-13-methylmyristate, 169
  occurrence, 28

8-Methylpalmitate, occurrence, 28

10-Methylpalmitate, occurrence, 28

14-Methylpalmitate, occurrence, 29

10-Methylstearate, *see* Tuberculostearate

Methylstearates, desaturation, 139–140, 151

12-Methyltridecanoate, occurrence, 29

Microsomes (Endoplasmic reticulum)
  desaturation, 135
  elongation of fatty acids, *120*, 121, *121*
  phosphoglyceride biosynthesis, 258
  thiokinase, 207, 211

Minerals, effect on fatty acid biosynthesis, 260–261

"Minor" fatty acids
  biosynthesis, 123, 129, 141, *141*, 157
  definition, 1, 3, 5
  "major" acids, relationship to, 3, 5, *6*
  occurrence, 2–9, *8*
  saturated, 3–4, *4*, *8*
  structures, 2–9, *4*, *6*, *8*
  unsaturated, 5–9, *4*, *6*, *8*
  "unusual" acids, relationship to, 9

Mitochondria
  electron transport, 272
  elongation of fatty acids, *120*, *121*, 122
  enzymes of, 207, *208*
  fatty acid biosynthesis, 122, 135
  fatty acids, 65–66, *66*
  lipid metabolism, 255, 257, 258
  lipids, 65–66, 268–269
  membranes, lipids in, 269
  oxidation of fatty acids, 203, 205–207
  structure, 268–269, *268*

Mitochondrial swelling, role of lipids, 272–273

Mnium, fatty acids, 7, *74*

Molecular species of lipids, separation, 289–292

Monoacyl phospholipids, *see* Lysophospholipids

Monodus subterraneus, extraction, 280

Monogalactosyl diglyceride
  as energy reserve, 275
  biosynthesis, 177, 183–188, *187*, 190, 257
  chloroplast development, 245–247
  chloroplast membranes, 265–267, *266*, 271, 272
  enzymic hydrolysis, 199–200, 259
  fatty acid interconversions in, 188, 190, 272, 274
  fatty acids, *91*, 92, *92*, 93, *94*, 95, *247*, 292
  identification, 289
  in "ageing", 257
  in electron transport, 271
  in Hill reaction, 271–272
  in photosynthesis, 250–252, 265
  in repair mechanisms, 277
  in senescence, 258–259
  in sugar transport, 249, 270, 271

Monogalactosyl diglyceride—*cont.*
  isolation, 285–288
  metabolism, 245–247, 249–252, 258, 259
  molecular species, 292
  quantitative determination, 295
  structure, 48, *48*
Monogalactosyl diglyceride, 6-*O*-acyl
  biosynthesis, 49
  isolation, 287
  occurrence, 49
  structure, 48, *48*, 49
Monogalactosyl monoglyceride, metabolism, 190
Monoglycerides
  in triglyceride biosynthesis, 179
  isolation, 282, 283
Monohydroxyacids
  biosynthesis, 168–172
  occurrence, 23–25
Monophosphoinositide, *see* Phosphatidyl inositol
Mosses
  fatty acids, 7, 73, *74*
  lipids, *67*, 73
Mung bean (*Phaseolus radiatus*), phospholipid metabolism, 244–245
Murraya koenigii
  seed composition, 62
  seed germination, 275
Mycobacterium phlei
  ACP, 112
  biosynthesis of branched-chain acids, 174
  elongation of fatty acids, 122
  fatty acid synthetase, *109*
  polyunsaturated fatty acid, 140
  unsaturated fatty acid biosynthesis, *130*, 134, 149
Mycocerosic acids, biosynthesis, 124
Mycolic acids
  biosynthesis, 174
  occurrence, 28
  structure, 28
Myosotis spp. *see* Forget-me-not
Myristate (Tetradecanoate)
  biosynthesis, 102–129
  desaturation, 135, 139
  elongation, 121, 132, *132*
  occurrence, 2, *3*, *8*, *62*, *66*, *69*, *71*, *72*, *74*, *75*, *77*, *79*, 92
  α-oxidation, 214
  ω-oxidation, 222
  structure, 2, *3*
Myristoleate (*cis*-9-Tetradecenoate)
  biosynthesis, 135, 139
  occurrence, 5, *8*
Myxosarcina chroococcoides, fatty acids, *78*

N

Naphthylalkanoates, β-oxidation in plants, 205
Narcissus bulb
  fatty acids, *65*
  fatty acids, effect of environment, 153
  fatty acids in individual lipids, 90
  lipids, 64
Nasturtium, seed fatty acids, 7
Navicula pelliculosa, fatty acids, *75*
Nervonate (*cis*-15-Tetracosenoate)
  occurrence, 7, *63*
  relationship to oleate, 5, *6*
Neutral lipids
  fractionation, 283–284, 288
  isolation, 281–282, 286
Nicotiana tabacum, *see* Tobacco
Nitella, fatty acids, *75*
Nitrogen, effect on fatty acid biosynthesis, 260, 261
Nitzschia closterium, fatty acids, *77*
Nonacosanoate
  biosynthesis, 108, 124, 220
  occurrence, 4, *72*
cis-9-*cis*-12-Nonadecadienoate
  biosynthesis, 145
  occurrence, *8*
10,13-Nonadecadienoate, biosynthesis, 145
Nonadecanoate
  desaturation, 139
  occurrence, *8*
cis-9-Nonadecenoate, biosynthesis, 139
cis-10-Nonadecenoate, desaturation, 145
Nonanoate, occurrence, *8*
Non-conjugated ethylenic acids
  biosynthesis, 157–160
  occurrence, 11–13, *11*
  structures, *10*, *11*
  "unusual" fatty acid class, 9

Non-esterified fatty acids, *see* Free fatty acids
Norlinolenate, *see* *cis*-8,*cis*-11,*cis*-14-Heptadecatrienoate

O

Ochromonas danica, fractionation of lipids, 288
Octacosanoate
  biosynthesis, 102–129, 108, 124
  occurrence, 4, *72*
cis-19-Octacosenoate, occurrence, 7
13,15-Octadecadien-9,11-diynoate
  biosynthesis, *165*, 166
  occurrence, 19
  structure, *18*
cis-13,17-Octadecadien-9,11-diynoate, *see* Bolekate
5,6-Octadecadienoate, *see* Labellenate
cis-5,*cis*-9-Octadecadienoate, occurrence, *11*, 12
trans-5,*cis*-9-Octadecadienoate, occurrence, *11*, 12
cis-5,*cis*-11-Octadecadienoate, occurrence, *11*, 12
cis-6,*cis*-9-Octadecadienoate
  biosynthesis, 141, *141*
  occurrence, *8*
cis-9,*cis*-12-Octadecadienoate, *see* Linoleate
cis-9,*trans*-12-Octadecadienoate
  as precursor of acetylenic acids, 166, *167*
  as precursor of conjugated unsaturated acids, 162, *162, 163*
  occurrence, *11*, 13
  β-oxidation, 204
  oxidation by lipoxygenase, 227, *227*
trans-9,*cis*-12-Octadecadienoate
  occurrence, 13
  β-oxidation, 204
trans-9,*trans*-12-Octadecadienoate
  as precursor of conjugated unsaturated acids, 162, *162, 163*
  occurrence, *11*, 13
  β-oxidation, 204
  oxidation by lipoxygenase, 227, *227*
trans-10,*trans*-12-Octadecadienoate
  biosynthesis, 160, 162, *162, 163*
  occurrence, 16

structure, *14*
cis-13,*cis*-16-Octadecadienoate, oxidation by lipoxygenase, 229
15,17-Octadecadien-9,11,13-triynoate
  biosynthesis, *165*, 166
  occurrence, 19
  structure, *18*
cis-9,*cis*-14-Octadecadien-12-ynoate
  biosynthesis, 167
  occurrence, 20
trans-11,*trans*-13-Octadecadien-9-ynoate
  biosynthesis, *165*, 166
  occurrence, 19
  structure, *18*
trans-11,17-Octadecadien-9-ynoate
  biosynthesis, *165*, 166
  occurrence, 19
  structure, *18*
9,11-Octadecadiynoate
  biosynthesis, *165*, 166
  occurrence, 19
  structure, *18*
1,18-Octadecandioate, biosynthesis, 221, 222–223
Octadecanoate, *see* Stearate
trans-3,*cis*-9,*cis*-12,*cis*-15-Octadecatetraenoate, occurrence, *11*, 12
cis-6,*cis*-9,*cis*-12,*cis*-15-Octadecatetraenoate
  occurrence, *8, 9, 68, 69, 75, 77, 79, 94*
  oxidation by lipoxygenase, *230*
  relationship to α-linolenate, *6*
  structure, *4*
cis-9,*trans*-11,*trans*-13,*cis*-15-Octadecatetraenoate, *see* α-Parinarate
trans-2,*cis*-9,*cis*-12-Octadecatrienoate
  biosynthesis, 160
  occurrence, 11, *11*
trans-3,*cis*-9,*cis*-12-Octadecatrienoate, occurrence, *11*, 12
5,6-*trans*-16-Octadecatrienoate, occurrence, 21
trans-5,*cis*-9,*cis*-12-Octadecatrienoate, occurrence, *11*, 12
cis-6,*cis*-9,*cis*-12-Octadecatrienoate, *see* γ-Linolenate
cis-8,*trans*-10,*trans*-12-Octadecatrienoate
  biosynthesis, 160, *163*

cis-8,*trans*-10,*trans*-12-Octadecatri-
enoate—*cont.*
  occurrence, 16
  structure, *14*
trans-8,*trans*-10,*cis*-12-Octadecatri-
  enoate, *see* Calendate
trans-8,*trans*-10,*trans*-12-Octadecatri-
  enoate
  biosynthesis, 160, *163*
  occurrence, 16
  structure, *14*
cis-9,*trans*-11,*cis*-13-Octadecatrienoate,
  see Punicate
cis-9,*trans*-11,*trans*-13-Octadecatri-
  enoate, *see* α-Eleostearate
trans-9,*trans*-11,*cis*-13-Octadecatri-
  enoate, *see* Catalpate
trans-9,*trans*-11,*trans*-13-Octadecatri-
  enoate, *see* β-Eleostearate
cis-9,*cis*-12,*cis*-15-Octadecatrienoate,
  see α-Linolenate
cis-13-Octadecen-9,11-diynoate
  biosynthesis, *165*, 166
  occurrence, 19
  structure, *18*
trans-13-Octadecen-9,11-diynoate, *see*
  Exocarpate
17-Octadecen-9,11-diynoate, *see* Isanate
trans-3-Octadecenoate, occurrence, *11*,
  12
cis-5-Octadecenoate
  biosynthesis, 159
  occurrence, *11*, 12, 13
trans-5-Octadecenoate, occurrence, *11*,
  12, 13
cis-6-Octadecenoate, *see* Petroselinate
trans-6-Octadecenoate, occurrence, 13,
  21
7-Octadecenoate, occurrence, 13, 73
cis-9-Octadecenoate, *see* Oleate
trans-9-Octadecenoate, *see* Elaidate
cis-11-Octadecenoate, *see* Vaccenate
15-Octadecen-9,11,13-triynoate
  biosynthesis, *165*, 166
  occurrence, 19
  structure, *18*
17-Octadecen-9,11,13-triynoate
  biosynthesis, *165*, 166
  occurrence, *18*
  structure, *18*

cis-9-Octadecen-12-ynoate, *see* Crepenynate
trans-11-Octadecen-9-ynoate, *see*
  Ximenynate
17-Octadecen-9-ynoate
  biosynthesis, *165*, 166
  occurrence, 19
  structure, *18*
9-Octadecynoate, *see* Stearolate
Octanoate, *see* Caprylate
7-(2-Octylcycloprop-1-enyl)heptan-
  oate, *see* Malvalate
8-(2-Octylcycloprop-1-enyl)octanoate,
  see Sterculate
7-(2-Octylcyclopropyl)heptanoate, *see*
  Dihydromalvalate
8-(2-Octylcyclopropyl)octanoate, *see*
  Dihydrosterculate
Oil-palm fruit
  kernel, fatty acid biosynthesis, 239–
    240, 243
  kernel, fatty acids, 62, *62*, 82
  kernel, lipid biosynthesis, 239–240,
    240
  kernel, lipid metabolism, 243–244
  kernel, lipids, 62
  mesocarp, fatty acid biosynthesis,
    239–240
  mesocarp, fatty acids, 62
  mesocarp, lipid biosynthesis, 239–240
  mesocarp, lipids, 61, 62
  mesocarp oil, optical asymmetry, *89*
  structure, 59, *60*
Olea europaea, *see* Olive
Oleate (*cis*-9-Octadecenoate)
  accumulation in seeds, 238–239
  autoxidation, 224, *224*, *225*, 226
  biosynthesis, 126, *132*, 134, *134*, 135,
    138, *138*, 139
  conversion to crepenynate, 166, *167*
  conversion to 10-methylstearate, 174
  conversion to sterculate, 175
  desaturation, 141–146, *141*, 188
  distinction from petroselinate, 12
  effect of temperature on accumula-
    tion, 152
  elongation, 121, 129, 261, 262
  genetic control of biosynthesis, 262
  in cutin biosynthesis, 192
  in erucate biosynthesis, 129, 261, 262

Oleate—*cont.*
  occurrence, 2, *3*, *8*, *62*, *63*, *64*, *65*, *66*,
    68, *69*, *71*, 73, 74, *75*, *77*, *78*, *79*, *91*,
    92, *94*, *247*
  α-oxidation, 214
  β-oxidation, 204, 207
  relationship to "minor" acids, 5, *6*
  structure, 2, *3*
Olive (*Olea europaea*)
  exocarp fatty acids, *86*
  exocarp lipids, 61
Omega-notation, definition, 5
Omega-oxidation, *see* ω-Oxidation
Onion, lecithin biosynthesis, 181
Onoclea sensibilis, fatty acids, *74*
Optical asymmetry, triglycerides, 85–
  89
Oxalacetate, *203*, *208*
α-Oxidation
  animals, 218–219
  biosynthesis of "unusual" acids and,
    160, 168, 175, 220
  function, 219–221
  germinating seeds, 213–214, 243
  leaves, 214–217
  mechanism, 216–217, *216*
  microorganisms, 217–218
  of acetylenic acids, *165*, 167
  of fatty acids, 213–221
  of phytanate in animals, 213
  of propionate in rust uredospores,
    212, *212*
  rate, relative to β- and ω-oxidation,
    217
  senescence, 259
  stereochemistry, 216
  systems, properties of, *215*
  wax biosynthesis, 253, 254
β-Oxidation
  biosynthesis of fatty acids and, 102,
    119, *120*
  biosynthesis of "unusual" acids and,
    160
  glyoxysomes, 207, 209–211
  mitochondria, 205–207
  of fatty acids, 201–213, *202*
  of fatty acids in plants, 205–213, *208*
  of unsaturated fatty acids, 204–205
  rate, 204, 217
  reversal of, 102, 122

senescence, 259
γ-Oxidation, of *cis*-3-enoates, 205
δ-Oxidation, of fatty acids, 221
ω-Oxidation
  of fatty acids, 173, 221–223
  rate relative to α- and β-oxidation,
    217
  stereochemistry, 223
  wax biosynthesis, 253
Oxidation of fatty acids
  by light, *see* Photo-sensitized
  by lipoxygenase, *see* Lipoxygenase
  non-enzymic, *see* Autoxidation
Oxirans, substituted
  acetylenic, 20
  acids, 22, 27
  biosynthesis, 172–173
Oxygen
  effect on fatty acid composition, 31,
    153
  necessity for desaturation, 130, 134,
    135, 142, 146, 158
  necessity for glyoxysomal β-oxida-
    tion, 209
  necessity for α-oxidation, *215*
Oxygenase, as model for desaturase,
  147, 148

P

Palm kernel oil, *see* Oil palm
Palm oil, *see* Oil palm
Palmitate (Hexadecanoate)
  as precursor of *trans*-3-hexadecen-
    oate, 158, 189
  as precursor of 2-hydroxypalmitate,
    168
  biosynthesis, 102–129, *131*
  desaturation to *cis*-5-hexadecenoate,
    159
  desaturation to palmitoleate, 134,
    139, 188, 189
  elongation, 121, *123*, 124
  incorporation into lipids, 158
  occurrence, 2, *3*, 4, *8*, *62*, *63*, *64*, *65*,
    66, 68, *69*, *71*, *72*, *74*, *75*, *77*, *78*, *79*,
    91, *92*, *94*, *247*
  α-oxidation, 214, 215
  ω-oxidation, 222
  structure, 2, *3*
  wax biosynthesis, 253–255

Palmitoleate (*cis*-9-Hexadecenoate)
  bacterial, 13, 128
  biosynthesis, 128, *131*, 134, 135, 139,
    159
  elongation, 121, 128, 160
  occurrence, 5, *8*, *63*, *65*, *66*, 68, *69*, *71*,
    73, *74*, *75*, *77*, *79*, *91*, *94*, *247*
  relationship to oleate, 5, *6*
  structure, *4*
Palmitoyl thiolesterase, 113–114
α-Parinarate (*cis*-9,*trans*-11,*trans*-13,
    cis-15-Octadecatetraenoate)
  biosynthesis, 160, *163*
  occurrence, 15
  structure, *14*
Parsley (*Petroselinum sativum*), seed
    fatty acids, 12, *64*
Pea (*Pisum sativum*)
  galactosyl diglyceride biosynthesis,
    185
  leaf, α-oxidation, 214–217
  leaf waxes, composition, 73
  phosphatidyl ethanolamine biosyn-
    thesis, 180
  phosphoglyceride metabolism, 244
  phospholipase D, 198
  propionate oxidation, 211
Peanut (*Arachis hypogaea*)
  germinating cotyledons, fatty acid
    biosynthesis, 124
  glyoxysomes, 209
  maturing cotyledons, fatty acid bio-
    synthesis, 124
  α-oxidation, 213, 214
  β-oxidation, 205, 206, 207, 209
  propionate oxidation, 211
  seed fatty acids, *63*, *86*
  seed, lipase, 193
  thiokinase, 101, 211
Pear (*Pyrus communis*)
  flavour component, 17
  mitochondria, fatty acids, *66*
Pentacosanoate
  biosynthesis, 108, 124, 220
  occurrence, 4, *72*
cis-9,*cis*-12-Pentadecadienoate, biosyn-
    thesis, 145
Pentadecanal, intermediate in α-oxida-
    tion of palmitate, 214, *215*, 216
Pentadecanoate

biosynthesis, 108, 213–218, 220, 243
  desaturation, 139
  occurrence, 4, *8*, *77*, *79*
cis-7-Pentadecenoate, biosynthesis, 139
cis-9-Pentadecenoate
  biosynthesis, 139
  occurrence, *8*
Perennial rye grass, *see* Grasses
Persea americana, *see* Avocado
Petroselinate (*cis*-6-Octadecenoate)
  distinction from oleate, 12
  occurrence, *11*, 12, 13, 21, *64*
  structure, *10*
Petroselinum sativum (*Apium petro-
    selinum*), *see* Parsley
Phaeodactylum tricornutum, fatty acids,
    77
Phaeophyceae, *see* Algae, brown
Phaseolus multiflorus, *see* Runner bean
Phaseolus radiatus, *see* Mung bean
Phaseolus vulgaris, *see* French bean
Phenoxyalkanoates,   β-oxidation   in
    plants, 205
Phosphatidic acid
  biosynthesis, 176–178, 241
  in glyceride biosynthesis, 178–179,
    187
  in mitochondrial swelling, 273
  isolation, 285
  metabolism, 241, 244
  occurrence, *61*, 66, *70*
  structure, 44, *45*
Phosphatidyl choline (Lecithin)
  biosynthesis, 180–182, 186, *187*, 188,
    241, 257, 258
  fatty acid composition, 90, *91*, *92*, 93,
    247
  identification, 289
  in "ageing", 257, 258
  in chloroplast development, 247
  in photosynthesis, 250–251
  in senescence, 258, 259
  involvement in fatty acid biosynthe-
    sis, 71, 144, 188, 250, 274
  isolation, 282, 285, 286, 288
  metabolism, 241, 244, 258, 259
  occurrence, *61*, 65, 66, *67*, 68, *70*, 71,
    76
  quantitative determination, 294–295
  structure, 44, *45*

Phosphatidyl ethanolamine
as lecithin precursor, 181
biosynthesis, 179–180, *187*, 241, 257
fatty acids, *92*, 93
identification, 289, 293
in "ageing", 257
in senescence, 258, 259
isolation, 282, 285, 286, 288
metabolism, 241, 244, 258, 259
occurrence, *61*, 65, 66, *67*, 68, *70*, 76
quantitative determination, 294–295
structure, 44, *45*
Phosphatidyl ethanolamine, *N*-acyl
biosynthesis, 241
isolation, 287
metabolism, 241, 244
structure, 44, *45*
Phosphatidyl ethanolamine, *N*-methyl,
in algae, 181
Phosphatidyl glycerol
biosynthesis, 182–183, *187*, 241, 257
fatty acids, 90, *92*, *247*
identification, 289
in "ageing", 257
in cation transport, 270
in chloroplast membrane, 266
in photosynthesis, 249, 250, 251
involvement in fatty acid biosynthesis, *156*, 158, 189, 274
isolation, 285, 286, 287, 288
metabolism, 241, 249, 250, 251, 257
molecular species, 93
molecular species, separation, 292
occurrence, 61, 66, *67*, *70*, 71, 76
quantitative determination, 294–295
structure, *45*, 46
Phosphatidyl inositol
biosynthesis, 183, *187*, 241
fatty acids, *92*
identification, 289
in mitochondrial swelling, 272–273
in photosynthesis, 249
isolation, 285, 286, 287, 288
metabolism, 241, 244, 249
occurrence, *61*, 66, *67*, *70*, 76
quantitative determination, 294–295
structure, *46*, 47
Phosphatidyl serine
biosynthesis, 180, *187*, 241
detection, 293

metabolism, 241, 244
occurrence, *61*, 66
structure, 44, *45*
Phosphoenolpyruvate, *203*, *208*
Phosphoglycerides
biosynthesis, 179–183, 185–190, 236–
241, 257, 258
enzymic hydrolysis, 195–199
fatty acids in different, 89–93
fatty acids within classes of, 93–95
in "ageing", 257
in electron transport, 272
in maturing seeds, 236–241
in membranes, 263–274
in repair mechanisms, 277
in wax ester biosynthesis, 191,
255
isolation, 283, 284, 287
molecular species, isolation, 292
quantitative determination, 294–295
separation from triglycerides, 282
structures, 44–48, *45*, *46*
Phospholipase A, 190
in wax ester biosynthesis, 191
occurrence, 196
role in mitochondrial swelling, 273
sites of attack by, *196*
Phospholipase B
occurrence, 196
sites of attack by, *196*
Phospholipase D
deactivation, 280
in phosphoglyceride biosynthesis,
180, 181, 182, 198
inhibition of, 197
occurrence, 197, 198
physiological importance, 198, 259
sites of attack by, *196*
stability, 199
substrate specificity, 197
Phospholipases, 195–199
classification, 196
4′-Phosphopantetheine
as binding site of ACP, *99*, 105
as binding site of CoA, *99*
as binding site of fatty acid synthetase complex, 111–113
structure, *99*
Photo-sensitized oxidation, of fatty
acids, *225*, 226

Photosynthesis, lipid metabolism during, 248–252
Photosynthetic tissues, *see* Algae, Chloroplasts, *and* Leaves
Phylogeny, 31
Phytanate (3,7,11,15-Tetramethyl-palmitate)
  occurrence, 29
  oxidation, 213, 219, 220
Phytoglycolipid
  fatty acids, 23, *24*
  isolation, 283, 287
  occurrence, 55-57, 66
  structure, 55–57, *56*
Phytol
  degradation, 213, 219, 220
  precursor of branched acids, 29
Phytoplankton, *see* Algae
Phytophingosine, structure, 54, *55*
Pine (*Pinus*)
  black, leaf fatty acids, 68
  β-oxidation in seedlings, 206
  Scots, lipase, 194
Pinus spp., *see* Pine
Pisum sativum, *see* Pea
Pityrosporum ovale, fatty acids, 25
Plant families, index, 36–39
Plant genera, index, 40–42
"Plant" pathway, unsaturated fatty acid biosynthesis, 135–137, *136*, 138, 156
Plant storage tissues
  fatty acids, *65*
  lipids, 64
Plants, *see* Higher plants, Algae *and under individual systematic or trivial names*
Plasmalogens
  occurrence, 50
  structure, *50*
Plocamium coccineum, fatty acids, 77
Poikilotherms, effect of temperature on fatty acid composition of, 153, 155
Polyacetylenes, biosynthesis, *164*, 167, *167*
Polyhydroxyacids
  biosynthesis, 173
  occurrence, *22*, 25, 26–27
Polyketides

biosynthesis, 166
cyclization, 164
structure, 17
Positional distribution of acids
  glycolipids, 93–95, 190
  phospholipids, 93, 190
  triglycerides, 81–89
Potato (*Solanum tuberosum*)
  fatty acid biosynthesis in "ageing", 255, 257
  fatty acids, 65
  glycolipid hydrolases, 200
  lipid metabolism in "ageing", 257
  lipids, 64
  α-oxidation, 214
  phospholipases, 196–197, 198
  thiokinase, 101
Pristanate (2,6,10,14-Tetramethyl-pentadecanoate)
  formation, by α-oxidation of phytanate, 219
  occurrence, 29
  β-oxidation, 219
Propionate
  as precursor of branched-chain acids, 123–125, 174
  as precursor of odd-numbered acids, 108, 220
  as "starter" for fatty acid biosynthesis, 108, 220
  carboxylation, 174
  elongation of fatty acids by, 123–125
  oxidation, 211–213, *212*
  production by γ-oxidation, 205
Prostaglandins
  biosynthesis and lipoxygenase, 234
  biosynthesis in animals, 142
Proteins, interaction with lipids, 263, 264, 265, *266*, 268, 269
Prototheca zopfii, oxidation of propionate, 212, *212*
Pseudomonas oleovorans, ω-oxidation in, 222
Pteridophyta
  fatty acids, 73, *74*
  lipids, 73
Puccinia graminis
  biosynthesis of epoxyacids, 172
  epoxyacids in, 27

Punicate (*cis*-9,*trans*-11,*cis*-13-Octa-
decatrienoate)
  biosynthesis, 160, *161*, *162*, 163, *163*
  occurrence, 15
  structure, *14*
Pyrulate (*trans*-10-heptadecen-8-
ynoate)
  biosynthesis, 165, 167–168
  occurrence, 18
  structure, *18*
Pyrus communis, see Pear

Q
Quantitative determination of lipids,
  294–295

R
Rape seed (*Brassica napus*)
  effect of nutrients on composition, 261
  effect of temperature on composition,
    153
  fatty acid biosynthesis, 129, 238–239,
    261, 262
  fatty acids, 7, 13, *63*, *86*
  genetic control of composition, 129,
    261, 262
  lipid biosynthesis, 238–239, 261
Ratsbane, fatty acids, 22–23
Reagents for detection of lipids and
  their derivatives, 283, 285, 289,
  292–293
Red algae, *see* Algae, red
Refsum's disease, failure of α-oxidation
  in animals, 219
Repair mechanisms, requirement for
  lipid, 276–277
Rhodomelia subfusca, fatty acids, *77*
Rhodophyceae, *see* Algae, red
Rhus spp., *see* Sumach
Ricinoleate (12-Hydroxy-*cis*-9-Octa-
decenoate)
  as intermediate in desaturation of
    oleate, 147, *150*, 152, 172
  biosynthesis, 169–172, 239–240
  dehydration to linoleate, 147, *150*
  elongation to lesquerolate, 129, *171*
  metabolism, 148, *171*
  occurrence, *3*, *22*, 24–25, *64*
  β-oxidation, 172
  structure, *10*

Ricinus communis, *see* Castor
Roots
  acetylenic acids, 19, 20
  branched acids, 29
  Rubredoxin, bacterial ω-oxidation, 222
Runner bean (*Phaseolus multiflorus*)
  glycosyl glyceride hydrolases, 199–200
  leaves, fatty acids in individual lipids,
    23, *24*, 90, *91*
  leaves, fatty acids within individual
    lipids, 93
Rust uredospores
  biosynthesis of epoxy acids, 172
  epoxyacids, 27
  oxidation of propionate, 212, *212*

S
Saccharomyces cerivisiae, nutritional
  requirements of mutants, 146
Saccharum officinarum, *see* Sugar cane
Safflower (*Carthamus tinctorius*)
  fatty acid biosynthesis, 143, 154, 240,
    244, 261, 262
  genetic control of composition, 261,
    262
  propionate oxidation, 211
  seed, fatty acids, 31
Santalbate, *see* Xymenynate
Santalum acuminitum, acetylenic acid
  biosynthesis, 166
Sapium sebiferum (syn. *Stillingia sebifera*)
  fatty acids, 17, 21
  tetra-acid triglycerides, *43*, 44, 290
Scenedesmus obliquus
  fatty acids, 7, *75*, 76
  lipids, 57
  phospholipase A, 196, 200
  sulpholipid hydrolase, 200
Seed
  classification based on fatty acid
    composition, 33–36, *34*, *35*
  effect of environment on composition,
    31, 152–154, 260–261
  fatty acid biosynthesis, 126–130, 138,
    169, 170–173
  fatty acids, 1–30, 63, 64, 236–241,
    260–261
  germination, 241–245, 274–275
  lipase activity, 242

Seed—*cont.*
  lipids, 62–64, 236–241
  lipoxygenase, 227
  maturation, 236–241
  α-oxidation, 213–214, *215*
  β-oxidation, 207
  taxonomy, 2, 33
Seedling development, 241–245
Senescence, lipid and fatty acid metabolism, 258–259
Separation of plant lipids, 281–295
  acetone fractionation, 282
  ethanol fractionation, 282
  preliminary bulk fractionation, 281–283
  separation of molecular species, 289–292
  separation of neutral lipids from complex lipids, 282
Simmondsia californica
  seed composition, 62
  seed germination, 275
Sinapis alba (Brassica hirta), lipase, 194
Skipped methylene structure, *see* Methylene-interrupted double bonds
Smut, ustilic acids from, 26
Solanum esculentum, see Tomato
Solanum tuberosum, see Potato
Soja hispida, see Soybean
Soybean (*Soja hispida*)
  desaturase, 134, 138
  effect of environment on composition, 154, 261
  fatty acid biosynthesis, 127, 154
  fatty acids, *63, 86*, 261
  lipoxygenase, 227–231, 234
  phosphoglyceride biosynthesis, 240–241
  phospholipase D, 198
  triglyceride biosynthesis, 179, 239
Sphagnum, fatty acids, *74*
Spherosomes, site of lipid storage, 207
Sphingolipids, structure and distribution, 23, *24*, 53–57
Sphingosines
  fractionation and analysis, 300
  structure, 53–55
Spinach leaf (*Spinacia oleracea*)
  ACP, 128

chloroplast lamellae lipids, *70*
fatty acid biosynthesis, *109*, 127, *130*, 149
fatty acids, *69*
fatty acids in individual lipids, 90, *92*
galactosyl diglyceride biosynthesis, 184, 185
metabolism of monoacyl glycerides, 190
monogalactosyl diglycerides, *94*
phosphatidic acid biosynthesis, 176–177
phosphatidyl glycerol species, 292
Spinacia oleracea, see Spinach
Spirulina platensis, fatty acids, 7, 33, *78*
Stability of lipids
  autoxidation, 223–226
  during chromatography, 293–294
  during extraction, 279–281
Stearate (Octadecanoate)
  biosynthesis, 102–129, *106, 123, 131*
  desaturation to *cis*-5-octadecenoate, 159
  desaturation to oleate, 134–140
  effect of temperature on accumulation in plants, 152
  elongation, 121, 124
  genetic control of biosynthesis, 262
  mitochondrial biosynthesis, 122, 135
  occurrence, 2, 3, 8, *62, 63, 65, 66, 69, 71, 72, 74, 75, 77, 78, 79, 91, 92*
  α-oxidation, 214, 217
  ω-oxidation, 221–223
  structure, 2, *3*
  wax biosynthesis, 254
Stearolate (9-Octadecynoate)
  biosynthesis, *165*, 166
  occurrence, 18, 19
  structure, *18*
Stearolic family of acetylenic acids
  biosynthesis, 164–168, *165*
  occurrence, 18, 19
  structures, *18*
Sterculate (8-[2-Octylcycloprop-1-enyl] octanoate)
  biosynthesis, 175
  inhibition of unsaturated fatty acid biosynthesis, 137, 143, 149, 155–156
  occurrence, 29, 30, *64*
  structure, *10*

Sterculia foetida, seed fatty acids, 29, *64*
Sterculynate (8,9-Methylene-8-octa-decen-17-ynoate), occurrence, 30
Stereospecific analysis of triglycerides, 83
Sterol esters
 isolation, 283, 284, 287
 occurrence, *61*, 66
 structure, 51, 52
Sterol glycoside esters
 biosynthesis, 191
 occurrence, *61*, 65, 66
 structure, 51, *52*
Sterol glycosides
 biosynthesis, 191
 in senescence, 258
 isolation, 285
 metabolism, 258
 occurrence, *61*, 76
Sterol glycosides, 6-*O*-acyl, *see* Sterol glycoside esters
Stillingia oil, tetra-acid triglyceride, 17, *43*, 44, 290
Stinging nettle (*Urtica dioica*), leaf fatty acids, 68, *69*
Stocksia brahuica, cyano lipid, 51
Streptomyces sioyaensis, α-hydroxyla-tion, 169, 174
Substituted acids
 biosynthesis, 168–173
 occurrence, 22–28
 properties, 23
 structures, *10, 22, 24*
 "unusual" fatty acid class, 11
Succinate, *203*, 207, *208*, 209
Succinate dehydrogenase, 203, *203, 208*, 209
Succinate thiokinase, *203, 208*
Sugar, biosynthesis from lipid, 201, *203*, 207, *208*, 209, 275
Sugar beet, chloroplast lipids, *70*
Sugar cane, leaf wax, composition, *72*
Sugar transport, lipids in, 270–271
Sulpholipid, *see* Sulphoquinovosyl di-glyceride
Sulphoquinovosyl diglyceride
 as energy reserve, 275
 biosynthesis, 185, *187*
 enzymic hydrolysis, 200, 259
 fatty acids, *92*, 93, *247*, 248

identification, 289
 in cation transport, 270
 in chloroplast development, 245–246, 247, 265, 267
 in chloroplast membranes, 265, *266*, 270
 in maturing leaves, 248
 in photosynthesis, 249, 250, 251, 265
 in senescence, 258, 259
 interaction with chlorophyll, 265, *266*, 267
 isolation, 285, 286, 288
 metabolism, 245–249, 250, 258
 occurrence, *61*, 66, *67*, 68, *70*, 71, 76
 quantitative determination, 295
 structure, 48, *49*
Sumach, dicarboxylic acids, 28
Sunflower seed
 fatty acids, 4, 27, 153
 genetic control of composition, 261–263
 lipoxygenase, 227
 wax, 4

T

Tarirate (6-Octadecynoate), occurrence, 13, 20–21
Taxonomy
 higher plants, 30, 33–42
 lower plants, 30–33, 80
TCA cycle, *see* Tricarboxylic acid cycle
Temperature, effect on fatty acid composition, 31, 152
Terpenoid esters
 occurrence, 53, 259
 structure, 53
Tetracosanoate, *see* Lignocerate
cis-15-Tetracosenoate, *see* Nervonate
cis-9,*cis*-12-Tetradecadienoate, biosyn-thesis, 145
1,14-Tetradecane dicarboxylate, *see* 1,16-Hexadecandioate
Tetradecanoate, *see* Myristate
cis-4-Tetradecenoate, occurrence, *11*, 12
cis-5-Tetradecenoate, occurrence, *8*
cis-7-Tetradecenoate
 biosynthesis, 139
 occurrence, 13
cis-9-Tetradecenoate, *see* Myristoleate

Tetragalactosyl diglyceride
  biosynthesis, 184
  occurrence, 49
9,10,12,13-Tetrahydroxybehenate,
  occurrence, *22*, 26
Tetrahydroxyheneicosanoate, occurrence, 27
Tetrahydroxytricosanoate, occurrence,
  27
Tetrahymena pyriformis, α-oxidation,
  219
3,7,11,15-Tetramethylpalmitate, *see*
  Phytanate
2,6,10,14-Tetramethylpentadecanoate,
  see Pristanate
2,4,6,8-Tetramethyloctacosanoate
  biosynthesis, 174
  occurrence, 28
Tetratriacontanoate
  biosynthesis, 102–129, 108, 124
  occurrence, *72*
Thiokinase
  acetic, 98, 101, 102, 126, 201, 211
  in maturing seeds, 210
  3-ketoalkanoic, 99
  malonic, 126
  mechanism, 98, *101*
  microsomal, 207, 211
  octanoic, 98, 201
  α-oxidation in absence of, 219
  palmitic, 98, 101, 143, 201, *202*, 206,
    207, 211
  properties, 98–101
  spherosomal, 207
Thiophorase, *100*, 101
Thylakoids, 265, *266*
Tiglate (2-Methyl-*trans*-2-butenoate),
  occurrence, 29
Tobacco (*Nicotiana tabacum*)
  chloroplasts, proportion of all leaf
    lipids, 66
  chloroplasts, lipids, *70*
Tomato (*Solanum esculentum*)
  leaf lipids, *67*
  lipoxygenase, 278
Torulopsis spp.
  fatty acid biosynthesis, 142
  fatty acids, 25
  ω-oxidation, 222
Trace elements, effect on fatty acid

biosynthesis, 261
Transacylases, *see* Acyl transferases
Transferases, *see* Acyl transferases *and*
  Transphosphatidylation
Transphosphatidylation, 181, 183
Traumatate (2-Dodecendioate), *see*
  Traumatic acid
Traumatic (2-Dodecendioic) acid
  biosynthesis, 173, 235, 276–277
  occurrence, 28
  structure, 276
Triacetic acid lactone, biosynthesis by
  fatty acid synthetase, 115
Triacontanoate
  biosynthesis, 102–129, 108, 124
  occurrence, 4, *72*
cis-21-Triacontenoate, occurrence, 7
Tricarboxylic acid (TCA) cycle
  absence of activity in young seedlings,
    206, 207, 209
  oxidation of acetate, 201, *203*, 207,
    208
Tricholoma grammopodium
  biosynthesis of acetylenic acids, 166
  fatty acids, 19
Tricosanoate
  biosynthesis, 108, 124, 220
  occurrence, 4, *72*
Tridecanoate, occurrence, *8*, *79*
cis-9-Tridecenoate, occurrence, *8*
Triester phospholipids, 46, *47*
Trifolium repens, *see* Clover
Trigalactosyl diglyceride
  biosynthesis, 184, *187*
  occurrence, 49, *70*
  structure, 49
Triglyceride
  as energy reserve, 201, 275
  biosynthesis, 178–179, *187*
  fatty acids, 1–42, 43, *63*, *64*
  hydrolysis by lipase, 193–195
  identification, 289
  isolation, 282, 283, 284
  maturing seeds, 236–240
  occurrence, *61*, 65, 66
  optical asymmetry, 85–89
  positional distribution of acids, 81–
    89
  species in oils, *87*
  stereospecific analyses, 83–85

Triglyceride—*cont.*
structure, 43
tetra-acid, isolation, 290
tetra-acid, structure, 43, 44
8,9,13-Trihydroxybehenate, occurrence, *22*, 26
9,10,18-Trihydroxy-12-octadecenoate
biosynthesis, 173
occurrence, *22*, 26
2,15,16-Trihydroxypalmitate (Ustilate B)
biosynthesis, 168
occurrence, *22*, 26
9,10,18-Trihydroxystearate
biosynthesis, 173
occurrence, *22*, 26
2,4,6-Trimethyl-2-tetracosenoate
biosynthesis, 124, 174
occurrence, 28
2,4,6-Trimethylundecanoate, occurrence, 29
Triticum vulgare, see Wheat
Tritriacontanoate
biosynthesis, 108, 124, 220
occurrence, *72*
Tropaeolum majus, see Nasturtium
Tuberculostearate (10-Methylstearate)
biosynthesis, 174
desaturation, 140
occurrence, 28
Tubers
fatty acids, 65
fatty acids in individual lipids, 90
lipids, 64
Tung oil, fatty acids, 15
Turnip (*Brassica rapa*)
root fatty acids, *65*
root lipids, 64

U

Undecanoate, occurrence, *8*
"Unusual" fatty acids, *see also*
Acetylenic acids
Branched-chain acids
Conjugated ethylenic acids
Non-conjugated ethylenic acids
Substituted acids
biosynthesis, 123, 156–175
definition, 1
occurrence, 9–30, *64*

oxidation, 213
structures, *10, 11, 14, 18, 20, 22, 24*
Urtica dioica, see Stinging nettle
Ustilago zeae, ustilic acids, 26
Ustilate A, *see* 15,16-Dihydroxypalmitate
Ustilate B, *see* 2,15,16-Trihydroxypalmitate

V

Vaccenate (*cis*-11-Octadecenoate)
biosynthesis, 113, *120,* 128, *131,* 133, 159, 160
conversion to lactobacillate, 174
occurrence, *11,* 13, 73, *74,* 133
Vegetable marrow, phospholipase D in seeds, 198
Vernolate (*cis*-12,13-Epoxy-*cis*-9-Octadecenaote)
as precursor of acetylenic acids, 166, *167*
as precursor of conjugated unsaturated acids, 161, *161*
as precursor of dihydroxyoleate, 173
biosynthesis, 173
configuration, 27
hydration, 173
occurrence, 19, *22,* 27, *64*
relationship to linoleate and crepenynate, 166
structure, *10*
Vernonia anthelmintica
hydration in, 173
lipase, 195
seed fatty acids, 27, *64*
Vitis spp., *see* Grape

W

Water melon (*Citrullus vulgaris*), seed germination, 243–244
Wax esters
as energy reserves, 62, 246, 275
biosynthesis, 190–191, 255, *256*
in cuticles, 71, *72*
structure, 52
Waxes (cuticular)
biosynthesis, 252–255, *256*
composition, 4, 23, 25, 71–73, *72,* 252

Wheat (*Triticum vulgare*)
  chloroplast fatty acids, *71*
  germ, acetyl-CoA carboxylase, 125
  grain, extraction of, 281
  grain, fatty acids, *63*
  grain, lipids, 62
  grain, structure, *60*
  leaf, cuticle, 71
  propionate oxidation, 211
Wound hormone, *see* Traumatic acid
Wyerone
  biosynthesis, 173
  structure, 21

X

Xanthophyceae, classification, *32*
Xanthophyll esters, 259
Xeranthemum annuum, epoxidation in, 173
Ximenynate (*trans*-11-Octadecen-9-ynoate)
  cis-analogue, 19

biosynthesis, *165*, 166
occurrence, 18, 19
structure, *18*

Y

Yeast
  acetyl-CoA carboxylase, 116–117
  effect of temperature on fatty acids, 153, 154
  fatty acid biosynthesis, 108–113, *109*, *110*, *112*, *130*, 134, 142, 147, 148
  fatty acids, 9, 24–26
  glycolipids, 24
  α-oxidation, 217
  ω-oxidation, 222
  termination reaction of fatty acid synthetase, 114–116

Z

Zea mays, *see* Maize